METHODS AND
MATERIALS
OF RESIDENTIAL
CONSTRUCTION

METHODS AND MATERIALS OF RESIDENTIAL CONSTRUCTION

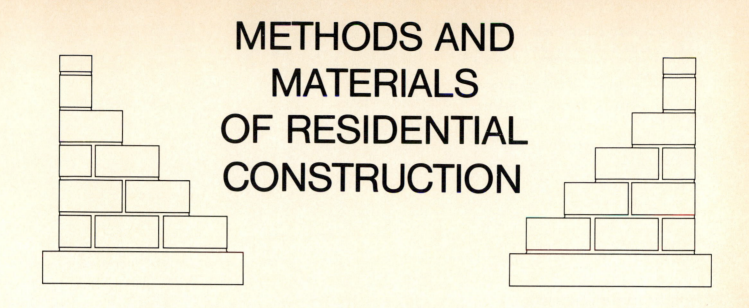

JAMES E. RUSSELL

PRENTICE - HALL, INC., Englewood Cliffs, New Jersey 07632

Library of Congress Cataloging in Publication Data

RUSSELL, JAMES E. (James Emmerson). (date)
    Methods and materials of residential construction.

    Includes index.
    1. House construction. 2. Building materials.
I. Title.
TH4811.R744 1984   690'.837   84-9917
ISBN 0-13-578881-1

Editorial/production supervision and
    interior design: *Gretchen Chenenko and Nancy DeWolfe*
Jacket design: *George Cornell*
Manufacturing buyer: *Anthony Caruso*

Printed in the United States of America

10 9 8 7 6 5 4 3

ISBN   0-13-578881-1   01

PRENTICE-HALL INTERNATIONAL, INC., *London*
PRENTICE-HALL OF AUSTRALIA PTY. LIMITED, *Sydney*
EDITORA PRENTICE-HALL DO BRASIL, LTDA., *Rio de Janeiro*
PRENTICE-HALL CANADA INC., *Toronto*
PRENTICE-HALL OF INDIA PRIVATE LIMITED, *New Delhi*
PRENTICE-HALL OF JAPAN, INC., *Tokyo*
PRENTICE-HALL OF SOUTHEAST ASIA PTE. LTD., *Singapore*
WHITEHALL BOOKS LIMITED, *Wellington, New Zealand*

Contents

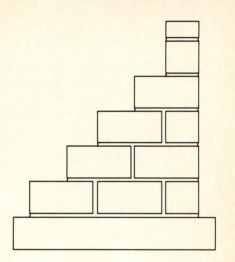

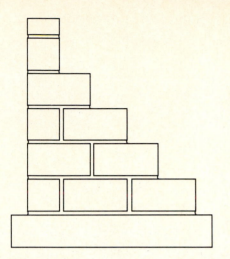

Preface

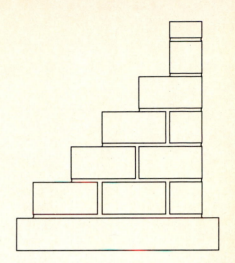

A teacher once told me that it takes 20 years to gain real proficiency in the construction field. I shrugged the statement off because 20 years sounded like an eternity then and because I knew that teachers always exaggerated the importance of their fields. I was much younger then. Those 20 years have whizzed by, and I'm still learning about construction.

Even a definition of residential construction is elusive. To the framing contractor, it may be framing; to the electrician, wiring; to the plumber, plumbing—and so forth. Specialization is absolutely necessary to building efficiently and well; at the same time, overspecialization (being aware of or concerned with only your particular specialty) is perhaps one of the chief reasons for much of the disharmony in our urban and suburban building. How, then, should we define residential construction? My own answer is: Residential construction is the art and technology of arranging houses so that they are in harmony with one another and with the land.

In keeping with my definition, I have attempted to show that residential construction involves the cooperative efforts of many building professionals. However, to study residential con-

struction, it is necessary to break down this elaborate process into increments convenient for study: into construction details, floor plans, window schedules, and so forth. I hope that in studying these details, presented by both text and drawings, the reader will be constantly aware of the larger picture, the process mentioned above. The appendixes included are extensive and will be handy for anyone involved in residential construction.

This text is intended for students at technical colleges and other schools, tradesmen, instructors, contractors, homeowners, manufacturers, and others interested or involved in the residential construction field. The book comes from many sources: from contractors, architects, interior designers, draftsmen, tradesmen, manufacturers, and others, and it is, or course, influenced by my own involvement in building a broad range of projects over many years. Thus, it would be impossible to thank properly all those who have been helpful to me in the writing of this book.

I hope this text serves its readers for years to come.

JAMES E. RUSSELL

The Building Business

1

THE GENERAL CONTRACTOR

The general contractor is usually responsible for seeing the house or houses to completion, from the beginning stages to occupancy. The general contractor may be one person, building a single house for a known owner (presold) or for an unknown future buyer (speculative housing). Or the general contractor may be a huge corporation building thousands of homes per year.

The general contractor may employ many specialists within the building business—heating and air conditioning people, plumbers, excavators, and many others—as well as supervisory personnel, or the general contractor may employ only a skeleton crew of supervisors and depend on subcontractors to do the majority of the hands-on work.

The amount of equipment owned or leased depends on the amount of work the general contractor does, how the company is organized, and similar factors.

Licensing. Both general contractors and subcontractors, as well as all the building trades specialists, are required by the states to maintain licenses. The exact requirements for these licenses vary somewhat from state to state. Generally, some sort of technical proficiency examination is required, plus other requirements to assure the public a degree of security.

Building codes. Building codes have evolved over the years to protect the public against unsafe or unhealthy structures. Local authorities are charged with enforcing the building codes. Some well-known codes are the National Building Code, the Southern Standard Building Code, the Uniform Building Code, the National Plumbing Code, and the National Electric Code. In addition to these code requirements, lending institutions and federal agencies may have their own requirements. It is critical that the contractor have a working knowledge of all the codes, particularly those that directly apply to the work being done, as code requirements often significantly affect cost and, if the work is not done to code or other requirements, may have to be redone at the contractor's expense.

The building site. Obviously, the building site must be secured before any building can be done. However, certain steps must be taken before the land is purchased to assure its suitability for the desired house or houses. The site must be zoned properly; zoning regulations assure that land is used in a manner that benefits the overall community—a factory, for example, is not suitable on property zoned for residential use. Similarly, it would not be wise for a builder to locate a house or houses too close to industrial land or land generally undesirable for residential use.

The building site must have a *clear title*. That is, when it is purchased, the owner must be assured that there are no claims against the land; to prevent this possibility, a *title search* should be performed, usually by the title company. Title insurance, available to the purchaser at a moderate fee, insures the purchaser against any claims that might develop later against the land.

Mortgages. Few Americans have the ready cash to purchase a house outright. Even if they did, it probably would not be the best use of their money. Lending institutions offer money to prospective homeowners to be repaid over a long period of time, typically 25 or 30 years. This type of loan is largely responsible for the massive growth of this country in the twentieth century. To protect the lender, a mortgage is held against the property until the loan is paid off; the mortgage gives the lender the right to reclaim the property should the payments not be paid. A loan from a lending institution directly to an owner is called a *conventional loan*.

To stimulate lending, and thus home building, the Federal Housing Administration and the Veterans Administration issue insurance that protects lenders against loan default. Both the FHA and the VA have requirements both of the lender and the buyer-owner. Obviously, VA loans apply only to veterans.

Bidding. Competitive bids are intended to lower the cost of construction through competition. It is typical for the owner to take three or more bids, if bidding is used. (See Appendix F.) Subcontractors may also submit bids to general contractors. In speculative housing, the general contractor or builder is the temporary owner and takes bids from the subs. If bidding is not done, building costs may be negotiated; that is, the cost of doing the various building components is simply discussed and agreed upon by the parties.

Real estate brokers. Small, speculative general contractors sometimes personally sell their own products. And sometimes, prospects seek out general contractors to build a home for them; these are called *presold* houses. Contractors may personally sell as many of their houses as they wish. However, most states require that persons *hired* to sell houses that do not belong to them be licensed. Thus, most general contractors engaged in speculative building choose to turn the marketing over to a real estate broker who is licensed by the state.

The sales contract. The sales contract is the initial instrument, or document, that the owner or broker uses to confirm a sale. The buyer signs the contract and usually is required to make an "earnest money" deposit. The exact amount of the deposit varies, but 5% of the contract price is not uncommon; thus, an earnest money deposit of $4,000 would be fairly typical for an $80,000 house. Should the buyer back out of the sale, for all but the most compelling reasons, the deposit is usually forfeited. If the buyer is serious about buying the property, the earnest money deposit should not be a major concern, because it applies toward the purchase price of the house if the sale is concluded.

TYPES OF BUSINESS OWNERSHIP

Businesses may be owned as corporations, sole proprietorships, and partnerships. Each type of ownership has ramifications that affect the contractor; thus, contractors entering the business for the first time should carefully study their personal and family situations, their professional abilities, and the type of business they will do (both immediately and in the future) before deciding on a particular type of ownership.

Corporations

Corporations are organizations of people that may legally act as a single entity. The corporation may buy and own land; build houses, apartments, and other structures; acquire loans; and, in general, perform all the activities that are spelled out in its charter (the charter lists all the business activities that the corporation will be involved in and is filed with the appropriate federal, state, or provincial government body).

Corporations are extremely flexible. They may be composed of three people or thousands of people. Corporations can be structured to gain most of the advantages of individual ownership or partnerships with few of their disadvantages. Corporations are easy and inexpensive to form (often under $500, at the time of this writing), in light of the benefits that may often be gained from them.

The corporation can be used by small contractors with few employees and, as business grows, can expand easily. Employees and stockholders may come and go with relatively little disruption of business activities.

As with the other legal formats for doing business, the corporation should be detailed in consultation with an attorney who is experienced in setting up and working with construction businesses (this attorney often becomes associated with the corporation).

Sole Proprietorships

Sole proprietorship (also called *individual ownership*) essentially means that one person is responsible for the operation of the business and anything that may happen as a result of that operation. One person is responsible for the bills and responsible for distributing the profits.

Many general contractors in business today started as workers in one of the trades: as electricians, plumbers, carpenters, and so forth, then took the next step as individual owners of their own subcontracting firms. In time, they became general contractors, still operating as individual owners.

There are some advantages to individual ownership, in small operations, but even in small operations they are largely a state of mind: The individual owner makes all the decisions, reaps the profits, sets the working hours, and generally runs the business in a highly personal, uncomplicated manner.

In fact, however, the disadvantages of individual ownership may outweigh the advantages. Sole proprietors tend to be small operators, "jacks of all trades," often working hard all day and then doing bookkeeping, working up ads, and performing other business-related activities at night. Small operators may have problems getting and keeping good employees, because employees tend to "graduate" to employers who offer them a better future.

During bad economic times (and even during good economic times), the larger contractors may receive better credit and lower-priced materials from the building supply houses.

If the business fails (a possibility whose consequences should always be weighed before entering a business), the individual owner's personal assets—as well as the assets of the business—may be seized to pay off debts.

Finally, should the individual owner die, the business often suffers due to the loss of the operational hand of the owner, and if the business suffers, so may the family of the deceased, because the business is so closely tied to personal property.

Partnerships

Partnerships consist of two or more people who have joined forces to conduct a business. The partnership agreement spells out what each partner's responsibilities are and how the profits are to be distributed.

A typical contracting partnership might be formed between a real estate broker and a person with a building background. The broker would typically be responsible for office management and sales, and the builder would handle construction and related duties; the contributions of capital to the business might be equal. (This is an example of a construction firm partnership arrangement, not necessarily a model.)

In larger firms there may be more than two partners: For example, an accountant, a builder, a real estate broker, an attorney, an architect, and still other professionals may join together in a partnership, to gain bigger jobs and more jobs because of their expertise.

Partnerships may be *general*, in which case each partner is equally responsible for debts, or the partnership may be *limited*. In limited partnerships, one or more of the partners operate the business and one or more partners are investors only. The investors may or may not have expertise in the business.

There are many variations in partnerships. In any partnership, a well-studied *partnership agreement* should be drawn up, with the aid of an attorney, and properly recorded. The agreement should include at least the following information:

1. Purpose of the partnership
2. Responsibilities of each partner
3. The manner in which the profits and losses will be distributed among the named partners
4. The names, addresses, and responsibilities of all persons associated with the partnership: partners, bankers, attorneys, advisers, etc.
5. The types of businesses the partners may engage in outside the partnership
6. The amount of capital or other monetary contributions each partner is to make
7. The date of the beginning of the partnership, its proposed period of existence, and the method and conditions of its ending
8. A method of handling disputes that may arise between or among the partners
9. A method of adjusting the partnership for continuance in the event of death or the inability of one or more of the partners to work (how and when the deceased partner's assets would be distributed; how and when the partner would be replaced; etc.)

It is impossible to overemphasize how important it is to work out a detailed partnership agreement with an attorney and have it properly recorded. In the excitement of starting a new business, it is easy to overlook the unfortunate realities: Partners sometimes die; friendships may end; pressures may cause individuals to hurt the business in unforeseen ways; other disturbing problems may affect the partnership. When problems do arise, a detailed partnership agreement may be indispensable to keeping the partnership afloat.

Typical advantages of a partnership include increased capital and the combined expertise of the partners. The better capital position helps in gaining credit. The expertise may bring larger and better jobs.

However, there are considerable disadvantages to partnerships that should be carefully weighed:

1. Partnerships can be slow and hard to get out of, once in them.
2. Undesirable partners (for whatever reasons) are difficult to remove from the partnership.
3. *Each* full (general) partner is responsible for the total liabilities of the partnership. Thus, should the partnership be unable to pay its bills, creditors are able to make claims against *any* of the partners' personal assets; this can result in one partner's having to pay the bills of the whole partnership.

The legal format within which a contractor chooses to work—corporation, sole proprietorship, or partnership—and the particular way each of these formats is structured should be carefully studied by the beginning contractor, in consultation with an attorney experienced in setting up construction businesses. It is perhaps impossible to offer any one format as the best solution for every contractor. Too many specific factors have to be analyzed. However, it probably could be said that the corporation is the most *versatile* of the various legal formats, for contractors of all sizes. Sole proprietorships usually work best for small operators, contractors with a helper or two who do small jobs, renovation, and repair, additions, garages, and so forth. Partnerships may work for two or more individuals who wish to combine capital and different but complementary skills (such as production, marketing, and financial planning skills). In any case, again, consult an attorney who has experience in setting up construction businesses before deciding.

INSURANCE AND BONDS

Many types of policies have evolved over the years to cover known and suspected areas of risk. These areas of risk include liability, fire, collision, upset, theft, worker injury, and other areas that can financially ruin a contractor. The method of cover-

age and the amount of coverage needed vary from policy to policy and with the needs of individual contractors. Each contractor has very specific insurance needs that should be met; the best coverage must be custom-tailored for each contractor, and this coverage varies with the type of work the contractor does. Arranging for proper insurance coverage, therefore, is a job for a skilled professional insurance or bond broker or counselor, working closely with the contractor. The best advice for students, contractors, or anyone building a project where workers, suppliers, or the public may be involved is to establish a good working relationship with a reputable insurance or bond professional; in practice, the relationship is very similar to the contractor's relationship with a banker.

Because of the variance in coverage and individual needs mentioned above, the following discussion is necessarily general and does not attempt to investigate insurance in depth; to do so would be beyond the scope of this text. Also, an intensive coverage of insurance is not warranted because it can be learned on the job, since most entrants to the contracting business will not start as owners but at some specialty or entry-level position.

Insurance coverage varies. It is important to note that the coverage should be itemized and the conditions of coverage fully detailed. For example, does the policy cover building materials such as siding, masonry, and roofing? *Where* does the coverage apply—on the job site only? During transit to the job? And what else does the policy protect against—fire? Storms? Theft? Vandalism? All of these events?

Many policies contain "extended coverage" or "all-risk" features, which generally enlarge and increase coverage. However, terms like *all-risk* are still relative. For example, even in so-called comprehensive policies, the contractor's liability and property damage insurance does not cover all types of damage that may occur. *Undermining* is a typical example.

Undermining sometimes occurs during the draining of excavations. A clue to the possibility of undermining is the continual flow of dirty, soil-laden water into the excavation from underground sources; this means that soil is being drawn from surrounding areas, perhaps areas with buildings or other property on them. Eventually, cave-ins may result, and the contractor is held responsible for them. A typical solution to this problem is the use of *well points*. Well points are pumps with pipes containing fine holes that allow only very small particles of soil to enter. The well points are located around the excavation such that water is removed before it reaches the excavation, and thus surrounding soil is not undermined. Other solutions include blocking the entry of outside water by grouting, freezing, or chemical action. In any case, insurance inspection should precede action, because a special endorsement to the policy will be necessary.

Jobs that for whatever reason may present liability hazards beyond the ordinary should be inspected by insurance company officials so that appropriate precautions can be agreed upon before construction begins. This may result in a higher rate for the particular job, but this is preferable to disagreements or expensive lawsuits with the insurance company should an accident occur.

A contractor's general liability and property damage insurance does not cover highway vehicles such as trucks, jeeps, cars, and pickups. These vehicles have special coverage at higher rates.

Adequate coverage should be provided for all participants on the job. Duplication of coverage, however, should be avoided, as insurance premiums are a considerable expense. For example, the owner, general contractor, subcontractors, and others need not all insure against the same risks. The job specifications typically spell out the phases of the job for which the various participants are responsible, and arriving at fair shares of the premium burden for the appropriate participants is usually not difficult. In some cases, the owner agrees to insure everyone, which, of course, influences the bidding price of the participants (lowers the bid), since they will not have to provide insurance. In other cases, the subcontractors may provide their own insurance, the general contractor may provide all coverage, or some combination of these approaches may be arrived at.

All the participants should be aware of their responsibilities, their risks, what coverage is being provided, and all the details of that coverage.

In the event of a just claim, the responsible participant—general contractor, subcontractor, or other—should cooperate in seeing that the claim is paid as quickly as possible. Conversely, *no* unjust claim should be paid, no matter how small. Busy contractors are inclined to pay small, annoying claims, even though they are unjust, to "get on with the work." This practice results in higher insurance rates and thus higher building costs, and a contractor who pays a small, unjust claim for expediency may then be open to many more similar claims.

In addition to the insurance coverages indicated earlier, contractors are usually required to carry workers' compensation insurance, which covers employees for job-related injuries. Clerical and office workers' premiums can be as low as 1%

of payroll, while those for hazardous occupations may range up to 20% of payroll. Typical rates for residential and light construction activities range between 4% and 8% of payroll. Obviously, contractors must factor these and other insurance costs into bid proposals.

Contract bonds work similarly to insurance and are typically required by various interested parties, as appropriate. The most typical bonds are bid bonds, performance bonds, completion bonds, and supply bonds.

Bid bonds are given to the owner by the contractor. These bonds assure the owner of serious bidding; they state that if the contract is awarded, the contractor will accept it and furnish performance and other bonds, as appropriate.

Performance bonds, as the name implies, guarantee that the work will be done. These bonds are given to the owner by the contractor. A maintenance bond covering workmanship and materials for a period of one year is typically included in the bond at the same cost; further maintenance bonds, if issued, cover maintenance for a specified period of time and are given to the owner by the contractor.

Completion bonds guarantee the owner that the job will be completed. These bonds are given to the owner by the contractor.

Supply bonds guarantee the owner that the manufacturers or distributors will deliver the contracted materials, equipment, and so forth, needed to build the job according to the contract for the work.

These four types are by no means all the bonds used in connection with the construction business; they are simply some of the more typical ones. Bonds are exchanged between developers and public agencies, public agencies and contractors, unions and contractors, and many other participants in construction activities. Bonds are simply guarantees from one party to another that money is available to see the job through, in the event that one or more of the participants does not or cannot perform the work contracted for.

In our discussion, the bonds are said to be "given" to one participant by another. Actually, nothing is literally given; bonds are cost items to those who issue them and are figured into the cost of doing the job.

WORKING DRAWINGS AND SPECIFICATIONS

Adequate and accurate working drawings and descriptions of materials and specifications are necessary to contractors, lending institutions, future owners, and building authorities. These documents speed the work, reduce confusion, reduce supervision, and contribute to accurate building.

The items discussed here are the *minimum* needed.

Plot Plan

The *plot plan* is typically drawn at a scale of 1 inch = 20 feet. The following information should be included, although the amount of information shown will vary with the job situation.

1. The lot and block number
2. A north arrow
3. Dimensions of the plot boundaries
4. Dimensions locating *all* improvements such as buildings, garage, walks, drives, steps, patios, fences, retaining walls, and pools
5. Dimensions and locations of all easements and location of building setback lines
6. The following elevations: first floor elevation of house, garage (or carport), any other buildings; finish curb elevations (or crown of street

elevations if there are no curbs); finish grade elevations at building corners. Additional elevations that may be necessary: any points or areas where special grading may be required (retaining walls, steep slopes or other dramatic grade changes, etc.); any other elevations that may be necessary to grade the plot properly for drainage or to construct the improvements at the desired elevation

7. Lot drainage (does the entire lot drain toward the street? Toward the back lot line? Some combination?)
8. Drainage swales, catch basins, and other drainage aids
9. Location of septic tank, distribution box, absorption field, etc., where an individual sewage-disposal system is used
10. Location of well, service line, etc., where an individual water-supply system is used
11. Location of any items on adjacent property that may affect the subject property (wells, sewage-disposal systems, drainage components, etc.) (See Figure 1-1.)

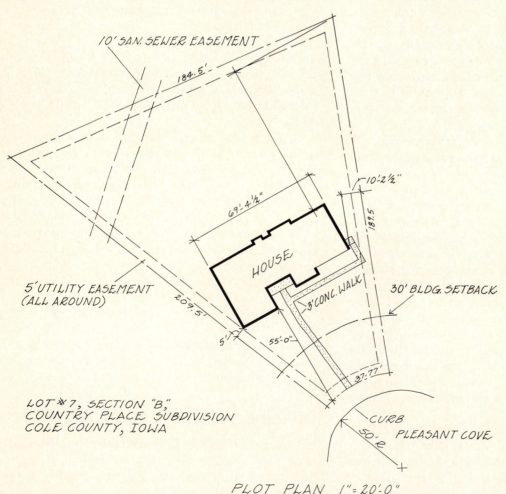

PLOT PLAN 1"=20'-0"

FIGURE 1-1 Typical plot plan for subdivision housing.

Foundation Plan

The *foundation plan* describes, dimensions, and locates the type of foundation that will be used and the materials that will be used to construct it. Concrete footings are typical. Foundation walls are concrete or, where weather permits, concrete block or masonry.

The foundation plan also describes the floor slab, where one is used. Typical floor slabs are 4 inches thick, reinforced with wire mesh, and laid over a 4-inch bed of gravel; a plastic membrane is usually placed between the gravel and the slab to prevent moisture from coming through the slab.

It is best that all details pertaining to the foundation be placed on the foundation plan; however, footing and other details are sometimes shown on the framing section.

The foundation plan is usually drawn at the same scale as the floor plan: ¼ inch = 1 foot. (See Figure 1-2.)

Floor Plan

The *floor plan* includes at least the following information:

1. A plan of all floors: basement, first floor, second floor, other. Plan of garage or carport, porches, patios, etc. Porch columns and piers are usually indicated on the floor plan.
2. Floor and ceiling framing members, including size, direction, and spacing.
3. Location of all wall partitions showing door location and direction of swing; window locations.
4. Plan view of kitchen cabinets; all shelving (closet, bathroom, kitchen, other); plumbing fixtures; appliances; etc.
5. Location with appropriate symbol of all electrical switches, outlets, thermostats, heating and cooling registers, other similar information.
6. Where space permits, door, window, electrical, and other schedular information. Kitchen and bathroom elevations also may be placed on the floor plan or on a special sheet.
7. For small houses and additions, heating and cooling systems, ducts, piping, registers, etc. (In this text, we will assume that a special sheet is used for the heating and cooling plan.) (See Figure 1-3.)

Exterior Elevations

The *exterior elevations* are drawn at a scale of ¼ inch = 1 foot.

The following minimum information is included:

1. Front, rear, and both side elevations; elevations of garage or carport; other elevations that may be required to give a clear picture of the building or buildings; all windows and doors (sizes may be indicated on the elevations or included in separate schedules).
2. Type of materials
3. Finish floor lines and grade lines at each building
4. The foundation system

One or more framing sections, which, as the

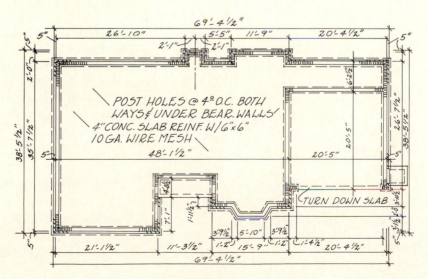

FIGURE 1-2 Example of foundation plan.

FOUNDATION PLAN ¼"=1'-0"

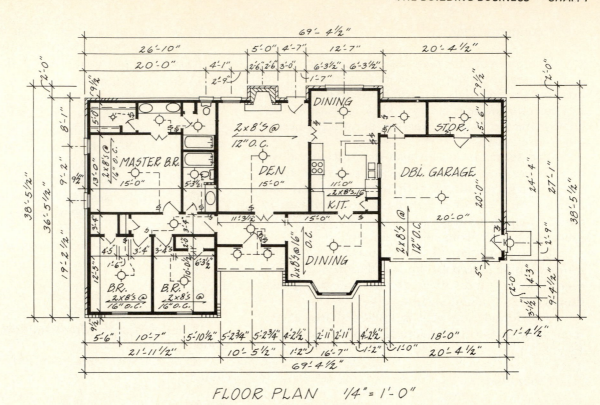

FLOOR PLAN   1/4" = 1'-0"

FIGURE 1-3 Example of floor plan (some detail omitted for printing clarity).

name implies, show how the framing members relate to each other, may be shown on the exterior elevation sheet or on a special sheet. (See Figure 1-4.)

Heating and Cooling Plan

The *heating and cooling plan*, or *mechanical plan*, is usually drawn on a special sheet at ¼ inch = 1 foot scale.

The following minimum information is included:

1. Graphic layout of the system, showing ducts, piping, registers, radiators, heating and cooling equipment, and other equipment
2. Sizes of all equipment
3. Location of thermostats
4. Heat loss and gain information (usually shown on a schedule placed on the sheet); model number of equipment, heating and cooling capacity, and similar information. (See Figure 1-5.)

Detail and Section Sheets

One or more special *detail and section sheets* are often required when space does not permit the details elsewhere.

Typical drawings included on detail sheets are: foundation footing sections and special foundation conditions such as grade beams, piers, and so on; wall sections, both interior and exterior; kitchen and bathroom elevations and details (see Figure 1-6); details and sections of stairs, steps, and chimneys; details of beams, roof trusses, and other structural components; details of other construction items that may not be clear from the other drawings (see Figure 1-7).

The minimum scale for these drawings is ¼ inch = 1 foot. Kitchen and bathroom elevations typically are shown at ⅜ inch = 1 foot.

Specifications and Descriptions

The *specifications* and *description of materials* vary with the particular job. The Federal Housing Administration requires completion of FHA Form

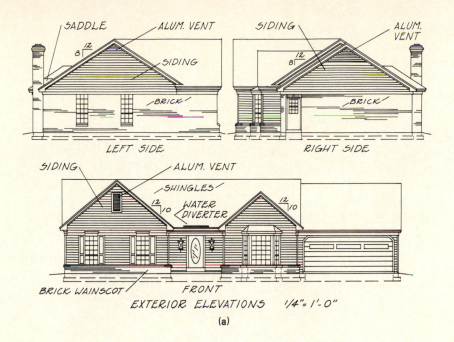

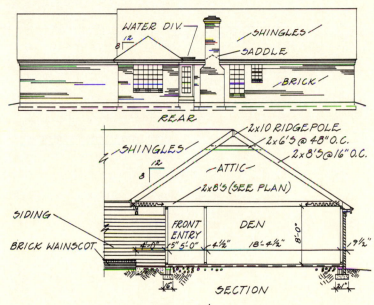

FIGURE 1-4 Exterior house elevations and framing section.

2005 for houses that will be federally insured. (See Appendix A.)

FHA Form 2005 is a good *minimum* framework, spelling out the building conditions and materials necessary for a decent, livable house. It is worthy of careful study; for more intensive study of specifications, refer to other texts that concentrate on residential construction specifications.

The most qualified professional to produce building plans is a licensed architect. The architect usually has completed a five-year course of instruction at an accredited college or university, then worked a three-year apprenticeship under a licensed architect, then passed a rigid state examination.

Instead of using architects, however, some builders rely on so-called plans services. The training that plans service personnel have varies. Often, they are people who have worked for architects specializing in residential designs and have accumulated enough experience to be technically qualified to produce good plans. Since the training varies so much, word-of-mouth recommendations are probably the best guide in selecting a plans service.

BUILDING PERMITS

Your building department probably has a procedure form for securing a building permit. Figure 1-8 shows one such procedure, obtained by mail from a city of about 500,000.

In your locality, the procedure may be more or less stringent than this one. Call the municipal building department and ask them to mail you the procedure for securing a building permit.

Appendixes F, G, H, and I contain standard documents frequently used in residential construction.

MAKE	GENERAL ELECTRIC	
MODEL NUMBER	WV942A100	
CAPACITY HEAT	108,280	BTU/HR.
CAPACITY COOL	42,000	BTU/HR.
TOTAL HEAT LOSS	65,504	BTU/HR.
TOTAL HEAT GAIN	32,167	BTU/HR.
BASED ON	6	IN. INSULATION IN CEILING
AND	3½	INCHES INSULATION IN WALLS
INSULATION ON DUCTS	2 IN. FOILBACK	
GRILLS	(not specified)	
REGISTERS	(not specified)	

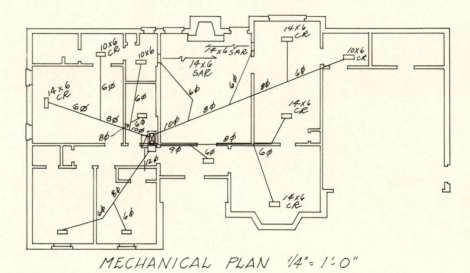

MECHANICAL PLAN ¼" = 1'-0"

FIGURE 1-5 Example of schematic mechanical plan.

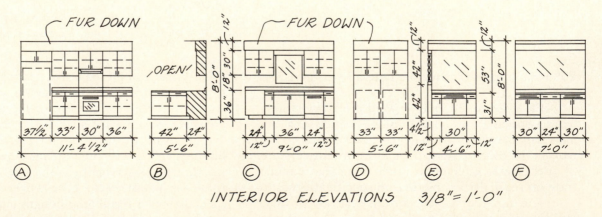

INTERIOR ELEVATIONS 3/8" = 1'-0"

FIGURE 1-6 Detail sheet for kitchen elevations.

DOOR SCHEDULE			
MARK	SIZE	THICK.	REMARKS
A	3'-0"x 6'-8"	1 3/4"	SEE ELEV.
B	2'-8"x 6'-8"	"	FRENCH
C	2'-8"x 6'-8"	"	1/2 GLASS
D	2'-8"x 6'-8"	"	FLUSH
E	2'-8"x 6'-8"	1 3/8"	6 PANEL
F	2'-0"x 6'-8"	"	"
G	1'-6"x 6'-8"	"	3 PANEL
H	2-1'-6"x 6'-8"	"	"
I	2-1'-4"x 6'-8"	"	"
J	4'-0"x 6'-8"	4 1/2"	CASED OP'NG
K	2'-8"x 6'-8"	"	" "
L	18'-0"x 7'-0"	—	GARAGE DOOR

WINDOW SCHEDULE			
MARK	SIZE	TYPE	REMARKS
1	3'-0"x 6'-0"	S.H	ALUM.
2	2'-8"x 6'-0"	"	"
3	2'-4"x 6'-0"	"	"
4	1'-6"x 6'-0"	"	"
5	2'-8"x 3'-0"	"	"
6	2'-0"x 3'-0"	"	"
7	6'-0"x 5'-0"	FIXED	"
8	4'-0"x 6'-0"	"	"

ELECTRICAL SCHEDULE			
SYMBOL	REMARKS	SYMBOL	REMARKS
	CEIL. FIX.		EXH. FAN
	SURFACE MT.	P.C.	PULL CORD
	RECESS LIGHT	W.P.	WATER PR'F
	FLUORES.		
	WALL RECEPT.		
	220 RECEPT.		
U.C.	UNDER CTR.		
	GARBAGE DIS.		

FIGURE 1-7 Sample door, window, and electrical schedules.

BUREAU OF CONSTRUCTION CODE ENFORCEMENT

RE: Code Enforcement Procedures for Securing a Building Permit

The following listed information and procedures are normally necessary when persons, firms, corporations, or others request permit applications for improvements of properties.

I. An applicant is required to have three (3) sets of complete plans with one (1) set of specifications, signed, sealed, dated, and drawn to scale by an architect or engineer who is licensed by the state. These plans should include the following basic information.

    A. Plot Plan
        1. Lot, scaled to size and dimensioned
        2. Subdivision information
        3. Location of proposed construction, ALL existing improvements, rights-of-way, easements, etc.
        4. Off-street parking area
        5. a. Existing and proposed grade elevations
           b. Proposed drainage runoff
        6. Curb cut sizes and location
    B. Footing/Foundation Plan
    C. Floor Plans
    D. Elevation Plans
    E. Sectional and Detailed Plans
    F. Roof Plans
    G. Electrical Plans
    H. Heating and Air Conditioning Plans
    I. Plumbing Plans
    NOTE: When consulting designers, architects, or engineers, the applicant should make absolutely sure that the architect or engineer verifies that Intended Use or Uses comply with all the zoning district regulations.

II. Once the applicant has the proper plans and is ready to apply for a building permit, correct procedure is as follows.

    A. Proceed to the City Building Department and make an application for a building permit. House number will be assigned at time of application.
    B. To make application for a building permit, the applicant must present three (3) sets of complete plans.
    C. The Building Department will process the three (3) sets of plans through the Central Plans Division for compliance of Code and Zoning Regulations. (This plan review is for Building Code, Electrical Code, Plumbing Code, Mechanical Code, Fire Department, and City Engineer's Office.)
    D. Full plan review normally takes five (5) to fifteen (15) working days, depending on the amount of work logged with the department and the needed information supplied on the plans.
    E. When plans and specifications are in Code and Zoning compliance, permits are authorized for issuance.

FIGURE 1-8 Typical procedure for obtaining a building permit.

Land Clearing and Land-Clearing Equipment

2

This chapter is a study of the methods and equipment used to clear large areas of land for development.[1] There has been a tendency in recent years toward larger building projects. There is an economy of scale in large projects, and with good planning by experienced professionals, there can be environmental gains also, because the impact of development on the environment is studied for larger regions. This broader view is to be applauded.

LAND CLASSIFICATIONS

In land clearing, several types of terrain and vegetation will be encountered, and these could be subdivided into countless subtypes. Most people, however, would agree to the following classifications for land-clearing purposes:

1. *Desert*. Devoid of vegetation above 2 or 3 feet (0.6 or 0.9 meters) in height, occasional trees in low spots where water accumulates or ground springs are present.
2. *Bush Country I*. Scattered bushes, usually thorny, woody vegetation with or without scattered cactuslike plants. When not overgrazed with livestock or wildlife, it is covered with annual or perennial grasses.
3. *Bush Country II*. Usually found in tropical

[1]Material on land clearing in Chapter 2 adapted from *The Clearing of Land for Redevelopment*, with permission of Caterpillar Tractor Co., © 1974. Photos reproduced with the permission of Caterpillar Tractor Co.

semiarid areas. Its characteristics are very dense thorny brush with scattered large hardwood vegetation and undergrowth so thick that neither annual nor perennial grasses can grow.

4. *Upland Woods*. Usually hardwoods and softwoods, deciduous and coniferous. Few, if any, vines entangle the treetops. All the trees will be approximately the same height, and there is usually only a single canopy (the uppermost branchy layer of a forest). This vegetation is limited almost exclusively to temperate climates.
5. *Tropical Rain Forest*. Characterized by a mixture of hardwoods and softwoods and a two- or sometimes three-storied canopy, with or without dense underbrush. Trees may grow beyond 15 feet (4.6 meters) in diameter. Often, the tops are heavily entwined with clinging vines. This type of vegetation requires good soil, year-round rainfall, and constant high or moderately high temperatures.

FACTORS TO CONSIDER IN DETERMINING METHODS AND EQUIPMENT

Determining exactly how certain variables increase or decrease the efficiency of specific methods and equipment is difficult. Success, however, is often a combination of being aware of the variables and applying common sense in equipping the job to meet them.

Many factors affecting production and cost stem from the type and density of vegetation. Among these are the number of trees, tree size, wood density, roots, vines, and undergrowth.

All of these variables can be determined by a *tree count*. To conduct this count, measure a straight line for any convenient distance, usually 328 feet (100 meters). Take tabulation along this line for a depth of approximately 16 feet (5 meters) on both sides. The tabulation should include the number of trees, their diameters, densities (hard or soft wood), and root system plus a description of any undergrowth and vines. The 328-by-32-foot (100-by-10-meter) area will produce a sample of about ¼ acre (0.1 hectare). This should be repeated two or three times for each area where the size or type of vegetation changes significantly. The tree counts need not be parallel.

More detailed methods of analyzing vegetative cover can be found in any good forestry textbook. However, these other methods are probably too complex for most land-clearing purposes.

Tree counting can be done by grouping trees into categories by diameter:

less than 12 inches (30 cm) (undergrowth)
12–24 inches (30–60 cm)
25–36 inches (61–90 cm)
37–48 inches (91–120 cm)
49–72 inches (121–180 cm)
over 73 inches (181 cm)

Trees can be counted as they are measured. The diameter of the tree is found, as shown in Figure 2-1, by measuring the trees at diameter breast height (DBH), or 4.5 feet (1.37 meters), above the ground (A). If a buttress (B) is present, the measurement should be at the top of the buttress where the trunk begins to run straight and true (C).

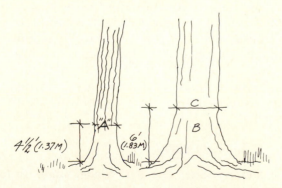

FIGURE 2-1　Measuring the diameter of trees.

Wood density (hardwood or softwood) and the root system (taproots or lateral roots) should be noted when measuring. Recording the presence of vines and the type of undergrowth should also be done at this time.

Generally, wet soil encourages trees and shrubs to develop wide lateral root systems near the surface. These can be grubbed out more economically than species with deep taproots found in dry areas.

Note the presence of high-climbing vines binding the tree tops together. This condition complicates the felling operation, regardless of the method used.

Diameter and density affect productivity (speed of land clearing). A hardwood tree 8.8 feet (2.7 meters) in diameter, in a case study, took 48 minutes to fell, while a softwood tree of nearly the twice diameter, 16.1 feet (4.9 meters), required only 7 minutes to fell. The same effect can be true for other variables.

Soils play a major part in the choice of methods and equipment, especially with regard to depth of topsoil, moisture content, and presence of rocks and stones.

Vegetation

In clearing operations, the soil is often disturbed in felling and piling the vegetation. In areas where there is a very shallow layer of topsoil and where planting will be done, it is imperative that this soil disturbance be kept to a minimum to ensure that the land will be conducive to planting after being cleared. Method and equipment choice will determine the degree of soil disturbance that can be expected.

Soil Conditions

The type of soil will also play a part when tree-felling production is considered. In sandy loam soils, roots often pop out of the ground when a tree is felled. At the other end of the soil scale is hard clay, which often holds roots so tightly that they have to be dug or uprooted with special tools or cut off at ground level and left.

Moisture content of the soil is another factor that affects the choice of method and equipment. In heavily wooded areas, sunlight is seldom in contact with the soil for long periods of time. This frequently causes the soil to be so damp that it will not support the weight of equipment in the felling and stacking operations.

If embedded rocks or stony outcroppings are

present in the area to be cleared, equipment that severs the vegetation at ground level is severely hampered in its operation, and maintenance becomes a problem. Other types of equipment could be used in these areas.

With all these factors to judge, some standard must be set so that underfoot conditions can be accurately described and understood. *Good* underfoot conditions exist when traction, flotation, and slope are not a problem for the tractor even after repeated passes in the same tracks. *Soft* conditions exist when more than one or two passes over the same track create moderate to severe impairment of machine performance. *Poor* conditions exist when the first pass is precarious and difficult.

Topography

Grade and terrain can greatly affect the normal operation of some equipment. Such factors as steep slopes, ditches, swampy areas, and the like can decrease production, increase maintenance costs, and thus influence the choice of methods and equipment considerably.

Rainfall and Climate

All phases of land clearing, from cutting to burning, can be affected to some degree by temperature changes and the amount of rain that falls during a clearing project. Rainfall and the resultant water table also affect flotation. Due to heavy rain, conventional track-type tractors will sometimes sink into the ground. Low-ground-pressure tractors should be considered.

End Use of Land

The end use of the land is an important factor when choosing the method and equipment. For example, if the land is to be used for building, highway, or dam construction, total removal of vegetation is necessary. On the other hand, if the land is to be planted, cutting the trees and brush flush to the ground or 3 to 4 inches (7.5 to 10.0 centimeters) below ground level is all that is required. If the land will be used for grazing, certain large trees may be left standing along with the stumps of other trees.

LAND-CLEARING MACHINES AND ATTACHMENTS

Machines and job conditions must be matched for each job site. What works well in one area may not be the best method in another area. Track-type tractors find application in almost every phase of

land development work. The D4E, D5B, D6D, D7G, D8L, and D9L track-type tractors are well suited in the initial stages of rough clearing and pioneering. The low-ground-pressure D4E, D5B, and D6D can help out in areas where underfoot conditions are extremely soft. Even in the final stages of land development work, track-type tractors, like the Special Application D4E, D5B, D6D, and D7G, are used in replanting or maintenance of vegetative growth.

Generally speaking, lower-cost clearing can be done with larger tractors if the amount of clearing involved is sufficient to merit the greater initial investment in the bigger machines. In any case, because of the need for constant shifting, a power shift transmission should be standard equipment on all tractors used in this application except in pull-type operations. In most applications, a Caterpillar Winch (D9L through D5B) should also be considered as standard equipment for at least one tractor per fleet.

Track-type tractors are by no means the only machines that should be considered. Imagination and resourcefulness can allow the adoption of other types of machines in specific applications. In fact, wheel loaders have been used in some areas for raking, stacking, and even grubbing operations.

In equipping land-clearing machines, consideration must be given to the various protective attachments available. For instance, the radiator, engine, and underside of the tractor should be well protected with perforated hoods, screens, crankcase guards, and other shields.

More important, though, cab guards should be considered a necessity in land clearing. (It is also interesting to note that production has been estimated to increase significantly when cab guards are used.)

Attachments

A familiarity with current land-clearing equipment is vital for determining effectively the proper tools for a clearing job. Substantial cost savings can be realized when the job is thoroughly planned and properly equipped. The following is a capsule description of land-clearing tools, their applications, and in some cases their advantages and disadvantages. No attempt is made to compare the attachments.

Rome K/G Blade

Description.　The K/G blade, made by Rome Industries (Cedartown, Georgia), is designed to apply the power and weight of a track-type tractor to a sharp cutting edge.

The blade angle is 30 degrees on all models. The blade is constructed of a special alloy steel. Its replaceable cutting edges and "stinger" can be re-sharpened with a small portable grinder. A guide bar is used to control the direction in which the trees fall—forward and to the right of the operator. (See Figures 2-2 through 2-4.)

Application. The K/G blade can be used for practically any type of land clearing, including clearing for reservoirs, lakesites, range lands, forest site preparation, agriculture, rights-of-way for pipelines or power lines, and large construction sites. It can also be used for building roads, firebreaks, and V-type drainage ditches. Stumping and windrowing are also possible with this tool.

FIGURE 2-4   Rome K/G blade on Cat D6C.

Advantages. This is a versatile attachment; the same tool can cut, pile, stump, and ditch. Cut trees normally fall in one direction. Given enough time, the K/G can fell any size tree.

Fleco V Blade

Description. The V blade, made by Fleco Corporation (Jacksonville, Florida), is available for Caterpillar track-type tractors through the D8L. It is equipped with a heavy-duty splitter, angled serrated cutting blades, and brush rack. V blades mount directly on the tractor trunnions and are available for cable or hydraulic control. The blade is in two sections and bolts together to form the working tool. The serrated blade and splitter are both hardened steel. (See Figures 2-5 and 2-6.)

FIGURE 2-2   Cat D8L/Medford cab guard, Rome K/G blade, and protection group.

FIGURE 2-3   Cat D8L/Medford cab guard, Rome K/G blade, and protection group.

FIGURE 2-5   Fleco D8L V tree cutter.

FIGURE 2-6   Fleco D8L V tree cutter.

FIGURE 2-7   Fleco D7G tree pusher.

Application. The V blade is useful in high-production clearing of trees, stumps, or brush that does not require removal of subsurface roots and stumps. Applications for V blades are for clearing dam sites, industrial sites, agricultural areas, jungles, road pioneering, and clearing rights-of-way.

Advantages. When properly matched to the job, the V blade is a highly productive tool because it can maintain a continuous path. Bolt-on design permits assembly or disassembly on the job. It also provides flexibility in moving the unit.

Disadvantages. Trees fall either to the right or to the left of the tractor, which can be a disadvantage depending on the particular job. It does not remove imbedded material such as roots and rocks. Stumps are usually not removed in one pass. the V blade is not recommended for piling or ditching.

Tree Pusher

Description. Two models of tree pushers are available from Fleco Corporation. They mount on a standard or angle bulldozer blade. One is attached with brackets to the top of the C frame or push arms and pinned to the top of the blade, so it will be raised or lowered with the blade. Another method of mounting is to pin it to the C frame or push arms so it can raise or lower independently of the dozer blade. A double drum-cable control is necessary for this unit. Both configurations are available for all track-type tractor models. (See Figure 2-7.)

The other tree pusher model is mounted

directly on the tractor trunnions. This model is available for all tractors except the smaller track-type tractors.

Application. Because it removes the entire tree, the tree pusher is a highly effective tool when the land-clearing project requires that no stumps be left in the ground. The tree pusher is extremely useful in chaining operations where the chain cannot fell a large tree or when help is needed in maneuvering the chain.

Advantages. The tree pusher provides a method for downing trees because the trees always fall away from the operator. Since the entire tree is uprooted, no stumps or roots have to be removed at a later time. It can significantly speed up a chaining operation.

Disadvantages. The effectiveness of this tool can be impaired if trees are too large, taproots are present, or traction is poor. Small-diameter growth usually needs to be removed by some other means. Holes left by the stumps sometimes need to be filled, depending on end use of the land.

Rakes

Fleco Corporation (Jacksonville, Florida) manufactures 12 different types of rakes, all of them specifically designed for certain jobs. All the rakes are interchangeable with standard tractor attachments unless otherwise specified. Fleco rakes are available for several models of Caterpillar equipment.

Multiapplication Rake

Description. Multiapplication (MA) rakes are available from Fleco Corporation and Rome Industries. They are designed to withstand extreme shock loads under the most severe clearing conditions.

Multiapplication rakes have teeth equipped with replaceable wear tips. A steel center plate is located in the rake frame to protect the tractor radiator. A high-brush rack is standard equipment. MA rakes interchange with the blade on a bulldozer arrangement. (See Figure 2-8.)

Application. The multiapplication rake is designed for all heavy-duty land clearing including small tree and rock removal. Applications are jungle clearing, road pioneering, dam-site clearing, industrial and agricultural clearing, clearing for rights-of-way, grubbing, stumping, rock work, and general-purpose clearing.

Advantages. This is a heavy-duty implement for land-clearing use in varying types of terrain. It is particularly effective in sandy soils.

Disadvantages. The multiapplication rake will not fell large trees.

Rock and Root Rake

Description. The rock and root rake is equipped with teeth made of manganese-carbon steel castings. The teeth are curved to get under rocks, boulders, roots, and the like. Replaceable wear tips are standard along with the center plate, which protects the radiator. Brush racks are available as optional equipment. (See Figures 2-9 through 2-13.)

Application. The rock and root rake should be used for general clearing, grubbing boulders, and spreading rock riprap. It is also useful in grubbing and piling trees, stumps, and debris.

Blade Rake

Description. The blade rake is mounted to the bulldozer blade by pinning it to the top of the blade with two pins. Models are available for both angle and straight dozers for most Caterpillar D4E

FIGURE 2-9   Fleco D8L rock and root rake with optional brush rack.

FIGURE 2-8   Rome multiapplication rake on Cat D6D.

FIGURE 2-10   Fleco D3 rock and root rake with optional brush rack.

FIGURE 2-11   Fleco D6D rock and root rake.

FIGURE 2-13   Fleco D7G rock and root rake with optional brush rack.

FIGURE 2-12   Fleco D3 rock and root rake with optional brush rack.

track-type tractors. Installation is accomplished by welding two sets of mounting brackets to the upper edge of the blade. The rake is constructed of high-strength alloy steel and fits flush with the blade.

Application. Blade rakes are designed for light grubbing, piling, and raking. They are not recommended for heavy-duty clearing such as removing trees, stumps, and rocks. They are particularly suited for tractor jobs that require both raking and blading.

Clearing and Stacker Rakes

Description. Clearing and stacker rakes are available from Fleco Corporation for Caterpillar wheel and track-type loaders. Stacker rakes are also available from Rome Industries for Caterpillar track-type tractors. The teeth of the clearing rake are shorter and less curved than those of the stacking rake. Both rakes are equipped with brush racks for greater load-carry capacity. (See Figures 2-14 and 2-15.)

Applications. Each rake is designed to perform a particular function. The clearing rake is strong enough for tree pushing, grubbing, stumping, and clearing and can also be used for boulders.

The stacker rake has extra height and longer curved teeth for jobs such as raking, piling, carrying, and loading downed debris. It is not, however,

FIGURE 2-14   Fleco 955L clearing rake.

FIGURE 2-15   Fleco 931 stacker rake.

FIGURE 2-17   Fleco 977L clamp rake.

designed for grubbing or raking large jungle-type trees.

Clamp Rake

Description. The clamp rake is available for Caterpillar track-type and wheel loaders. With the regular bucket, it requires an extra hydraulic valve arrangement. The clamp rake retains all the basic features of the clearing rake for track-type loaders with the added feature of two built-in hydraulically operated clamps. (See Figures 2-16 and 2-17.)

Application. The clamp rake is designed for clearing and grubbing trees, stumps, and boulders. It can also be used for raking, carrying, loading, and piling debris.

FIGURE 2-16   Fleco 977L clamp rake.

Wheel Root Rake

Description. The pull-type wheel root rake is available from Rome Industries. It was designed specifically to follow up root plowing to remove unwanted roots. It leaves an area clear, ready for disk harrowing or agricultural operations.

The wheel root rake is secured to the tractor drawbar. The wheel assembly consists of two widely spaced steel drum-type wheels for flotation and stability. Two high arches run from the wheel assembly to the rake beam to give added brush capacity.

Application. The wheel root rake is basically intended as a land-clearing implement for semiarid areas.

Advantages. With its 21-foot (6.4-meter) width, the wheel root rake rapidly combs unwanted roots out of the soil, leaving them in windrows or piles, as desired.

Disadvantages. The wheel root rake cannot remove the large roots that are frequently found in rain-forest areas. Soil, roots, and stumps must be loosened before using in order to avoid serious bending or breakage.

Wake Rake

Description. The pull-type wake rake, available from the Darf Corporation (Edenton, North Carolina), has spinning wheels that literally sweep the topsoil clean of all light debris. Attached to the drawbar of a track-type tractor, it can clear the soil at speeds up to 5 miles per hour.

Application. The wake rake is useful in final cleanup for picking up small loose pieces of stumps, roots, limbs, and unburned residue.

Heavy-Duty Clearing Rake

Description. The heavy-duty clearing rake is a solid weldment of high-strength, lightweight steel. Features of this rake are recessed mounting points for good balance, a high frame that carries a large load, and wear tips that can be replaced with a hammer and punch.

Application. The heavy-duty clearing rake is available for most Caterpillar track-type tractors. It can be used for medium to heavy-duty land clearing, including the removal of trees and stumps. This rake is especially well suited for clearing mountainous terrain.

Rake Summary

Care should be taken in determining the application and the correct rake to fit that application. Picking the wrong rake can be a costly mistake because of the high wear and low production rates.

Rakes are almost universally recommended for repiling burned or burning material. Since the ash residue and soil can sift through the teeth, cleaner, hotter burns can be achieved. A summary of the rakes discussed and their application is presented below:

PRIMARY RAKES

Type of Rake	Application
Multiapplication rake	All heavy-duty land clearing including tree and rock removal
Rock and root rake	Clearing and grubbing boulders, spreading rock riprap, and grubbing and piling tree stumps and debris
Blade rake	Light grubbing, piling, and raking
Wheel root rake	Follow-up to the root plow for removal of unwanted debris
Heavy-duty clearing rake	Used in difficult land-clearing conditions

UTILITY RAKES

Type of Rake	Application
Clearing rake (for track-type and wheel loaders)	Tree pushing, grubbing, stumping, and clearing trees and boulders
Stacker rake (for track-type and wheel loaders)	Raking, piling, carrying, and loading downed debris
Clamp rake (for track-type and wheel loaders)	Clearing and grubbing trees, stumps, and boulders; raking, carrying, loading, and piling debris
Wake rake	Cleanup operations

Fleco Skeleton Rock Bucket

Description. The Fleco Skeleton rock bucket is designed so that small rocks and soil will sift out of the load through openings in the sides, back, and bottom. This heavy-duty bucket is made entirely of alloy steel. It is equipped with Cat-built tips, adaptors, and pins as standard equipment. This bucket is available for most Caterpillar wheel and track-type loaders.

Application. The Skeleton rock bucket is used to size, separate, and handle rock.

Advantages. By filtering fine material out of the load, this bucket eliminates the need for screening equipment. Eliminating fine material during loading provides maximum utilization of handling equipment.

Disadvantages. The Skeleton rock bucket cannot be used to handle or load small rocks, gravel, or soil.

Fleco Detachable Stumper with Splitter

Description. The Fleco detachable stumper is a one-piece manganese carbon-steel casting for use on the C frame of angle dozers. The curved face facilitates penetration in hard soil and cradles the stumps for removal. For larger stumps, the detachable splitter is available as an attachment. It is welded to the left side of the standard stumper. The detachable stumper with or without splitter is available on special order for most Caterpillar track-type tractors.

Application. This implement is designed for

ing and tree-removal operations. It can used as a utility ditcher.

Advantages. Power of the tractor is concentrated at one focal point. As a specialized tool, the Stumper does one job well.

Disadvantage. The stumper is limited in its use.

Root Plows

Description. Root plows made by Fleco Corporation mount on the tractor trunnions and are complete with overhead lift frame, sheaves, hydraulic group, and auxiliary trunnions. Spreader boxes are flanged in two places to speed assembly and disassembly of the unit. Both vertical standards are equipped with replaceable wear shins. Bolt-on cutting edges are reversible and hard-surfaced. Optional depth of the root plow is controlled by a depth-setting arrangement with seven or more optional adjustments. This root plow is available for most Caterpillar track-type tractors.

Rome Industries' root plows consist of a rear trunnion-mounted frame with a horizontally mounted knife-type moldboard. This attachment moldboard is pulled through the ground by a tractor at a depth of 8 to 18 inches (203 to 457 millimeters). Wedge-type adjustment allows the operator to make quick and easy moldboard settings. The throat clearance from this moldboard to the crossbeam allows brush and debris to flow through the plow easily. The Rome root plow is also available for most Caterpillar track-type tractors. (See Figure 2-18.)

Application. The root plow is designed for killing brush and growth by undercutting vegetation at the crown or bud ring. Large roots are forced to the surface by fins welded to the horizontal blade. Root plows also shatter hard surface crusts and hardpan, resulting in better water retention as well as preparing a good seedbed. An ideal use for root plows is in the restoration of range lands or irrigated crop land like that found in semiarid regions.

Advantages. Cuts roots below the bud ring. Depth is easily set and controlled.

Disadvantages. Will not work well in sandy or wet soils. Not applicable to large trees.

FIGURE 2-18 Rome/Holt root plow with over-the-cab hydraulics.

Tool Bar Root Plow

Description. The root plow for a tool bar mounts on the 4½-by-7½-inch (114-by-190-millimeter) tool-bar beam on the Cat D4E through D6D track-type tractors. Depth of cut is controlled by adjustment of the tool-bar screw jacks. This is the only difference from the standard root plow. Applications, advantages, and disadvantages are the same.

Fleco Rolling Choppers

Description. Fleco rolling choppers are available as a single unit or three combinations. The drum of the chopper, which is normally filled with water to increase its weight, has welded-on cutting blades that can penetrate 6 to 10 inches (152 to 254 millimeters) deep. Multiple-drum choppers have swivel assemblies that connect the drums. Optional equipment available includes a spring-loaded tongue to reduce shock loading, a hinged tongue that swings up to reduce the overall length of the chopper for transporting, reversible blades with backup plates, and roller bearings. The choppers are available for the Caterpillar D4E through D9L track-type tractors. (See Figures 2-19 and 2-20.)

Application. The rolling chopper offers a fast, economical method of controlling undesirable growth. While operating the tractor in second and third gear, the chopper blades can fracture and shatter growth. Soil crust is furrowed and loosened with minimum topsoil disturbance. Applications

FIGURE 2-19 Fleco D7G rolling chopper.

FIGURE 2-20 Fleco D7G rolling chopper.

include preparation of forest sites, brush control on rights-of-way, clearing for reservoirs, preparation of seedbeds, and pasture renovation and maintenance.

Advantage. Properly matched to the application, this implement can be a very inexpensive clearing tool.

Disadvantage. Large-diameter trees cannot be chopped with this tool.

Rome Disk Harrows

Rome Industries manufactures two basic types of harrows—gang and offset. Many configura-

tions are available for each type. Descriptions applications of each harrow will be given individually. (Harrows that are used primarily for seedbed preparation will not be discussed.) Since there are so many harrows to discuss, the advantages and disadvantages of harrows in general will be given first.

Advantages. Harrows offer a fast and economical method of mixing organic matter with soils, leveling stumped areas, and speeding decomposition of embedded material.

Disadvantages. Harrows cannot be used to bring fines to the surface as does a moldboard or disk plow. Eradication of noxious grasses is difficult.

Rome TACH Harrows

Description. This harrow is very similar to the Rome Series TRCH except that it is designed for use with lower-horsepower tractors. It features extra-high gang-carrier construction to prevent clogging.

Models are available with 10, 12, and 16 disks. Hydraulic angling control may be used for all models.

Boxed angle design imparts strength and rigidity to the TACH. Tapered roller bearings and overhead knife-type scrapers are standard.

Rome Series TACH harrows provide level, clog-free plowing and thorough mixing. (See Figure 2-21.)

Application. The TACH harrows provides

FIGURE 2-21 Rome TACH 10-32 harrow and Cat D6D.

thorough mixing and aeration to a maximum depth of 9 inches (228 millimeters). It is particularly useful for incorporating heavy trash and litter into the soil.

Rome MR Harrow

Description. The Rome "Master," as this series is commonly called, comes with 28- or 30-inch (711- or 762-millimeter) notched disk blades, in groups of 8, 10, or 12 blades for the bush and bog configuration. A tandem model with 16 blades is available. Standard equipment includes mechanical angling control, scrapers, and weight box (bush and bog only). Tapered roller or white hard iron bearings with heavy-duty gangs for severe applications are available on all models.

Heavy-duty gangs, hydraulic angling control (8-disk models only), and thicker blades are available as options. From 40 to 130 drawbar horsepower is required, depending on the model desired and the width of cut of that particular harrow.

Rome TRH Disk Harrow

Description. The series TRH has groups of 14 to 28 disks for a cut of 7¾ to 15 feet (2.2 to 4.6 meters). Standard disks are 30 inches (762 millimeters) in diameter and ⅜ inch (9.3 millimeters) thick. Optional blades are 30 by ½ inches (762 by 12.7 millimeters), 32 by ⅜ inches (812 by 9.3 millimeters), and 32 by 1½ inches (812 by 38 millimeters). Special high-tensile alloy-steel axles are available and are recommended for severe conditions such as forest-site preparation and newly cleared land. Depending on the size of harrow used, 65 to 175 drawbar horsepower is needed.

Application. This series is built for heavy service in breaking new ground and forest-site preparation. It is in between the TAH and TRCH harrows in weight per disk and toughness of application.

Rome TRCH Harrows

Description. Rome Series TRCH harrows have a high ratio of weight per disk, high clearance, and mechanical, hydraulic, or cable angling controls. Width or cut ranges from 7 to 12 feet (2.1 to 3.7 meters) with 10 to 16 disks that are 36 inches (914 millimeters) in diameter and ⅜ inch (9.3 millimeters) thick. A 36-by-½-inch (914-by-12.7-millimeter) version is also available. From 85 to 175

drawbar horsepower is required, depending on the model used.

Application. The TRCH harrow is used for plowing, pulverizing, and mixing vegetation into the soil in one operation. It is designed especially for difficult soil conditions requiring good penetration, deep tillage up to 13 inches (330 millimeters), and extra-high frame clearance to prevent clogging.

Rome TYMH Harrows

Description. The TYMH is available in only one size, which gives a 10-foot (3-meter)-wide cut with 10 disks that are 50 inches (1270 millimeters) in diameter and ¾ inch (19 millimeters) thick. It has hydraulic controls and a high-clearance design. Drawbar horsepower required is 165 to 200.

Application. The design of the TYMH is for penetration down to 19 inches (482 millimeters) and maximum trash-handling capacity. It is ideal for deep plowing and tillage through heavy trash and crop residue.

Rome TYH Harrows

Description. In one design configuration, the TYH has 16 disks that are 36 inches (914 millimeters) in diameter and ⅓ inch (8.4 millimeters) thick, with a cutting width of 9½ feet (2.9 meters). Standard equipment includes replaceable hardened bushings on all pivot points and weight box. Required drawbar horsepower is 150 to 180 (See Figure 2-22.)

FIGURE 2-22 Cat D8H pulling Rome TYH 16–36 harrow.

Application. The TYH is designed for the most severe land-clearing jobs requiring equipment of maximum strength. It is built for clearing land, breaking newly cleared land, or wherever both deep penetration and maximum strength of equipment are necessary.

Rome TRW Harrows

Description. The TRW comes equipped with wheels for transportation and depth control. It has an 11- or 13-foot (3.4- or 4.0-meter) cutting width, with either 20 or 24 disks that are 30 inches (762 millimeters) in diameter and ⅜ inch (9.3 millimeters) thick. Required drawbar horsepower is 95 to 155.

Application. The TRW harrow is used for plowing, pulverizing, and mixing vegetation into the soil in one operation.

Fleco Hydraulic Tree Shear

Description. The tree shear utilizes hydraulic action to shear softwood trees up to 30 inches (762 millimeters) in diameter and hardwood trees up to 22 inches (558 millimeters) in diameter in less than a minute. A "kicker" mounted on the shear throws the log butt up so that the treetop hits the ground first. The shear is available on Caterpillar track-type tractors, track-type loaders, and wheel loaders. (See Figures 2-23 through 2-25.)

Application. Primarily designed for harvesting trees for pulpwood, the tree shear could be used to cut trees in any location, wherever it is economical.

FIGURE 2-23   Fleco 910 tree shear.

FIGURE 2-24   Fleco 910 tree shear.

FIGURE 2-25   Fleco 931B tree shear.

Advantages. Can fell trees quickly when applied correctly. Provides directional felling, which in turn speeds choker setting and skidding. Shears trees at ground level to provide increased wood volume.

Disadvantages. Capabilities are limited by tree size and composition. Cuts only trees; if underbrush has to be removed, another implement must be used.

Rome Series SS and SH Grapple Shears

Description. The Rome grapple shear is designed to fell, skid, and bunch with one machine. It features in-line directional felling with virtually no wood fracture. It will shear trees up to 20 inches (508 millimeters) in diameter, leaving stumps nearly flush with the ground. Models are available for use with either hardwood or softwood.

The Rome grapple shear uses a guillotine shearing principle to provide maximum shearing speed and efficiency. The straight-line cut provides positive directional control of felling. Simple controls provide for ease of operation.

The shear is mounted on the front of Caterpillar track-type and wheel loaders.

Application. The main purpose of the grapple shear is harvesting salable timber.

Advantages. The grapple shear provides rapid felling and bunching of trees that are to be hauled off for use as lumber, plywood, or pulpwood.

Disadvantage. Felling capability is limited to trees of 20 inches (508 millimeters) in diameter.

Ideas to Consider When Equipping Clearing Machines

After the machines and clearing tools have been selected, they must be properly equipped to handle various clearing operations effectively.

In operations such as felling large vegetation and piling where frequent direction changes occur, a power-shift transmission is recommended. In constant-speed operations such as felling smaller vegetation, root plowing, and harrowing, a direct-drive transmission is desirable. If a tractor is to be used in more than one operation, its predominant use should dictate the transmission choice. Under a constant-load operation, a direct-drive transmission is preferable.

In large vegetation, a land-clearing tractor is subject to falling trees and severe impact loads.

In a clearing operation, tractors and operators must be well protected. A heavy-duty cab guard is essential. It should withstand any vegetation that might fall on the tractor and have support and deflector bars extending forward to protect the engine compartment.

Heavy screening should be installed around the cab to protect the operator from brush. Inside-mounted household-type screens can further protect the operator from painful and harmful insect stings. Other desirable protection includes extra-heavy radiator, crankcase, track-guiding, and fuel-tank guards.

Provision for adequate engine cooling is a must

FIGURE 2-26   Cat D8K/Remco cab guard with optional screens.

FIGURE 2-27   Rome feller-buncher on screened Cat 930.

in any type of clearing operation. Special attachments such as perforated steel hoods and side panels have been designed with this in mind. Perforated steel side panels consisting of hinged inspection doors allow cleaning of debris around the engine and the radiator core. A reversible blower fan may be advantageous, depending on job conditions. An air-breather extension running from the hood to the operator's compartment is recommended, especially when working in leaves or trash. Such a modification will ensure that the air breather is not plugged and that the engine is receiving an adequate supply of air. The hydraulic system, including cylinders, lines, and tank, should be well protected with guards. Each machine should also be equipped

FIGURE 2-28 Cat 931B and Rome SJ100 shear.

FIGURE 2-29 Rome A-frame grapple on Cat 518.

FIGURE 2-30 Cat 518 stretch frame with Rome CG-405C 118-inch grapple.

FIGURE 2-31 Rome/Holt towed rake and Cat D8 tractor equipped with Rome hydraulic protection group and grille guard.

with a fire extinguisher or water tank for putting out small fires that may occur in the engine compartment.

METHODS OF LAND CLEARING

Land clearing can be accomplished by the following methods, each of which will be discussed in this section.

1. Complete removal of trees and stumps by physically uprooting and moving them to piles for disposal by burning or other means

2. Shearing the vegetation at ground level with sharp cutting blades and piling into windrows or piles for burning. The stumps and roots can also be removed, left in the soil to decay, or shattered in subsequent operations by root plowing or harrowing.

3. Knocking all vegetation down and crushing it to the ground for later burning in place

4. Plowing and chopping the vegetation into the top 6 to 8 inches (152 to 203 millimeters) of the soil in a once-over plowing operation. This allows the vegetation to decay.

For each of these general methods for clearing land there are several types of equipment that may be used. This section will deal with the equipment in these specific applications and will discuss the advantages and disadvantages of each.

The first and second methods in the list are similar in that the woody material is not only knocked down but is also removed from its site of growth and piled for burning. These two types of clearing differ in that uprooting depends on overpowering the vegetation with massive brute force, while shearing operates on the principle of shaving the material at ground level with a sharp cutting edge.

Methods 3 and 4 are similar in that the felled material is not piled. They are different in that the organic material of one is disposed of by burning and the other by decaying in the soil.

It should be noted that burning is not listed as an initial land-clearing method. It is difficult to get a good burn and control it in any type of vegetation without some previous preparation. Burning also removes organic matter that might help improve soil fertility. Burning, nevertheless, has been used effectively in many areas as a method of disposal for knocked-down brush left in place or windrowed. Burning as a method of disposal will be discussed in detail later in this chapter.

Uprooting the Vegetation by Brute Force

Several types of equipment are used in this kind of clearing operation. These may include bulldozer blades, rakes, knock-down beams, or chains drawn between two or more large track-type tractors.

Bulldozing

Early attempts to clear land mechanically involved simple modifications of earthmoving techniques using standard earthmoving equipment—the crawler tractor with its ordinary straight or angling bulldozer. Bulldozer blades are still used throughout the world even though it has been repeatedly shown that 30% to 40% more land can be cleared in a specified time with specialized land-clearing tools.

The bulldozer blade can be more economical in intermediate-size areas of upland woods and bush country when the size of the area to be cleared does not warrant purchase or rental of a specialized tool. Bulldozer blades are available for all sizes of track-type tractors and many wheel-type farm tractors. The most economical power-unit size will vary with the amount and type of vegetation and the amount of land to be cleared.

Generally, the bulldozer blade is not an efficient land-clearing tool. When larger trees that cannot be pushed over are encountered, they must be dug out of the ground—a costly and time-consuming operation. Small trees and bushes are so limber that they bend down and the dozer blade passes over

them, or else they break off, leaving stubs protruding above the soil surface that might have to be removed later. Valuable topsoil may be removed in many instances when a bulldozer blade is used to windrow brush. Since a ball of dirt is often left on the roots of the tree when it is uprooted, subsequent burning is made more difficult.

Raking

As previously shown, various types of rakes are being manufactured and sold for clearing land of both trees and rocks. Rakes have the advantage of permitting the soil to pass between the teeth as it is pushed through the soil, ripping out and pushing the rocks, stumps, brush, and other vegetation. Rakes work best in extremely sandy soil such as that encountered in northern Florida and southern Georgia. Rakes sometimes do not work well in clay soils or wet soils because of clogging between the rake teeth. When this happens, the rake is in effect converted into a bulldozer blade. In most cases, a good operator can overcome this effect.

Rakes are used successfully and are almost universally recommended for repiling burned or burning material. The ash residue sifts through the teeth, and cleaner, hotter burns can be achieved.

Tractors used in the burning operation should always be equipped with a blower or reversible fan (and fire extinguishers of at least 5-gallon capacity) to prevent the ash and live sparks from being sucked under and into the tractor. This is necessary both from the standpoint of safety and operator comfort. Also, a blower fan promotes better burning by forcing large volumes of air into the fire.

Tree Pushing

Both bulldozers and rakes depend on the brute force of the tractor to accomplish their objectives. This principle is in part applied with the tree pusher. It is a structure extending above and forward from the tractor, giving the tractor added leverage in pushing over larger trees. The size of tree that can be felled is dependent on the size and weight of the tractor, leverage, tractive conditions, the soil, and the tree's root system. Tree pushers are ideal when used in combination with some device for cutting the roots around the larger trees. This allows the trees to be pushed over with greater ease. They are also used in conjunction with chaining to lift the chain higher on larger trees for increased leverage or to assist the chain against larger trees.

Even though the tree pusher may be considered a highly specialized tool with limited applications, it has proved to be very efficient in many areas when working with other felling methods.

Chaining

The chain is dragged behind two crawler tractors. The outside tractor travels along the edge of the uncleared area. The inside tractor travels through the uncleared area, avoiding any large vegetation to be left or vegetation it cannot knock down. The distance between the two tractors will vary with the size of the tractors and the vegetation. The tractors should be close enough to allow travel in an almost continuous forward direction. The chaining passes should be made as long as possible to minimize tractor maneuver time.

When chaining areas of unknown terrain, a walker may be required to go in front of the inside tractor to warn of any obstructions or depressions that might hinder tractor operation.

Two passes in opposite directions may be required to uproot smaller vegetation as found in semiarid areas. The need for this will vary with later clearing operations and the end use of the land.

Chain clearing is most economical in arid or semiarid types of vegetation where only limited or no undergrowth occurs. Population of woody species of all sizes should not exceed 1,000 plants per acre (2,500 per hectare). However, experience has shown that chaining can be used in all sizes of vegetation. The upper limit in size and density will vary with the size of the tractors used and the width of the area chained.

Chaining can be very difficult when extremely heavy undergrowth conditions are present because they reduce the operator's visibility. This in turn impedes the operator's ability to maneuver the tractor around larger trees.

The terrain should be well drained, level to gently sloping, and without large gullies, stone outcroppings, or other obstructions that prevent free passage and maneuverability of the tractors and chain.

Root Plowing

The root plow is another tool for removing vegetation below the soil surface. It is designed to kill brush and growth by undercutting vegetation at the crown or bud ring. Large roots are forced to the surface by fins welded to the horizontal blade. Root plows also shatter hard surface crusts and hardpan, which results in better water retention and prepares a good seedbed.

Root plows are available for mounting on tractor trunnions or tool bars. The vertical standards are equipped with replaceable wear shins, and bolt-on cutting edges are reversible. The root plow is normally operated in the "float" position, and depth is controlled by a depth-setting arrangement with several adjustments. It operates at depths from 8 to 20 inches (203 to 508 millimeters) and is available in a number of sizes from several manufacturers.

One advantage of the root plow is that it cuts the vegetation below the bud ring, killing brush that would normally resprout if cut at ground level. Since depth is easily set and controlled, it is easy to operate and does an effective job.

The main disadvantages of the root plow are that the size of vegetation it can handle is limited, and it does not work well in sandy soils.

Grubbing

A variation of the root plow, generally smaller and mounted on track-type or wheel tractors, is the grubber. The grubber is for use in medium to light brush of average density where the tractor can move from one plant to another. The grubber can be a highly efficient tool in areas suitable for its application.

Cutting the Vegetation at Ground Level

This method varies from those previously discussed in that the vegetation is cut off at or slightly above ground level, leaving the stumps in the ground to decay or to be removed later. The tools and their use for this type of clearing probably vary more than any other method. They range from the use of a band ax or machete to the use of a large tractor with a cutting blade.

The main advantage of this method is that if the stumps can be left in the ground, the initial clearing cost is decreased considerably. It is especially effective in larger vegetation where the brute force of the tractor is not enough to uproot the vegetation without digging around it. The topsoil is left undisturbed (important in areas where it is thin), and the stumps will rot in place. This method is not, of course, suitable for areas to be under buildings or elsewhere where firm bearing soil is required.

Hand Clearing

Clearing with hand tools is probably the oldest and most widely used method of clearing. Hand tools are adequate for small areas that do not warrant investment in mechanical equipment. Their economical use in larger areas will be affected by the availability of labor versus capital and the degree of clearing desired.

Single- or double-bitted axes can be used to cut most top growth. They become less efficient in very small or very large growth. Axes can also be used as an aid in grubbing roots. They must, of course, be properly sharpened to be effective.

Machetes can be used on smaller stems and branches that are commonly cut with an ax. They are especially effective in the underbrushing that precedes tree felling during the hand clearing of jungle areas. Machetes can be sharpened with a whetstone or file.

Brush hooks are useful in cutting small vegetative growth. The brush hook is swung like a scythe. It is sharpened by grinding with an abrasive wheel or emery.

Grub hoes and mattocks can be used to chop off small brush near ground level or to dig out small roots. They are not effective in large vegetation.

Power Sawing

In larger vegetation, chain saws are more efficient than axes or similar hand tools. Most chain saws are available in lengths from 1 to 5 feet (0.3 to 1.5 meters). They are generally powered by two-stroke gasoline engines. They leave a stump about ground level that hinders later clearing and land-use operations. Chain saws are most economical in intermediate to large vegetation in small to medium-size areas and in pruning felled material.

Sickle Mowing

Light brush, with a stem diameter of less than 1.5 inches (3.8 centimeters), can be cut with regular farm-tractor mowers adapted for heavy-duty operation. These mowers have short and heavier sickle bars equipped with stub guards and extra hold-down clips. The tractor is run in low gear to give a high sickle speed in comparison to the forward speed. Larger-diameter brush can be cut, but continuous forward direction is not usually possible. This equipment is usually most economical in intermediate-size areas.

Blade Shearing

Perhaps the most efficient land-clearing tool for medium and large vegetation in intermediate to large areas is a shearing blade on a track-type tractor. The shearing-blade principle differs from the bulldozer principle of land clearing in that the total horsepower of the tractor is applied to a sharp cutting edge. The shearing blades are usually equipped with a stinger or wedgelike projection. This allows larger trees to be split in one or more successive passes before actually felling the trees with the cutting edge. By cutting and splitting the trees, much larger trees may be felled with a given size tractor. Later, burning is also faster and cleaner because the tree has been split and there is no root-ball on the butt of the logs.

Another advantage of shearing blades over bulldozer blades in land clearing is ease of operation. Clearing blades are equipped with a flat sole at the bottom of the ground to prevent digging in. This permits faster operation and less operator fatique because the operator is not constantly manipulating the controls to keep the blade from digging into the ground.

Two types of shearing blades exist: angle blades, like the K/G manufactured by Rome Industries, and the V-type blade manufactured by the Fleco Corporation.

The K/G blade is available for most track-type tractors. Blade angle is 30 degrees, and the blade can be operated either by cable or hydraulic control. The replaceable cutting edges and stinger should be resharpened with a portable grinder daily. A guide bar is used to control the direction the tress fall—forward and to the right of the operator.

This type of blade is a versatile attachment in that it can cut, pile, stump, and ditch. Cut growth normally falls in one direction, and given enough time, the K/G blade can fell any size of tree.

The disadvantage of the K/G blade is that it does not remove embedded material such as roots and rocks in one pass. When stumps are to be removed, the blade requires another pass over the area with the blade tilted. Rocks and stony outcroppings can severely damage the cutting edge of the blade and should be avoided.

The V blade is available for track-type tractors through the D8L size. It is equipped with a heavy-duty splitter, angled serrated cutting edge, and brush rack. This blade mounts directly on the tractor trunnions and is available for cable or hydraulic control. The blade is in two sections and bolts together to form the working tool.

When properly matched to the job, the V blade is a high-production tool that maintains a continuous path. The bolt-together design permits assembly or disassembly on the job and provides flexibility in moving the unit.

In some jobs with the V blade, it is a disadvantage to have trees fall to both the right and the left of the tractor. The V blade, like the angle blade, does not remove embedded material such as roots and rocks, and it does not usually remove stumps in one pass. Unlike the angle blade, the V blade cannot pile or ditch. This is especially important in smaller jobs where only one tractor is used for the entire clearing operation.

Where land use requires the immediate removal of the stumps left in the ground by shearing blades, a second operation is performed. The stumps may be shattered with very heavy duty offset disk plows, removed with the stinger of the Rome K/G blade, or grubbed out with a rake or stumper.

The use of shearing blades of whatever type should be limited to heavier clay and loam soils relatively free of stones. Under extremely sandy conditions, the clearing can best be done with rakes or other clearing tools.

All tractors with shearing blades should be equipped with heavily constructed cab guards for protection against falling trees. In certain areas, an insectproof compartment should be provided to protect the operator.

Tree Shearing

Another land-clearing tool, primarily designed for harvesting trees for pulpwood, is the tree shear. This tool utilizes hydraulic action to shear softwood trees up to 30 inches (760 millimeters) in diameter or hardwood trees up to 22 inches (558 millimeters) in less than a minute. A "kicker" mounted on the shear throws the log butt up so that the treetops hit the ground first.

The tree shear can fell trees quickly at ground level and provides directional felling. The disadvantages of the tree shear are that its capabilities are limited by tree size and composition, and it only cuts trees. If underbrush has to be removed, another implement must be used.

Knocking the Vegetation to the Ground

Knocking the vegetation to the ground is an inexpensive way of clearing land, but unfortunately its successful application is limited to specific situations. It is highly effective in clearing bush-type vegetation and smaller upland woods where the diameter of the woody vegetation does not exceed 6 to 8 inches (152 to 203 millimeters). In rain-forest areas, smaller vegetation may be knocked down while larger vegetation is uprooted.

Since much of the vegetation is not uprooted and is sometimes chopped up, it is normally not piled. It may be burned in place at a later time, left in place to decay, or incorporated into the soil. If it is to be burned in place, the vegetation must be sparse enough to allow native grasses to grow intermittently because the grasses are required to provide fuel for the burn.

Several types of equipment are used to knock the vegetation down and leave it or chop it up into smaller pieces.

Rotary Mowing

Rotary mowers pulled by farm tractors can be used in vegetation up to 4 inches (102 millimeters) in diameter. They have one or more revolving blades, rotating around a vertical shaft and powered by tractor power takeoff. These blades sever the vegetation at or near ground level and shred it into small pieces. Rotary mowers are available in several sizes and are most efficient in small to medium-size brush in intermediate-size areas. Their use is not recommended in hilly terrain or where rocks and stones are present. Extreme tractor-tire damage can be expected when recrossing areas moved if vegetation is of the type that leaves spikes in the ground.

Flail-Type Cutting

Flail-type rotary cutters, which are also tractor-drawn, have cutting knives that rotate around a horizontal shaft to knock down and shred small brush at or near ground level. These cutters are also available in a wide range of sizes as power takeoff units for both wheel-type and track-type tractors. They are most efficient in small to medium-size brush in intermediate-size areas.

Chopping

In larger areas, this type of clearing is generally done with rolling choppers. The drum of the chopper, which can normally be filled with water to increase its weight, has welded-on cutting blades that can penetrate 6 to 10 inches (152 to 254 millimeters) deep. These blades cut, fracture, and shatter growth. Soil crust is furrowed and loosened with minimum topsoil disturbance. The woody vegetation is left in a thick flattened mat.

Rolling choppers are available in a number of combinations, both self-propelled and pulled by track-type tractors. They come in several weights per length of cut. The tractor-drawn models are available in single, tandem, and dual combinations in a wide range of sizes.

The main disadvantage of this tool is that large-diameter trees cannot be chopped. Another problem is that for the vegetation to be burned in place, it must be left for a long period of time to dry thoroughly.

Many thousands of acres are cleared every year simply by the use of a tillage implement pulled behind a tractor. The implement cuts and chops the material into the upper 6 to 10 inches (152 to 254 millimeters) of the soil. This method is limited to situations where the vegetation does not exceed 3 or 4 inches (75 to 102 millimeters) in diameter. There should be adequate rainfall to promote rapid decomposition of the material chopped into the ground. The soil surface should be free of large protruding stumps and of stones or stone outcroppings that would limit the effectiveness of the disks in chopping the woody vegetation.

Incorporating the Vegetation into the Soil

Moldboard plowing. Pull-type, semimounted or full-mounted moldboard plows drawn by wheel or track-type tractors can be used to turn and cover small brush if the soil is not excessively hard, sticky, or rocky.

For all but the lightest plowing, plows with heavy-duty frames should be used.

Jointers and coulters should be removed, and covering wires may be added to help cover the trash. Plowshares should be kept sharpened for best performance. Plows are available in a wide range of sizes from many farm equipment manufacturers.

Disk plowing. Tractor-drawn disk plows can also be used effectively in covering small brush. They can be used on soil that is dry, hard, sticky, or rocky, where moldboard plows are generally not recommended. However, disk plows do not normally cover brush as thoroughly as moldboard plows. Variations of the standard disk plows have been developed that incorporate a "stump-jump action" whereby each disk is independently adjustable and spring-loaded to permit it to ride over any fixed obstruction it may encounter. They are available in sizes that can be pulled by track-type or wheel tractors.

Offset Harrowing and Gang Harrowing

For slightly larger brush, heavy-duty offset and gang harrows pulled by track-type tractors equipped with bulldozer blades can be effective land-clearing tools. The vegetation is bent over with the dozer blade and then cut and chopped into the soil by the harrow. In smaller brush, the dozer blade is not needed.

Offset harrows are available in many types and sizes. Care should be taken to select the type and size most suitable to the application for which it is intended.

Heavy-duty offset and gang harrows will not invert the soil layer and vegetative matter but will chop up this material and mix it throughout the soil profile, unlike the moldboard or disk plow.

All of these harrows and plows can be used to incorporate previously shredded or uncut vegetation into the soil. Larger stumps may damage conventional moldboard plows. Newer moldboard plows, with protective devices that allow each bottom to swing back when it hits an obstruction, can be used in areas with larger stumps. Under these conditions, the heavy-duty harrows are usually more effective because of their chopping action.

Once the vegetation has been felled, it must usually be disposed of in some manner. The method of disposal will be determined by a number of variables including type and size of vegetation, end use of land, rainfall, and terrain. Since disposal is often an expensive operation, care should be taken to select the most economical method.

Disposal of Vegetation after Initial Clearing

Leaving in place. If the variables affecting disposal permit elimination in place, this may be the most economical method available. With small vegetation that is chopped up during initial clearing, the vegetation may be left to decay or be incorporated into the topsoil. With larger vegetation, the end use of the land may permit scattered vegetation to be left in place. However, larger vegetation that cannot be incorporated into the soil must usually be disposed of in some other manner.

Burning in place. Burning vegetation in place after it has been knocked down can be an effective means of disposal when conditions permit. These conditions include fuel for burning in the form of grasses or small brush, a dry enough climate to allow burning, and an end use of the land where some larger trees can be left on the ground.

All brush can be burned more cleanly if a high proportion of the fuel is dry. For this reason, all brush should be prepared in some manner before it is burned. Crushing of light or medium material will usually cause it to dry enough during clear weather. A period of a few weeks may be necessary for proper drying.

Heavy jungle brush with many large trunks should often be left to dry 1½ to 2 months or longer before burning. Dryness is indicated when the bark on the larger trees cracks and the foliage has dried and begins to fall. (Dry foliage is, however, an excellent fuel and should be utilized, if possible.) It is sometimes necessary to plant grass in the felled areas and to delay burning until this grass has grown and dried sufficiently to serve as a fuel.

In small vegetation, chemical spraying can sometimes effectively prepare the vegetation for an efficient burn. Two chemical treatments can be used to prepare brush for burning: contact herbicides that kill only leaves and twigs and systematic herbicides that kill part of the stems as well as the leaves and twigs.

Contact herbicides should be used when the interval between spraying and burning is brief. The contact chemicals can be applied during clear weather only a few days before burning. Systematic herbicides can be applied at the height of plant growth in the spring to prepare the brush for burning the next fall or the following spring.

If weather conditions are suitable, burning can normally be accomplished during any season of the year. However, the season can affect the efficiency of the burn and the condition of the seedbed afterwards. In areas with wet and dry periods, burning may be possible only during the dry season, after the vegetation has had an opportunity to dry out. Care must be taken to ensure that surrounding areas are not so dry that the burn cannot be controlled. In areas with four seasons, where control is a problem, late spring and late fall are generally the best burning periods because the surrounding areas are higher in moisture content.

Weather conditions. The most important factor affecting burning is the weather. The burn should take place only after a reliable forecast of favorable weather. Unstable weather conditions should be avoided. In jungle areas, at least one week of continuously dry weather is desirable before the actual burn takes place. In bush areas, it is important to burn only in weather under which spot fires will not be started by embers flying across the control lines. Extreme caution should be exercised in setting burns in heavy winds.

Control lines. To control the in-place burn effectively, it is desirable to locate fire lines or belts around the boundaries of the area to be burned. In rain-forest areas, these belts may be of unfelled vegetation. In bush areas, they are previously cleared land. It is also necessary to have escape roads cut in the area to be burned for the safety of the burners.

Lines may divide the area into units as small as 10 to 20 acres (4 to 8 hectares) for difficult control conditions or as large as 2,000 acres (800 hectares) where control is not a problem. If possible, each unit should be a "natural burning unit" within which topographic features will aid the effectiveness of burning and the control of fire. If the units cannot be determined by terrain, they should be nearly square or rectangular, with the long axis parallel to upslope drafts or to the prevailing wind.

The units may be burned in a way to aid in the effective burning of the entire area. The choice of units may be determined in conjunction with the felling procedure.

Ignition techniques. In-place firing is normally carried out by burners who go through the area on foot with kerosene-soaked strips that are thrown into the felled brush. Closely spaced ignition points help build up the intense heat needed to obtain a clean burn.

To ignite an area, the firing crew starts at the upslope or downwind side of the area and sets closely spaced fires along the line. When the line can be held, the firing is speeded up until the many small fires pull together and burn as a single intense fire over a few acres. The firing progresses downslope or against the wind at the speed required to keep a small acreage burning, until the entire burning unit has been covered.

The spacing of ignition points and the number of people required to burn an area depend on vegetation and burning conditions. To jungle areas, it has been found that the maximum area one person can cover in one burn is about 10 acres (4 hectares). Ignition points may vary from a few feet (meters) to 100 feet (30 meters) apart, depending on conditions.

In areas of larger vegetation, it may be necessary to prune, stack, and reburn, according to clearing specifications, all timbers that are to be removed. This should begin as soon as it is possible to reenter the area. Front-mounted rakes can be used, or the vegetation can be cut into lengths suitable for carrying and stacking around large buttresses.

Safety recommendations. It should be remembered that burning can be safely conducted only by people who thoroughly understand fire behavior. They should observe the following recommendations:

1. Make sure the brush within the area to be burned is more burnable than the woody fuels outside the area.

2. Determine what weather conditions will allow efficient burning of the prepared area and yet will assure continuous control of the fire.

3. Burn areas large enough for an efficient burn but small enough that the burn is under control at all times.

4. Prepare a complete plan for the burn, have it approved, and follow it rigidly.

5. Delay burning until all desired conditions are met, regardless of the cost of rescheduling.

6. Provide for the safety of the crews at all times.

Piling

If vegetation is not disposed of in place, it is normally piled, then burned or left to decay. The equipment used for piling will often be determined by what equipment is used for felling. The angled shearing blades and clearing rakes discussed previously are quite suitable for piling. In small areas, piling of light material may be done by hand.

As mentioned earlier, one advantage of the K/G blade for clearing is its multiple application as both a felling and a piling tool. The cutting blade is equipped with a flat sole, which allows it to float on top of the ground without disturbing the topsoil. The stinger on most cutting blades can be used to lift the vegetation partially, making it easier to push. Because of the floating and lifting action of the blade, very little dirt is carried to the pile with the vegetation. This facilitates burning.

Clearing rakes also can be used for felling and piling. Since their principal felling application is in sandy or rocky type soils, they are often used for piling in those conditions. They are especially effective in removing some surface roots as they pile. Again, rakes are not recommended for wet or clay-type soils, which tend to build up between the teeth, causing a bulldozing effect. The resulting soil mixed with the vegetation may hinder later burning. Types and sizes of rakes were discussed in the section on machines and attachments.

Disposal of Piles

The method for disposal of piled brush and the location of the piles is determined by the size,

type, and density of the brush; the climate; rainfall; and the size and type of equipment used.

Leaving in low-lying areas. When swamps or other low-lying areas are present, it is sometimes desirable to pile brush in these areas. By leaving the brush to decay, further disposal is eliminated.

The main problem with this method of disposal is that drainage is often adversely affected. Insect and rodent breeding and the threat of disease may also be problems if such methods are used.

Piling in well-placed windrows. When no suitable low-lying areas are available and burning is not practical, vegetation is sometimes piled in well-placed windrows and left to decay. In areas of level terrain, the windrows may be parallel, facilitating piling and later operations. In hilly areas, the windrows may be left on the contour to facilitate piling and to help control erosion. The distance between windrows may be from 100 to 200 feet (30.5 to 61.0 meters). It will vary with the size and density of vegetation and the size of equipment used.

Piling in windrows for burning. In many situations, the best way to dispose of the vegetation is by burning. A clean and economical burn requires careful planning and close supervision. The windrows must be piled as compactly as possible, parallel to the prevailing winds, with a minimum of soil mixed with the vegetation.

Weather is the most important factor affecting the burning of piled brush, just as it is when burning in place. Unstable weather conditions should be avoided, and at least one week of continuously dry weather is necessary before the actual burn takes place.

The vegetation should be allowed to dry until the bark on the larger trees cracks and the foliage has dried and begins to fall. When vegetation dries in a pile, it will go through a period of heating. This is often a good time to burn the material, if other conditions permit, due to the added internal heat.

The number and timing of the ignition sets will vary with the type and size of the vegetation, the size of the pile, and the moisture content. Since the brush will be repiled to complete the burning, ignition timing is not so critical as with in-place burning. However, it is important that sets be close enough to ensure a complete initial burn in order to minimize repiling costs.

In certain clearing conditions, if the vegetation is very difficult to burn, it may be necessary to aid the burning with forced air and fuel or a combination of the two. The combination of air and diesel fuel allows burning of green material even during inclement weather. It should be remembered that the use of a burner is no substitute for adequate drying time. Costs will be increased and the burn may be less complete if conditions require the use of a burner.

Brush burners equipped with four-cycle engines, airplane-type propellers, and self-priming fuel pumps supply air at a high velocity and a continuous fuel spray for starting and maintaining fires. They are highly mobile and are a safe method of promoting a more complete burn of materials.

Repiling. After the initial burn, repiling should start as soon as the heat has subsided enough to permit a crawler tractor with a blower-type fan to approach the fire without damage. The piles should be stoked and the fire kept alive until the woody material is completely consumed by the flames. The initial heat should be used to get larger logs burning while the smaller material is piled on to keep them burning.

Windrows should be cut into segments as soon as possible, making round piles of burning debris. When one pile has lost most of its heat, it should be pushed to another pile to maintain the greatest possible concentration of heat and burning material and achieve quicker and more complete burns.

The best tools for repiling are tractor-mounted rakes, which allow the dirt and ashes to sift through the teeth. Shaking the rake up and down with the controls while moving forward with a load helps remove dirt. Dropping the load and picking it up again just before pushing it into the windrow also helps to sift the dirt and ashes through the teeth.

Other Methods of Disposal

Other methods of disposal include utilizing cut material for firewood and charcoal, hauling to other areas for disposal, and burying. All three methods are generally limited to small localized areas where only a small amount of vegetation is involved. The demand for firewood and charcoal is small in most areas. Burying or hauling the vegetation away is too costly for most land-clearing operations. A more economical method can usually be found.

Final Cleanup

Regardless of the method of felling and disposal used in clearing the land, loose pieces of stumps, roots, and limbs will be left on the ground.

There will also be some unburned debris left after even the most efficient burns.

This material can be disposed of in many ways. In many areas, the cheapest and most practical method is to pick up this residue by hand, then pile, burn, or haul it away in carts or trucks.

Mechanical tools used for final cleanup include rakes and spring-tooth chisels or tines mounted on wheel or track-type tractors. Rakes can be used in sandy soils to pick up the debris and allow the soil to sift through the teeth. A recent innovation for this type of operation has been the wake rake developed in Australia. The principle of this tool is literally to sweep the topsoil clean of all light debris at speeds up to 5 miles per hour. The final windrows in heavy material can be up to 5 feet (1.5 meters) high. Most mechanical tools are not recommended in wet or clay soils because of their bulldozing effect. The debris is piled, buried, hauled away, or burned.

Other Operations

Some land-clearing jobs may include further operations as a part of the specifications or contract. In areas with vegetation left in place, this may be no more than hand-seeding the area with grasses or legumes. In other areas, it may include leveling, plowing, and seeding.

Minor leveling operations can be done with bulldozers, shearing blades, disk harrows, or even chains. It is recommended that more extensive leveling be carried out under a separate operation because it is not generally considered land clearing as such and requires different types of equipment.

Plowing may be considered a part of the complete land-clearing job. Disk harrows are used to mix organic material into the soil. They split and chop embedded roots and stumps, speeding decomposition of these materials. They are also used to help prepare a seedbed for planting. A number of types and sizes of disk harrow are available from various manufacturers. They should be consulted to determine the size and type for a particular application. It is, however, imperative that the selected tool be of extra-heavy design and quality to withstand the loads and shock treatment it will be subject to.

After the first plowing, it is often desirable to plant a leguminous crop such as soybeans, velvet beans, or alfalfa on the newly cleared land. The legume plants fix nitrogen from the air, making it available in the soil for use by bacteria that aid in the decomposition of the roots. Legumes also shade the ground and inhibit resprouting of the chopped roots while decomposition takes place.

The need for any or all of these operations is determined by each specific clearing situation. No attempt is made here to recommend which, if any, should be carried out.

Rocks

Perhaps the most universal method of mechanical rock clearing is a crawler tractor with front-mounted heavy-duty rakes. Another type of rake is available on a track-type loader. The latter has the advantage of being able to pry out boulders and load them into some type of conveyance.

DETERMINING COST AND PRODUCTION

As stated before, it is extremely difficult to established specific rules of thumb for selecting land-clearing equipment. Too many variables are involved. Such factors as the end use of land, underfoot conditions, type of vegetation, and specifications directly influence the selection of equipment for any specific clearing job. The final decision must be based on good judgment and common sense.

It is equally difficult to estimate with a reasonable degree of accuracy the total production, per-unit production, and cost per acre (hectare) for a land-clearing operation. One excellent source of information both for machine costs (depreciation, repairs, etc.) and local costs (labor, fuel, etc.) is the equipment dealer servicing the project.

In addition to job-cruise samples, all possible data should be gathered from aids such as:

Aerial photo maps
General topography maps
Water table maps
Yearly rainfall figures

Before beginning production and cost calculations, all pertinent information should be acquired and tabulated as follows.

Acquire Known Information

1. Total acres to be cleared _____
   (acres)

2. Time available for job (in years) _____
   (years)

3. Time available in hours per year per tractor [machine availability[2] (%) X hours worked per day X working days per year]: _____
   (hours/year/tractor)

4. Acquire all information possible from each job cruise. Tabulate each one separately for later use. Use Table 2-1 as a guide.

TABLE 2-1

Tree Diameter	1'-2'	2'-3'	3'-4'	4'-6'	Above 6'
Number of trees per acre					
Percent that are hardwoods					
Vines present? (yes or no)					
Description of root system					
Description of undergrowth					
Description of soil					
End use of land					
Debris disposal method					
Soil conservation to be practiced					
Grade and terrain					
Water table conditions					
Rainfall					
Underfoot conditions					

5. Select general method for clearing land. _____

6. Decide on machine(s) to be used. _____

_____

_____

[2]Efficiency due to weather, transportation, repairs, etc., expressed in percent. These figures may be obtained from the equipment dealer.

7. Calculate local owning and operating (O & O) cost for each machine.

Machine	*O & O Cost*
_____	_____
_____	_____
_____	_____
_____	_____

8. Determine implement(s) or attachment to be used and calculate local owning and operating (O & O) cost for each.

Attach-ment	*O & O Cost*
_____	_____
_____	_____
_____	_____
_____	_____

9. Estimate working speed (in miles per hour) for these operations:

	Machine	*Speed*
Harrowing	_____	_____
Root plowing	_____	_____
Rolling chopper	_____	_____
Chaining	_____	_____

Calculating Production and Costs

Now that the preliminary material has been gathered, it is possible to proceed with the calculations necessary to arrive at an estimate of production and costs. (They are estimates because much of the material that goes into the calculations are estimates. Consequently, the final figure will only be as accurate as the numbers used.)

To eliminate confusion, the following section has been organized in this manner:

Cutting	Production
	Costs
	Additional calculations
Piling	Production
	Costs
	Additional calculations
Harrowing	American Society for Agricultural Engineers (ASAE) formula for production
	Fleco formula for production
	Cost

Root plowing, rolling chopper	Procedures
Burning	Production
	Costs
Chaining	Production
	Costs
Conclusion	Total costs

Each step under production and costs that is used to determine a final figure has a letter designation. Some steps may be eliminated or used as necessary. For example, if root plowing, raking, and burning are to be done, the cutting, chaining, and harrowing calculations can be eliminated. A number used as a reference indicates acquired information; a letter or double letter indicates the actual calculation.

Cutting

Production

A. Using Rome's formula for cutting, calculate total time (in minutes) for cutting per acre:

_____
(minutes/acre)

B. Divide 60 (minutes) by A:

_____
(acres/hour)

C. Divide total number of acres to be cleared by B (see E_1):

_____
(hours required for cutting)

D. Divide C by total number of years for the job:

_____
(hours/year)

E. Divide D by number of hours per year per tractor (item 3 under "Acquire Known Information"):

_____
(number of tractors required)

E_1. If more than one tractor is required (E), divide C by E to get actual hours required for cutting per machine: _____

(hours required for cutting machine)

Costs
F. Machine owning and operating cost (O & O cost × number of machines): $ _____ /hour

G. Attachment owning and operating cost (O & O cost × number of attachments): $ _____ /hour

H. Add F and G: $ _____ /hour

I. Multiply H by C for the total cost of the cutting operation: $ _____

Additional Calculations
Divide I by

total number of acres (to find cost per acre),
total number of hours (to find cost per hour),
total number of weeks (to find cost per week),
or any other cumulative total such as months, years, etc.

Piling

Production
J. Using Rome's formula for piling, calculate total time (in minutes) per piling per acre: _____

(minutes/acre)

K. Divide 60 (minutes) by J: _____

(acres/hour)

L. Divide total number of acres to be cleared by K (see N_1): _____

(hours required for piling)

M. Divide L by total number of years for the job: _____

(hours/year)

N. Divide M by number of hours per year per tractor (item 3 under "Acquire Known Information): _____

(number of tractors required)

N_1. If more than one tractor is required (N), divide L by N to get actual hours required for piling per machine: _____

(hours required for piling machine)

Costs
O. Machine owning and operating cost (O & O cost × number of machines): $ _____ /hour

P. Attachment owning and operating cost (O & O cost × number of attachments): $ _____ /hour

Q. Add O and P: $ _____ /hour

R. Multiply Q by L for the total cost of the piling operation: $ _____

Additional Calculations
Divide R by

total number of acres (to find cost per acre),
total number of hours (to find cost per hour),
total number of weeks (to find cost per week),
or any other cumulative total such as months, years, etc.

Harrowing

Production—ASAE Formula
S. Speed (in miles per hour) of machine pulling implement: _____

(miles/hour)

T. Multiply S by width of cut (in feet) of the implement: _____

(acres/10 hours)

U. Divide T by 10: _____
                                                                            (acres/hour)

V. Metric equivalent:
Multiply U by 0.405: _____
                                                                            (hectares/hour)

W. Divide total number
of acres to be harrowed
by U to determine hours
required for one complete
harrowing: _____
                                                                            (hours)

X. Multiply W by num-
ber of harrowings for
total hours required for
harrowing: _____
                                                                            (hours)

Production—Fleco's Formula
Y. Working width (in
feet) of implement: _____
                                                                            (feet)

Z. Multiply Y by 10%
and add this to Y: _____
                                                                            (feet)

AA. Multiply Z by speed
(in miles per hour) ma-
chine with implement: _____
                                                                            (acres/10 hours)

BB. Divide AA by 10: _____
                                                                            (acres/hour)

CC. Metric equivalent:
Multiply BB by 0.405: _____
                                                                            (hectares/hour)

DD. Divide total number
of acres to be harrowed
by BB to determine hours
required for one complete
harrowing: _____
                                                                            (hours)

EE. Multiply DD by num-
ber of harrowings for total
hours required for harrowing: _____
                                                                            (hours)

Costs
FF. Machine owning and
operating cost (O & O
cost X number of ma-
chines): $ _____ /hour

GG. Implement owning
and operating cost (O & O
cost X number of imple-
ments): $ _____ /hour

HH. Add FF and GG: $ _____ /hour

II. Multiply HH by EE
or X for the total cost of
harrowing: $ _____

Again, any cumulative total can be divided
into II to arrive at a per-unit cost.

Root Plowing, Rolling Chopper

For calculating production or cost for either
of these operations, use ASAE's or Fleco's production formula for harrowing and the cost calculations immediately following them.

Burning

Production
JJ. Estimate production
of machine[3] with imple-
ment in hours/acre: _____
                                                                            (hours/acre)

KK. Estimate acres to be
burned: _____
                                                                            (acres)

LL. Multiply JJ by KK
to get approximate hours
required for burning: _____
                                                                            (hours)

Costs
MM. Machine owning and
operating cost (O & O
cost X number of ma-
chines): $ _____ /hour

---

[3] This is the machine that will be tending the fire; production
will usually be between 0.5 and 1 hour/acre.

NN. Attachment owning and operating cost (O & O cost × number of attachments): $ _____ /hour

OO. Add MM and NN: $ _____ /hour

PP. Multiply OO by LL for subtotal cost of burning: $ _____

QQ. Multiply gallons of fuel to be used times cost per gallon and add to PP for total cost of burning: $ _____

Any cumulative total can be divided into QQ to arrive at a per-unit cost.

Chaining

NOTE: No analyses have been given for the tree pusher, stumper, tree shear, and chaining. No known method will allow the computation of production and costs for these methods. However, for a very rough estimation of production and costs when chaining, the following analysis is possible.

Production
RR. Machine speed (in miles per hour):

_____ (miles/hour)

SS. Multiply RR by working width of chain (in feet)

_____ (acres/10 hours)

TT. Divide SS by 10 for total acres chained per hour:

_____ (acres/hour)

UU. Divide total number of acres to be chained by TT to determine total number of hours required for chaining:

_____ (hours)

Costs
VV. Machine owning and operating cost (O & O cost × number of machines): $ _____ /hour

WW. Chain owning and operating cost (O & O cost × number of chains): $ _____ /hour

XX. Add VV and WW: $ _____ /hour

YY. Multiply XX by UU for the total cost of chaining: $ _____

Conclusion

Now that the individual production and cost phases have been estimated, the complete expenses incurred for land clearing can be tabulated. Enter the cost for each operation chosen on its respective line and add. This is the total estimated cost for the clearing operations used.

Reference Letter	Operation	Cost
I	Cutting	$_____
R	Piling	$_____
II	Harrowing	$_____
	Root plowing	$_____
	Chopping	$_____
QQ	Burning	$_____
YY	Chaining	$_____
	TOTAL COST	$_____

Any cumulative total can be divided into the total cost to arrive at a per-unit cost. For example: Total cost divided by number of acres gives cost per acre; total cost divided by number of hours gives cost per hour.

Because this discussion centered on land clearing, no additional operations (road construction, drainage, leveling, etc.) were taken into account. Costs for these must be calculated separately.

SPECIFICATIONS

The manner in which specifications are written greatly affects the outcome of a land-clearing project. Poorly written specifications can make the difference between success and failure.

All specifications and supplementary information must be written in such a way that the contractor or governmental project engineers can clearly understand exactly what the job requirements are, what the job limitations are, and what is expected by the contracting agency or owner.

If specifications are not written clearly and concisely, the contracting agency should expect to pay more per acre. If contractors do not understand the job, they will understandably bid high to protect against loss.

Points to remember when writing specifications:

1. State all terms clearly and concisely.
2. Specify job requirements exactly. Use dimensions, examples, and comparisons to illustrate what is desired.
3. Define the end use of land clearly and exactly.
4. Knowing that all variables mentioned affect the total time requirement, specify a job's completion date realistically.
5. Avoid writing specifications so rigidly that they force the contractor to use a less efficient method.
6. Specifications should be written for each operation required and with as few marked "as directed" as possible.
7. Job requirements should not call for more work to be done than is absolutely necessary.
8. If total clearing is necessary, ensure that the work to be done is really land clearing and not earthmoving. For example, if gullies, streams, and ditches are to be made usable, a major earthmoving project (rather than land clearing) might have to be done.
9. When determining disposal methods or requirements, remember that tree size and rainfall may limit burning—thus, disposal may be done correctly and economically by just piling the refuse on unusable land.
10. If logging of salable wood is to be done, it should be done before or after the actual clearing.

The following materials can be helpful to the contractor:

Aerial photo maps of the entire area to be cleared
General topography map
Water table map
Yearly rainfall figures
Map showing access routes to and throughout area
Average tree count per acre—including the number of trees per acre, percentage of hardwoods, soil conditions, root systems present,

and the size ranges of the trees—tabulated in a clear and concise manner

LAND-CLEARING FORMULAS

Cutting and Windrowing

The following is Rome Plow Company's formula for determining either cutting or windrowing time per acre (in minutes) for a tractor equipped with a shearing blade such as the K/G.

To estimate tractor hours per acre on a specific land-clearing job, apply the factors shown in Table 2-2, together with data obtained from the job cruise in the field, in this formula:

$$T = B + M_1 N_1 + M_2 N_2 + M_3 N_3 + M_4 N_4 + DF$$

where

T = Time per acre in minutes

B = Base time required for each tractor to cover 1 acre of light material where no trees need splitting or individual treatment

M = Minutes required to cut or pile each tree in each diameter range at ground level

N = Number of trees per acre in each diameter range, obtained from field cruise

D = Sum of diameters in feet of all trees per acre above 6 feet (180 cm) in diameter at ground level, obtained from field cruise

F = Minutes per foot (30 cm) of diameter for trees above 6 feet (180 cm) in diameter

In cutting, certain factors will affect the production time:

If stumps are dug out after cutting to ground level, add 50% to diameter-range times (M, F).

If trees and stumps are taken out in one operation, add 25% to diameter-range times (M, F).

If over 75% hardwoods, add 30% to total time (T).

If under 25% hardwoods, subtract 30% from total time (T).

If there are more than 600 trees under 1 foot (30 centimeters) in diameter per acre, add 100% to base time (B).

If there are less than 400 trees under 1 foot (30 centimeters) per acre, subtract 30% from total time (T).

If heavy vines are present, add 100% to base time (B).

TABLE 2-2 PRODUCTION TABLE

		Diameter Range				
Caterpillar Track-Type Tractor[a]	*Base Minutes (B)*[b]	*1–2 ft. (30–60 cm) (M₁)*[c]	*2–3 ft. (60–90 cm) (M₂)*	*3–4 ft. (90–120 cm) (M₃)*	*4–6 ft. (120–180 cm) (M₄)*	*Over 6 ft. (180 cm), per ft. (30 cm) (F)*[d]
Cutting (K/G Blade)						
D9L	39	0.1	0.4	1.3	3.0	1.0
D8L	45	0.2	1.3	2.2	6.0	1.8
D7G	62	0.5	1.8	3.6	11.0	3.6
D6D	100	0.8	4.0	8.0	—	—
Windrowing[e]						
D9L	97	0.08	0.1	1.2	2.1	0.3
D8L	111	0.1	0.5	1.8	3.6	0.9
D7G	135	0.4	0.7	2.7	5.4	—
D6D	185	0.6	1.2	5.0	—	—

[a]Current models (power shift, when applicable) working terrain under 10% grade with good footing and no stones, in average hard-soft wood mix. Tractor and tools in proper operating condition.

[b]Minutes required to cover acre of light material where *no trees* need splitting or individual treatment.

[c]M_1, M_2, M_3, M_4 refer to minutes required to cut or pile trees in each diameter range at ground level.

$$M_1 = 1\text{-}2 \text{ ft. } (30\text{-}60 \text{ cm}) \text{ trees}$$
$$M_2 = 2\text{-}3 \text{ ft. } (60\text{-}90 \text{ cm}) \text{ trees}$$
$$M_3 = 3\text{-}4 \text{ ft. } (90\text{-}120 \text{ cm}) \text{ trees}$$
$$M_4 = 4\text{-}6 \text{ ft. } (120\text{-}180 \text{ cm}) \text{ trees}$$

[d]Number of minutes needed per foot of diameter to cut or pile trees over 6 ft. (180 cm) in diameter. Therefore, a D9L cutting an 8 ft. tree requires

$$1.0 \text{ min./tree} \times 8 \text{ trees} = 8 \text{ min.}$$

Piling time is as follows:

$$.3 \text{ min./tree} \times 8 \text{ trees} = 2.4 \text{ min.}$$

[e]Assuming 200 ft. between windrows.

In windrowing, other factors will affect production time:

Add 25% to total time (T) when piling grubbed vegetation.

Subtract 25% to 50% from total time per tractor when using three or more tractors in combination.

For burning, allow between 30 and 60 minutes tractor time per acre.

When working in dense, small-diameter brush with few or no large trees or when vines are entangled within the felled material, reduce piling base by 30%.

Estimating Hourly Production (Constant-Speed Operation)

The American Society of Agricultural Engineers' formula for estimating hourly production of a constant-speed operation is:

$$\frac{\text{Speed (mph)} \times \text{width of cut (feet)}}{10} = \text{Acres/hour}$$

When production calculations are required for cutting, piling, and stumping, the production table (with formula) can be used. When performing harrowing, root plowing, and other activities, production can be estimated with the above formula. This formula allows for 82.5% efficiency.

The graphs in Figure 2-32 show the speeds, production per hour, and width of cut that can be accomplished in a constant-speed operation.

Fleco Corporation has a formula that is used to figure approximate acreage covered by an agricultural implement in an hour. This formula is the same as the ASAE formula except it adds 10% to the working width of the implement in feet, then multiplies this figure by the speed of the tractor in miles per hour. This figure is in turn divided by 10 to obtain acres per hour in a 10-hour day.

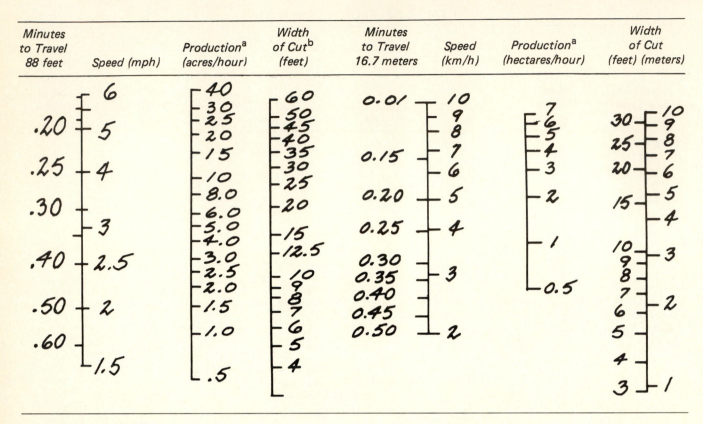

Minutes to Travel 88 feet	Speed (mph)	Production[a] (acres/hour)	Width of Cut[b] (feet)	Minutes to Travel 16.7 meters	Speed (km/h)	Production[a] (hectares/hour)	Width of Cut (feet) (meters)

[a]Based on 82.5% efficiency.

[b]When width of cut exceeds 60 feet (Graph 1) or 10 meters (Graph 2), use a multiple of the width of cut and increase production proportionately.

FIGURE 2-32 Speed, production, and width of cut in a constant-speed clearing operation.

This formula is used for Fleco's rolling choppers and root plows. The 10% factor is an allowance for losses due to turning and other procedures. The efficiency of this formula is approximately 90%. Neither Fleco Corporation nor Rome Industries claims absolute accuracy in these formulas because of the many variables that increase or decrease production.

Estimating the Number of Machines Needed

To find the number of machines required for each phase of land clearing, use this formula:

Number of machines needed

$$= \frac{\text{Hours/acre* X number of acres}}{\text{Hours scheduled to complete the job}}$$

*Average machine production for all operations in hours per acre.

Equipment Selection

Refer to Figure 2-33 for guidelines on equipment selection by size of area, vegetation, and method.

LANDSCAPING

Landscaping is a very general term. It means different things to different people. *The need of the ultimate user* is the most important consideration for the builder. But the builder usually depends on landscaping professionals for the plans and specifications; therefore, a brief description of these professionals will be helpful.

The *landscape architect* is the most qualified person a builder can use to evolve designs, produce working drawings, and supervise the implementation of the landscape plans. In most states, only a

LIGHT CLEARING Vegetation up to 2 in. (5 cm) Diameter)

	Uprooting Vegetation	Cutting vegetation at or above ground level	Knocking the vegetation to the ground	Incorporation of vegetation into the soil
Small areas 10 acres (4.0 hectares)	Bulldozer Blade Axes, Grub hoes and Mattocks	Axes, Machetes, Brush Hooks, Grub Hoes and Mattocks, Wheel-mounted Circular Saws	Bulldozer Blade	Moldboard Plows Disc Plows Disc Harrows
Medium areas 100 acres (40 hectares)	Bulldozer Blade	Heavy-duty Sickle Mowers (up to 1½"-3.7 cm diameter) Tractor-mounted Circular Saws; Suspended rotary mowers	Bulldozer Blade, Rotary Mowers; Flail-type Rotary Cutters; Rolling Brush Cutters	Moldboard Plows; Disc Plows Disc Harrows
Large areas 1,000 acres (400 hectares)	Bulldozer Blade Root Rake, Grubber, Root Plow, Anchor Chain drawn between two crawler tractors, rails		Rolling Brush Cutter; Flail-type Cutter; Anchor Chain drawn between two crawler tractors; rails	Undercutter with Disc; Moldboard Plows; Disc Plows; Disc Harrows

The most economical size area for each type of equipment will vary with the relative cost of capital equipment versus labor. It is also affected by whether there are alternate uses for equipment such as using tractors for tillage.

INTERMEDIATE CLEARING Vegetation 2 to 8 in. (5-20 cm) Diameter

	Uprooting Vegetation	Cutting vegetation at or above ground level	Knocking the vegetation to the ground	Incorporation of vegetation into the soil
Small areas 10 acres (4.0 hectares)	Bulldozer Blade	Axes, Crosscut Saws, Power Chain Saws, Wheel-mounted Circular Saws	Bulldozer Blade	Heavy-duty Disc Plow; Disc Harrow
Medium areas 100 acres (40 hectares)	Bulldozer Blade	Power Chain Saws, Tractor-mounted Circular Saws	Bulldozer Blade Rolling Brush Cutter (up to 5 in-12 cm Diameter), Rotary Mower (up to 4 in-10 cm dia.)	Heavy-duty Disc Plow; Disc Harrow
Large areas 1,000 acres (400 hectares)	Shearing Blade Angling (Tilted) Bulldozer Blade, Rakes, Anchor Chain drawn between two crawler tractors Root Plow	Shearing Blade (Angling or V-type)	Bulldozer Blade Flail-type Rotary Cutter, Anchor Chain	Bulldozer Blade with Heavy-duty Harrow

LARGE CLEARING Vegetation 8 in. (20 cm) Diameter or Larger

	Uprooting Vegetation	Cutting vegetation at or above ground level	Knocking the vegetation to the ground
Small areas 10 acres (4.0 hectares)	Bulldozer Blade	Axes, Crosscut Saws, Power Chain Saws	Bulldozer Blade
Medium areas 100 acres (40 hectares)	Shearing Blade Angling (Tilted), Knockdown Beam, Rakes, Tree Stumper	Shearing Blade (Angling or V-type), Tree Shear (up to 26 in. (65 cm) softwood; 14 in. (35 cm) hardwood), Shearing Blade—Power Saw Combination	Bulldozer Blade
Large areas 1,000 acres (400 hectares)	Shearing Blade Angling (Tilted), Knockdown Beam, Rakes, Tree Stumper, Anchor Chain with Ball drawn between two crawler tractors	Shearing Blade (Angling or V-type), Shearing Blade—Power Saw Combination	Anchor Chain with Ball drawn between two crawler tractors

FIGURE 2-33 Equipment selection guidelines.

person licensed by the state may use the title *landscape architect*. A landscape architect has passed a rigorous state examination and has probably completed four or five years at an accredited institution offering a course of study in landscape architecture. The intensity of the training is similar to that for the profession of medicine. Landscape architects are trained for total landscape design of home sites, apartment complexes, office buildings, parks, city planning—any element of landscape design. In practive, landscape architects tend to specialize in residential work or one of the other major categories of landscape work.

Landscape architects usually prefer to work by contract, typically the American Society of Landscape Architects' Standard Form of Agreement between Owner and Landscape Architect (for contractors building speculative housing, the contractor is considered the owner unless the house is presold and the landscaping is handled by separate contract). The contract simply spells out what is to be done, what the landscape architect's responsibilities are, what the fee will be, and how the payments will be made. There are, however, a wide range of possible agreements that contractors may enter into with landscape architects.

Many contractors have the mistaken idea that they cannot afford a landscape architect. In fact, they often save money by using a landscape architect.

But a landscape architect is not the only choice available to contractors. Through the years, many of the landscaping professions have merged and crossed responsibilities. The final choice the contractor makes about what kind of service is needed depends on the contractor's own capabilities, the type and size of job that is being done, and the amount of responsibility the contractor is capable and willing to assume.

Landscape designers perform parts of all the duties traditionally performed by landscape architects. Most of these designers do their jobs well. But the title *landscape designer* carries no legal requirements for qualification—anyone may use the appellation. Landscape designers are often employed by nurseries, landscape contractors, and sometimes contractors, if the contractor does enough work to justify the salary. A contractor may also permanently employ a landscape architect.

Landscape contractors traditionally execute the landscape plan for the owner or contractor according to the landscape architect's working drawings and plans; the landscape architect may be employed to supervise the landscape work and disperse payment, or this function may be left to the contractor.

Landscape contractors are often capable of preparing the site for implementation of the landscaping plan. They may dig drainage and irrigation lines, build retaining walls and terraces, and build fences and walks and drives; they sometimes even provide storage facilities for materials and equipment. What they do depends on their individual and company capabilities and on what they are contracted to do.

There are no legal educational requirements for landscape contractors; their skill and ability come from experience. But, as mentioned earlier, the landscaping professions have merged considerably, and the landscape contractor who has

prospered may employ a landscape architect or landscape designer. There is another variation on this tendency to merge: the landscape architect who adds a landscape contracting business to his services; this is sometimes called the "design-build" concept. In the design-build concept, the landscape architect takes the project from the design stage all the way to completion.

Nurserymen and landscape nurserymen traditionally grow plantings for wholesale use by landscape contractors, general contractors, and other professionals. They may also sell retail. Often, nowadays, nurseries are "warehouses" for plants bought from some distant regional growing center; this practice is all right if the nurseryman knows plants and is not just a broker; reputation is important here. Nurserymen also may be landscape architects, landscape contractors, or both. But, as with landscape contractors and landscape designers, there are no legal requirements for the education of nurserymen—experience is the teacher.

Which expert or combination of experts does the contractor choose? The answer depends on the complexity of the job, on the contractor's capability, and most important, on what the eventual owner wants. Large landscaping contractors are often unsuitable for small jobs. Similarly, small landscape outfits may not have the equipment or qualified labor to handle big jobs. One thing is certain: The contractor should not attempt to talk a landscape professional into doing a job that the professional shows no enthusiasm for. If, for example, the contractor needs an all-service design-build professional, a firm whose members appear afraid to get their hands dirty should not be used. Conversely, if design is especially important, a landscape contractor weak in design capabilities should be avoided, or the design function should be contracted for separately.

In any case, the contractor should be careful not to "overlandscape." If, for example, the eventual owner is unknown—that is, if the contractor is building speculatively—only the basic landscaping should be done. Some owners love to garden and maintain their landscaping plants; others abhor this. There is a tendency nowadays for low-maintenance, hardy plantings (includes trees, shrubs, and all plants) that require little time, expense, and labor from the owner. Even watering the lawn is too much for some owners, and thus automatic sprinklers have become popular. Lawn maintenance companies have sprung up that will do all the lawn maintenance chores—for a price.

Thus, from a return-on-investment standpoint,

it is wise not to overlandscape or overbuild when the eventual owner is unknown. However, good basic landscaping, in addition to performing its functional requirements (discussed later), also creates what real estate marketers call "curb appeal"; that is, good basic landscaping may make the difference as to whether a prospect drives past the house or stops to have a look inside.

There is also a timing aspect to the value of landscaping. Landscaping is not as important to value in the winter as it is in the spring, when it becomes a visible asset. A rule of thumb that some professionals use to establish landscape budgets is that landscaping costs should be approximately 10% of the cost of the overall development. Thus, a house and lot designed to sell for $80,000 would include $8,000 for landscaping. This figure includes both plantings and "hard" landscaping items such as terraces, drives, walks, garden walls, fences, and so forth. This is, of course, only a rough guide.

The Basic Landscaping Design and Planting Plan

Landscaping, like other building projects, is composed of three phases: design, installation, and maintenance. Too often the third phase is not given thorough consideration. Clients may approach the project with preconceived notions about the design and insist on plants that are not suitable for the project and the local weather and soil conditions. Poorly installed plants may die; if they die before the warranty is up (typically one year), the builder and/or landscape contractor is responsible for replacement. If they die after the warranty period expires, the homeowner is responsible for them. Properly designed and installed landscapes should provide the contractors with a fair profit, and no one should be burdened with excessive replacement costs.

The general contractor should always know the provisions of the warranties of subcontractors, such as landscape contractors. And, should the landscaping be done by separate contract between the owner and the landscape contractor, the general contractor is often asked for recommendations; if so, the general contractor should advise the owner to go over the warranty agreement carefully. The owner often feels that the general contractor is responsible for the successful outcome of the project whether this is legally true or not; contractors, especially small ones who depend largely on word of mouth for business, can avoid hard feelings, and sometimes lawsuits, by simply being sure

the owner is made aware of the advantages and pitfalls of handling portions of the job by separate contract.

The builder usually inherits the landscaping parameters from others—from developers or planning consultants and landscape architects who plan subdivisions, planned communities, and even larger developments. The individual builder's responsibilities are usually much more segmented, being concerned with building lots where the house, garage, drives, walks, and other physical components are already located.

Thus, builders are more often concerned with the final stages of relatively small numbers of lots than with areawide considerations of drainage, zoning, and similar factors that greatly influence what the final landscape will be.

Regardless of the scale of the project—one house lot or a thousand—the design process is very similar. Obviously, it is simpler to design landscaping for ten houses than for a thousand. Nevertheless, communities are made up of individuals, whether they live in condos, single-family houses, or apartments, and the final landscape must take the individuals' needs into consideration when analyzing physical site possibilities. A grasp of the design process of landscaping, or at least an appreciation of it, is essential to builders.

First of all, the best landscape designs are created with the house interior in mind. In a sense, the landscape is an extension of the house interior. For example, no one would want a game court or other high-activity area adjacent to a room used as an office or library where quiet is required. Similarly, it would not be good planning to locate a swimming pool where users would have to enter the house through a formal, carpeted area such as a living room. Such planning, obviously, is very basic. But if you continue to consider the relationship between the interior spaces and the landscape, you will see that the house and the landscape can become one flowing, continuous space, and you gain a sense of increased space. This may be because many houses are built as a collection of small cubicles stuck together: the living room, the dining room, the kitchen, the bathroom, and so forth. When you open these spaces up more to each other and then relate those spaces to the landscape (and the uses of the landscape), all the spaces relate to each other and to the outside, and you gain a feel for the whole rather than just the room or the part of the landscape you happen to be in.

This feeling of increased space can be aided by the use of materials. Brick walks leading to a patio, for example, may be used in nearby areas (if appropriate) in the house interior. Brick makes good entryways, halls, and kitchen floors and can be used for vertical surfaces too. Blending interior and exterior materials creates in users an enlarged sense of space with more subtle distinctions of usage between the areas, inside and out. To the builder, this can mean happy buyers and more sales, at little or no extra cost. Design time is relatively clean, compared to houses that sit for months before they sell.

Sun, wind, drainage, and soil are some of the major influences on landscape design and landscape components.

Sun affects the placement and design of houses as well as plants. Trees offer shade, where it is needed: patios, terraces, conversation and sitting areas, eating areas; they also may be used to shade houses. Trees are one of the best natural landscaping plants and may, of course, be located for visual effect elsewhere on the lot if shade is undesirable at or near the house. Each plant has soil and shade requirements: Some plants prosper in sunny areas with rich soil, others in shady areas with poor growing soil; there are many variations of plant soil and shade requirements. Protection from the wind is sometimes necessary for certain activities—sports, eating, sitting areas, and so on. There are many types of wind screens: hedges, low-growing trees, fences, other natural or built landscape components. Drainage must be considered when locating the house; in general, "positive" drainage is desirable. Positive drainage means positioning the house and other built components so that runoff is directed away from the components to some pre-planned disposal point such as a catch basin, drainage swale, or gutter. Plants are dramatically affected by drainage (*drainage* here means the ability of the soil to soak up water—sandy soils are better at this than clayey soils); as with the sun, plants require specific drainage conditions to prosper and to avoid excessive maintenance. Good drainage, to typical lawns, usually means that the lot will shed water after a heavy rain and not retain puddles in low spots.

Builders usually work with developers, planners, architects, and market analysts in determining what type of houses (in terms of size, style, and cost) are built in a given area. Architects typically design the houses and show their locations on the lot along with the drives, walks, and so forth.

Thus, when the typical builder enters the picture, at least two important elements of landscaping have been determined: lot drainage and circulation. *Circulation* here refers to the location of the house relative to the street and the location of drives and walks relative to the house. But arrangement of the remaining natural and built landscaping elements affect the visual quality of the lot and house, as well as lot usage and circulation between the lot and the house.

Each area of the lawn should be studied for usage relative to the house, and appropriate plants and materials must be selected to support those activities. Sports areas, for example, should be located where damage to the house or injury to the players is least likely: away from glass doors and power poles, for example. Grass is an excellent ground cover for vigorous sports like football. But grass requires regular, careful maintenance.

For less vigorous owners, especially the elderly, who do not need grass for its functional value and do not want to spend the time and energy maintaining it, other ground covers are available. Sandwort, for example, grows to a height of about 2 inches, then stops. You do not have to mow it. It prospers in open sun or light shade, needs good drainage and sandy, moist, slightly acid soil. In the spring, this ground cover produces a small, white bloom. It is always green, except in very cold areas, where it turns a kind of silver. Sandwort will grow in most areas of the country.

The foregoing is one small example of the thought processes involved in thinking through land usage, circulation, and selection of complementary landscaping elements. Part of the landscaping design process is a kind of matching game: function (desired usage, needed circulation patterns, and so on) is matched with landscape elements, both natural and built, that make the desired function easier and more pleasant for the user.

For each element of the landscape, similar questions will be asked: What does the owner want out of this project? Where are the *possible* locations for the project (the game court, the patio, the privacy screen, ground covers, etc.)? Which location is best, considering its relationship to the other activities on the lot and to the house? Which plants, combination of plants, or combination of plants and built elements will best aid the function, look good, and work well with the house in increasing the sense of wholeness in the overall design?

As an example of this kind of questioning, consider a fence. A fence may be built of wood, metal, fiberglass, brick, stone, adobe, even glass. But each fence for each project has a material or two that best meets the need of the owner in terms of function and budget. First, ask what is the most important function of the fence. Is it privacy? If so, then a high board fence will work. But privacy in what area? If shielding only the patio is important to the owner, why spend money to fence the whole lot? Only if the entire lot needs to be fenced in should the fence go around the lot lines. If the budget is flexible and the owner wants a feeling of prestige in addition to privacy, a brick or stone fence (wall, actually) is a good solution.

Light, wind, and sound should be considered when building a fence. Any solid fence shuts out considerable light, especially where there are numbers of trees. If light is desired, some kind of perforated fence design should be used. Perforations are typically accomplished in wood fences by alternating the boards on each side of the fence, creating spaces that let light and wind through. Brick fences and walls may be handled similarly by omitting certain bricks, thus creating a pattern that allows air and light to pass through.

No fence does a really good job of blocking sound. Most people experience a psychological relief from sound by not being able to see the source; a fence that blocks the view of a busy street, for example, may afford the owner some sound relief. Fences and walls do stop some sound, but they will not significantly reduce the sound level of busy streets and similar noise sources. Thick masonry walls block more noise than thinner, less substantial materials such as wood. The best sound protection, however, comes from high earth berms; unfortunately, this solution is usually not possible in urban and suburban locations. Trees, shrubs, and other plants, like fences, offer psychological relief from noise but do not significantly block noise.

If security is the main function for a fence, chain-link is one of the more economical solutions. Wood and plastic inserts are available, where privacy is desired. And the fences may be bought vinyl-coated in several colors, typically dark green and dark brown, which helps reduce the utilitarian appearance of these fences. Open-wire fences are the most secure because they do not afford intruders a place to hide, once inside the fence. No fence will keep out someone determined to get in; security fences are most valuable for reducing casual trespassing (such as neighborhood children) and for keeping young children and pets inside. Chain-link fences can now be bought with square posts, which look better for residential use than the typical round ones. There are other wire fences that can be used for the purposes described in this section. And all wire fences may be made aesthetically pleasing by paying close attention to the frame design; wire fences with wooden framing, for example, can be both attractive and economical.

Return to the landscape questioning process, using fences as the subject. What is the function of the fence? What is the best location? What is the best material that will meet the function and harmonize with the overall feeling and aesthetic factors of the house and site? Can the apparent solution be built within the budget?

Against such questioning, evaluate realistically what can be achieved by the fence. Fences do not stop burglars; some fences even aid theft. Fences do not offer significant protection from noise. Fences are not cheap.

Well-designed fences can be a significant aesthetic feature. Fences are excellent circulation devices, routing people where you want them to go. Fences protect landscaping and other property from casual trespassers. Fences help keep young children and pets inside the lot. Fences are good privacy screens and can be designed as wind screens (especially useful in coastal or similar open areas where wind is a nuisance to outdoor activities).

The person who actually puts the landscape design on paper may be a specialist, such as the architect, landscape architect, landscape designer, or one of the other specialists, or the contractor. Also, the contractor may work with the future owner of the house and lot. In practice, there are still other influences on the final landscape design and planting plan.

The plan itself varies in presentation. It may be a formal plan showing grades, types of plants, and all the technical information necessary to implement the desires of the owner. Or it may be simply a sketch with an accompanying listing of plants and their specifications for installation.

Grading, Earthwork, and Drainage

By the time the typical residential builder acquires a lot or lots in a subdivision, much grading and earthwork has been done. Thus, most builders are more interested in the final aspects of landscaping, such as finish grading, sodding, laying walks, drives, and so forth (these items are discussed elsewhere in the text). However, construction activities overlap. Therefore, it will be useful to contractors to know something about the early site activities.

The land is almost always reshaped from its natural state. Drainage must be assured; surface water must be routed away from houses and other activity areas; parking areas, streets, and other public areas must be located where drainage may be accomplished; all the built elements of the overall site must be located such that drainage can be accomplished for the individual houses and community facilities without causing drainage problems to neighboring property. Erosion must be prevented. These are typical considerations of area site planning, and the early grading and earthwork reflect these considerations.

Wherever possible, trees and other desirable vegetation are left intact. To do so, the site and grading plan must utilize natural grades as much as possible, because trees and plants are quite sensitive to grade change. It may be necessary to build retaining walls, tree wells, and similar devices to maintain existing grades near trees and shrubbery.

Grading, both preliminary and final, is an aesthetic skill: The shape of the land on the built areas must harmonize with the shape of the land that is left in its existing state, so that the houses, streets, and other built elements look like they belong there.

All the grading, in addition to the aesthetic and functional considerations, must be done economically. Excessive cuts and fills and retaining walls, catch basins, underground drainage systems, and similar drainage aids must be kept to a minimum if the project is to be economically feasible (many projects are *not* economically feasible).

Thus, building sites for houses are evolved with a number of factors in mind: drainage, activity requirements, aesthetics, and cost.

The developer may have graded the site entirely by cutting. In this method, the vegetation on all the home sites is cut down as required to facilitate good drainage and address the other considerations mentioned earlier. In doing so, the soil uncovered is usually stable and well compacted. However, erosion may result if seeding, sodding, or other steps are not taken to prevent it. Also, cutting removes topsoil, which is expensive and must be replaced before planting can be done.

Building sites may be graded entirely by fill. This method is very expensive and is usually done only when other methods are not possible; small areas may, however, be graded in this manner without too much expense. When fill is used, strict engineering practices must be followed to avoid serious problems with foundations, slabs, and similar building components. Topsoil must be provided when using this method also.

On typical large sites, grading is usually accomplished by a near balance between cut and fill methods. It is important to builders to know what kind of soil is in place on the particular lots on which they will build and to know what additional grading and earthwork, soil compaction, or topsoil replacement, if any, will have to be done before the lots can be built on.

For a residential building, the general contractor is typically responsible for final or finished grades. Careful planning and grading are required to ensure that the subgrades (the grades upon which the final materials rest) are dug to the exact depth required for the finish materials; the finish materials may be sod, concrete drives and walks, or similar built or natural materials.

Before finish materials are laid, other soil preparations may have to be made. On slopes, for example, it may be necessary to scarify the subgrade to ensure a bond between the subgrade soil and the finish topsoil. Special soils may have to be imported for specific plants; it is, however, advisable to avoid plants that do not prosper in the subject locality and will consequently require extra maintenance. Different building elements, of course, require different soil compaction; obviously, the heavier the building or built element, the more stable and compacted the soil will have to be. Where existing soils cannot achieve the needed compaction, more stable soil must be imported to the site and installed by appropriate engineering means.

Almost any lot can be graded so that a house can be built on it. But a good lot will have been graded so that the house looks good on the lot, like it belonged there. The lot will drain well and there will be little more preliminary grading to do; thus, the house can be built economically—within the budget.

The good aspects of a grading plan are difficult to itemize; but there are some simple indications of poor preliminary grading: If the site will need an extraordinary number of steps, the grading plan may be suspect; the same criticism applies to overusing retaining walls, terraces, catch basins, and other drainage aids. Slopes should be away from houses, if this can be accomplished without undue cost. Extreme unbalances in cut and fill quantities (is the lot almost entirely cut or almost entirely fill?) often indicate a poorly conceived grading plan. Steep slopes limit usage, complicate drainage, and make building more difficult; very level lots are hard to drain and require underground drainage pipes and similar drainage aids. The best lots have gentle, easy-to-drain slopes. Lots that have been stripped of all or most of their vegetation must have that vegetation replaced—at a significant cost. The existence of these types of grading conditions do not always rule a lot out, but they do make it suspect.

It is impossible—and undesirable—to reduce subdivision grading and drainage plans to a set of rules and formulas. However, there are several typical plans that often are used; these plans are shown in Figure 2-34.

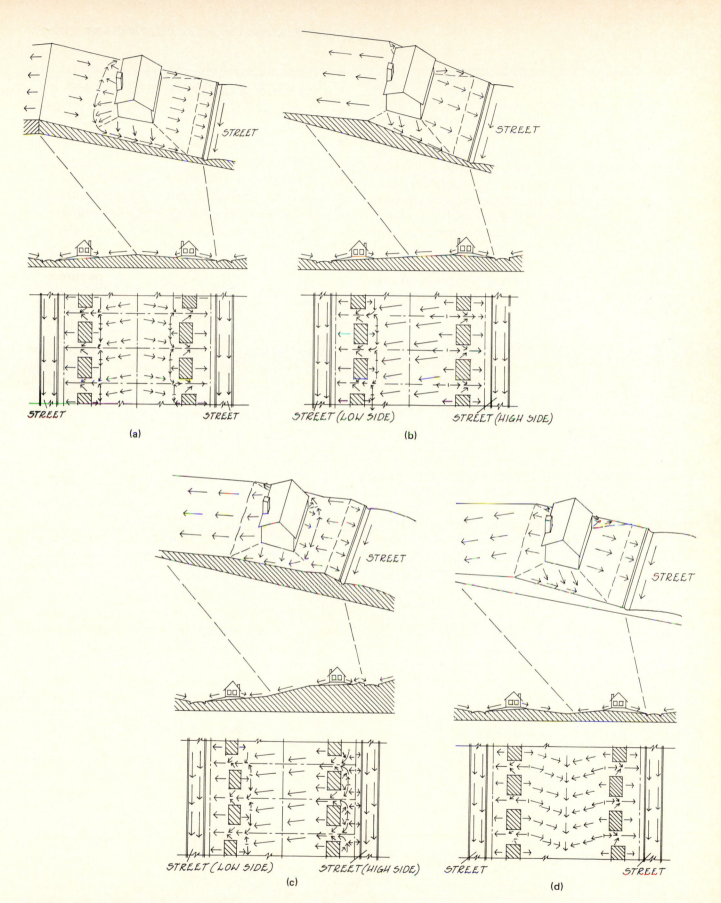

FIGURE 2-34 Block grading plan showing (a) drainage from ridge along rear lot lines to street; (b) drainage across a gentle cross-slope; (c) drainage across a steep cross-slope; and (d) drainage down valley along rear lot lines.

STREET

STREET

STREET (LOW SIDE)

STREET (HIGH SIDE)

(a)

(b)

STREET (LOW SIDE)

STREET (HIGH SIDE)

STREET

STREET

(c)

(d)

Landscaping Aids for Grade Changes

Builders and developers hope to achieve good drainage and acceptable lot grades by grading alone, manipulating the surface of the lots. They want to shape the lots by grading alone because this is the cheapest solution and thus helps increase profits or allows more flexibility in pricing the house.

When acceptable grades cannot be achieved by grading alone, when grading results in slopes that are too steep for desired homeowner activities, or when the slopes are too steep to be walked safely or create drainage or erosion problems, grading must be supplemented. The supplements are typically terraces, steps, ramps, retaining walls, and baffles.

Terraces. *Terraces* are generally a series of level or nearly level grades within a steeper overall slope. Terraces are, in effect and appearance, giant steps; the treads of these steps may be several yards from one elevation to the next, or dozens of yards, or even more, depending on lot size. (See Figure 2-35.)

The risers of these steps are actually minor slopes, formed most cheaply by grading. When walking up or down a terraced slope, some provision must be made to negotiate the elevation changes at the minor slopes or risers; the provisions are usually steps, stepped ramps, or both. In cases where the differences between the terrace elevations are too great to allow a graded slope, retaining walls must be used.

The terraces themselves may be surfaced with a variety of materials, depending on usage. In general, the cheapest solution is grass or some other hardy ground cover, if no specific activity is to be planned for. Otherwise, each activity has its own best set of materials.

The following slopes are typical in many areas of the United States for the respective uses. These relationships are not exhaustive and may vary somewhat locally, but they should give a feel for when terracing may be necessary and what the range of slopes is for some typical activities.

- mowed grassy slopes: maximum 3:1
- unmowed grassy slopes: maximum 2:1
- patios and sitting areas: maximum 2%, minimum 1%
- rear or "casual" walks: maximum 8%, minimum 1%
- main entry walk: maximum 4%, minimum 1%
- parking areas: maximum 5%, minimum 0.5%
- ramps: maximum 10%, minimum 1%
- grassy recreational areas: maximum 3%, minimum 2%

Steps and ramps. Steps and ramps function to allow passage of people and equipment between grades of different elevations in safety and comfort. That statement may seem heavy for a building component as apparently simple to locate and construct as a set of steps or a ramp. They are simple to build, but careful attention must be paid to their design.

Consider the following set of basic guidelines for designing *steps*:

1. Typical riser-to-tread relationships are as follows: 7-inch riser, 11-inch tread; 6½-inch riser, 13-inch tread; 6-inch riser, 15-inch tread; 5½-inch riser, 16-inch tread; 5-inch riser, 17-inch tread; 4½-inch riser, 18-inch tread; 4-inch riser, 19-inch tread. Avoid risers of less than 4 inches or more than 7 inches; avoid treads of less than 11 inches. Where possible, a good outdoor riser-to-tread relationship is a 6-inch riser with a 15-inch tread.

2. Keep the riser-to-tread relationship the same throughout the flight (a flight is two or more steps).

3. Do not taper treads and risers. For example, a 6-inch riser should measure 6 inches across the full width of the steps; a 15-inch tread should measure 15 inches across the width of the steps.

4. Pitch outdoor treads to drain—approximately ⅛ inch per foot.

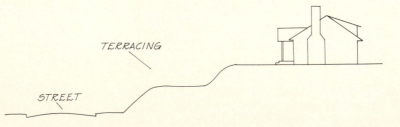

FIGURE 2-35 Example of terraces used like giant steps to soften a steeper overall slope.

5. Steps should be the same width as the walks they join.

6. For safety, where possible, steps that join walks or drives at right angles should be set back at least 2 feet from the walk or drive.

7. Landings between flights of steps should be the same width as the steps.

8. Where a series of step flights are connected by walks or ramps, the lengths of the walks or ramps between the flights of steps should be kept the same, if possible.

9. Stair railings are recommended for flights greater than 30 inches.

These guidelines are rudimentary, not conclusive.

Outdoor stairs may be planned with simple instruments: wood stakes, a line level, and perhaps a plumb bob for measuring vertical distance (see Figure 2-36). If outdoor stairs are built into the slope, the amount of cut should approximate the amount of fill, as shown in Figure 2-37. Whether built into the grade or spanned over the ground (usually determined by the steepness of the grade), a variety of materials and combinations of materials may be used (see Figure 2-38).

Ramps perform the same function as steps, transporting people across elevation changes, but ramps are more convenient in areas where wheeled equipment must be used or where the handicapped or the elderly will travel.

Also, the topography may call for a ramp: where the change in elevation to be transcended is too low for steps (where the risers would be less than 4 inches, for example) but too steep for a normal walk. Generally speaking, walks that slope approximately 5% or more are considered ramps.

Stepped ramps are ramps used in series. Stepped ramps may transcend elevation changes with a single riser, several risers, or flights, depending on the slope to be negotiated; obviously, gentle slopes require fewer risers than steeper slopes.

Topography, weather conditions, type of materials used, users, and usage are typical considerations in determining what ramp slopes are appropriate, within the range of physical possibilities. Typical ramp slopes range between 5% and 10%: A 6-inch rise in a 10-foot run gives a 5% slope; 8 inches, a 6.6% slope; 10 inches, an 8.3% slope; and 12 inches, a 10% slope. (See Figure 2-39.) A 5% ramp of broom-finished concrete might be called for by an unaided user in a wheelchair, while a 10% ramp would work for the nonhandicapped. Generally, however, if the grade is more than 8.3%, more comfortable walking results if a ramp is combined with steps.

Ramps and stepped ramps require somewhat more demanding specifications than do walks with steps. The following guidelines are basic, not conclusive:

1. Steps between ramps should be 3 inches high or be three flights of two risers each. All risers should be the same height, all treads the same depth.

2. Risers should not exceed 6 inches.

3. Ramps between flights should be the same length.

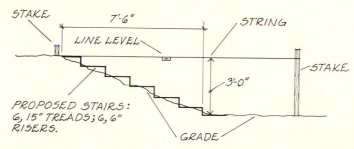

FIGURE 2-36 Outdoor stair plan.

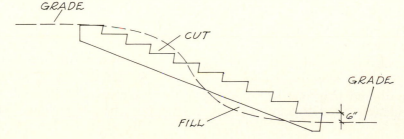

FIGURE 2-37 Outdoor stairs built into a slope.

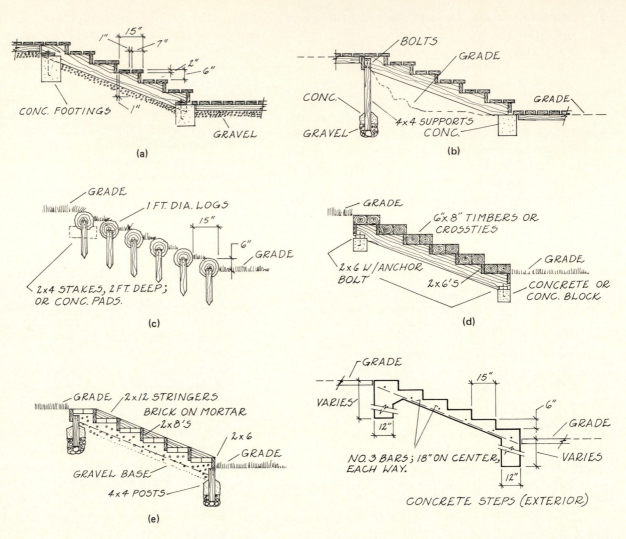

FIGURE 2-38 Materials that can be used for outdoor stairs.

4. Where ramps are connected by single risers, the ramp length should allow two comfortable foot paces (approximately 6 feet long). The finish grade should be ⅔ the height of the risers (see Figure 2-40). Where ramps are connected by flights of two risers, allow for three comfortable paces (approximately 9 feet long). The finish grade should follow a line connecting the bottom of the upper riser in each flight (see Figure 2-41).

5. Ramps and stepped ramps should be the same width as the walks they join.

6. The approaches to ramps should be clear, level, and approximately 5 feet long to allow for wheelchair and baby carriage maneuvers and to provide for general safety.

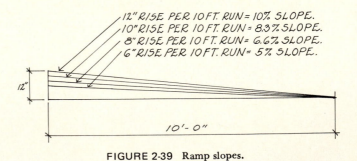

FIGURE 2-39 Ramp slopes.

7. Handrails, when used, should extend 18 inches beyond the top and bottom of the ramp.

8. Ramps should be at least 3 feet wide, wider where heavy traffic is anticipated or where the ramp will be used by service people.

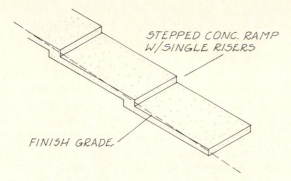

FIGURE 2-40   Stepped concrete ramp with single risers.

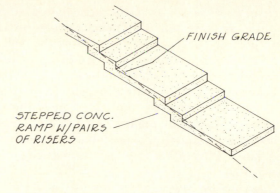

FIGURE 2-41   Stepped concrete ramp with pairs of risers.

9.   Plantings should be located so as not to cast shadows on ramps, which can increase ice buildup in winter.

10.   Low curbs at the sides of ramps are desirable where wheeled equipment will travel.

11.   Ramps should be well illuminated to ensure night safety; illumination is particularly important at ramp approaches.

Retaining walls.   *Retaining walls* may be called for when the slope of a lot is too steep to be handled by a series of terraces, at the property line of a steeply sloping lot, or in similar situations. In general, retaining walls should be avoided if other acceptable solutions are available because retaining walls are expensive to build and must be built with considerable care.

The height of the retaining wall and the angle of the earth on its high side are two very critical factors in designing a retaining wall. The higher the wall, the more structural strength will have to be built into it. The more steeply the earth slopes up from the high side of the wall, the more structural strength will be required. Surcharges increase the structural requirements further (surcharges are additional forces the wall may encounter, such as vehicles, buildings, and driveways located near the wall).

In addition to these factors, important considerations in building retaining walls include drainage, soil, and weather conditions. A well-designed retaining wall uses the most economical material that is compatible with the overall design of the project in a manner that meets the structural needs.

There are four common types of retaining walls: the gravity retaining wall, the cantilever retaining wall, the counterfort retaining wall, and the buttressed retaining wall.

Gravity retaining walls utilize their own weight and positioning to achieve stability. Typically, the fill side of the wall slants down from the top, into the fill. The side opposite the fill is usually vertical. (Note: Most retaining walls are built with a slight slant toward the fill side, not for structural purposes but to offset the visual impression that the wall is leaning toward the lower side, which occurs when the wall is built perfectly vertical.) The great width of gravity walls, at the bottom, and their weight keep them in place against the resistance of the earth. Gravity walls are usually masonry, but they may be built of stone, concrete, or any substantial, heavy material.

Gravity walls often require no steel reinforcement, but because of the massive amount of masonry used, they can be expensive. Where large amounts of native stone or suitable materials are available, gravity walls may be used economically. Also, gravity walls require less skilled labor than walls requiring steel reinforcement. (See Figure 2-42.)

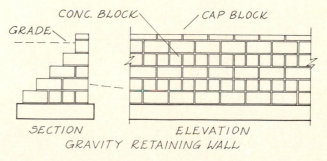

FIGURE 2-42   Gravity retaining wall.

Cantilever retaining walls are shaped something like an inverted T. Earth, resting on the footing portion, helps hold the stem of the wall in place. Cantilever retaining walls are often of reinforced concrete or masonry and may be veneered with other materials, if desired. Because these walls have steel reinforcement, they are much thinner than gravity walls and require less material. They do, however, require professional design, and skilled labor is required to construct them. (See Figure 2-43.)

Counterfort retaining walls use a series of vertical supports (posts or columns) and spanning panels. A typical example of this system would be reinforced concrete posts with concrete panels spanning from post center to post center; steel cables attached to the fill side of the posts would run down, at an angle, to a concrete foundation, where they would be anchored. The steel cables would hold the posts in place, in turn holding the panels in place. The panels would have enough structural strength to withstand the earth fill. The cables in this system are in tension. (See Figure 2-44.)

Buttressed retaining walls are similar in construction to counterfort walls. However, the buttressed wall, as the name implies, uses structural buttresses to maintain structural stability. The structural buttresses, located opposite the fill side of the wall at periodic intervals, keep the wall from toppling over; these buttresses are in compression. (See Figure 2-44.)

Drainage is important to all retaining walls; in areas subject to frost heave, it becomes especially important. The typical method of drainage is to build a series of drainage holes ("weep holes") through the wall near the base. These holes prevent the buildup of water on the fill side of the wall. To protect the holes from stoppages, wire or other filters may be used on the weep holes on the fill side; gravel is required around the weep holes to aid drainage and is recommended all along the fill side of the wall.

Where weep holes cannot be used, a drainpipe with periodic entries may be laid at the base of the wall on the fill side. The entries should be protected from stoppages similarly to weep holes. The drainpipe runs to some acceptable disposal point.

In areas subject to frost heave, the soil on the fill side of the wall should be checked. If the soil is largely clay, the wall may have to be lined over its entire fill-side surface with gravel or with a sandy soil; clays are particularly susceptible to frost heave and, in extreme cold, can topple the best retaining walls if precautions are not taken.

Retaining walls often require building permits, especially those over 3 feet tall. Engineering consultation or design is usually necessary to avoid damage to the wall or injury to people. The cost of engineering services is small, since builders can easily spend more on guesswork design than they would for economically designed walls that do not waste steel reinforcement or costly materials.

Figure 2-45 shows some the many materials that can be used in building retaining walls. Timber retaining walls are simple to build and long-lasting if the wood is treated properly. Wall *a* uses timber ties at each joint all along the bottom course. Vertical steel reinforcement is used at 3 feet on center horizontally. Wall *b* uses all vertical members set in concrete. Wall *c* uses vertical supports at 8 feet on center, set in concrete. The horizontal members are spaced about 1 inch apart for drainage.

Low retaining walls, such as wall *d*, may be built of solid masonry without reinforcement if the earth at the top is relatively level. For a more finished appearance, a masonry veneer may be applied, as in wall *e*. Stone wall *f* is built on a reinforced

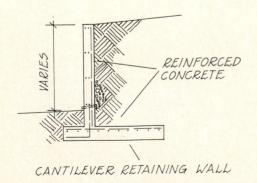

FIGURE 2-43   Cantilever retaining wall.

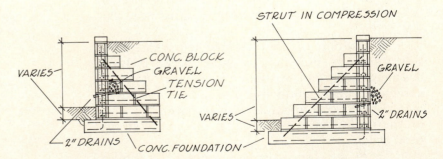

FIGURE 2-44   Retaining walls: (a) counterfort wall—note the steel tie in tension; (b) buttressed wall—note the strut in compression.

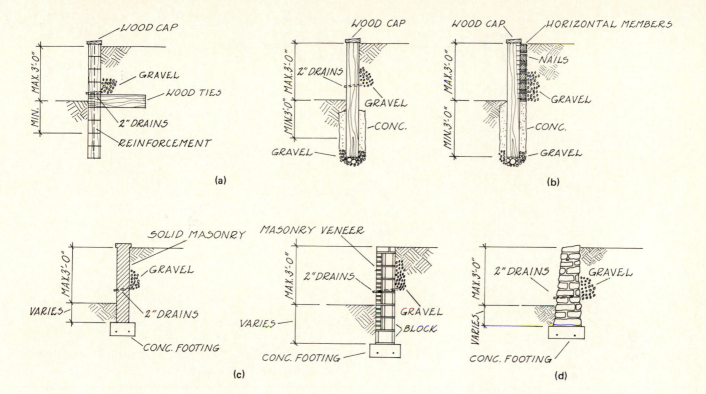

FIGURE 2-45 Materials used in retaining walls.

concrete footing. The drains have gravel at the back to prevent clogging.

Baffles. *Baffles* are materials used on slopes to slow down or divert the flow of water, preventing erosion. Any material suitable to the use and to overall design may be employed: masonry rubble, stone, broken concrete, crossties, and many others. Effective baffles with an acceptable appearance also may be made from combinations of these materials.

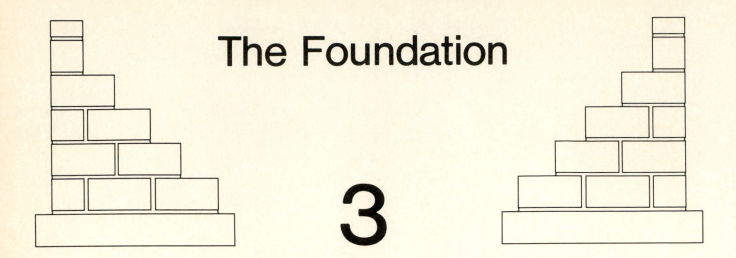

The Foundation

3

STAKING

Most builders enter the development picture only after all the major site clearing, grading, and preliminary site work has been done. Thus, the builder is usually concerned only with locating the house or houses to reference points such as the street, property lines, and so forth that have already been established.

The foundation is the first building component to be staked out. The foundation then itself becomes a reference for the construction that will be done over it. Typically, a corner is located by measuring from the property lines or other reference points. The first corner is marked with a stake and is sometimes called the *hub*. From this point, the remainder of the foundation may be located. The layout is largely a matter of maintaining square corners.

One simple, practical method of maintaining square corners is to use a large 3-4-5 right triangle. This triangle may be constructed with materials readily available on the job, such as one-by-fours or similar members, and may be any convenient size, 3 feet by 4 feet by 5 feet, or any multiple thereof—6 feet by 8 feet by 10 feet, for example. The triangle is placed at the hub, and string is then used to measure distances; the triangle assures a square corner. All the corners may be located and staked out in this manner. As the building progresses, more and more reference points become available, and the staking job gets easier and easier; the drive, for example, may be located quickly and

easily relative to the foundation wall, once the foundation wall is located. (See Figure 3-1.)

When the foundation is staked out, it should be thoroughly checked for accuracy before any concrete or masonry work begins. One fast, simple check is to measure the diagonals across the various rectangles formed by the stakes. Most buildings are composed of rectangles, and the diagonals of rectangles should be equal; if they are not equal, one or more corners are not square (90 degrees). Individual corners may be checked by measuring along the legs of the 3-4-5 right triangle, then checking to be sure the hypotenuse is the proper length; for example, measuring 6 feet along one side of the foundation stake out line, then 8 feet along the other side, the hypotenuse of the triangle formed should be 10 feet. If it is not 10 feet, the

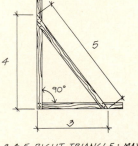

3-4-5 RIGHT TRIANGLE; MAY BE
ANY CONVENIENT SIZE: 3FT. x 4FT. x
5FT.; 6FT. x 8FT. x 10FT.; ETC.

FIGURE 3-1 A 3-4-5 right triangle, used to obtain square corners in trench work, footings, foundation walls, and similar projects.

corner is not square. Measuring major diagonals is another check; if they are equal, the corners are square. (See Figure 3-2a.)

A level or transit level may also be used to stake out a building. A fairly wide range of levels and transit levels is available. But to the residential builder, these instruments usually function simply to provide straight lines and square corners. The instrument is placed over a corner stake and sightings taken to other corners or similar locations.

When the foundation is staked out, *batter boards* are erected. Batter boards serve to maintain the points located earlier. The corners of the building, for example, are notched onto the batter boards; the strings may then be taken on and off, as desired, and the reference points will not be lost. The batter boards should be located far enough away from the footing trenches so as not to be damaged by excavation or other machinery. (See Figure 3-2b.)

COMPONENTS OF THE FOUNDATION

The most common foundation system in residential construction is composed of the *foundation wall*, upon which the exterior walls of the house rest, and the *footing*, which is the final transfer point between the house weight and the bearing soil beneath.

Footings

Footings distribute the weight above to the ground below. Consequently, the footings are wider than the foundation wall that rests on them. The foundation wall, in turn, is usually wider than the exterior wall that rests on it.

The size of the footing, and the reinforcement it has, if any, depends on the amount of weight it must bear. This weight is both "dead" and "live." The *dead load* does not change: It is the weight of permanent loads, such as the walls, ceilings, framing materials—all the materials assembled to make up the house. *Live load* changes: Comprised of people, furniture, the force and weight of sleet, snow, rain, and wind, and other variables, it is predictable only within certain limits. The footings must be strong enough to handle these loads, and the soil underneath—on which all the weight of the house including the foundation finally rests—must be able to accept the load.

In residential construction, the footings are almost always concrete. Other materials—such as brick and stone—have been used. But concrete is used much more, for a number of reasons: It has tremendous compressive strength, which, when combined with steel reinforcement, makes it able to support any house; it may be installed quickly, levelly, without the troublesome joints of masonry

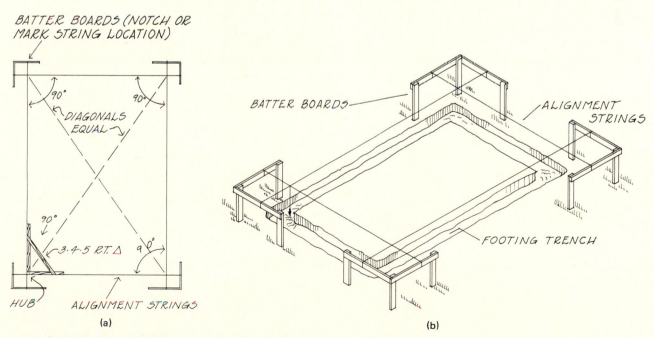

FIGURE 3-2 (a) Checking squareness of individual corners. (b) Batter boards and alignment strings used in keeping footing trenches level and square.

and stone; and it is predictable—concrete has been tested and observed in use for a long period of time—which simplifies design.

In general, little design is required for house footings, because local building practices are usually highly standardized. The footing must rest on soil capable of withstanding the weight of the house without settlement significant enough to cause the footing or the house to crack; this can often be accomplished by digging the footing trenches down to undisturbed soil—soil that has not been dug for a long time and is free of organic matter like tree limbs, brush, and grass.

In any case, the footings must be installed below the frost line, which is the depth to which the soil freezes for the particular region. Some areas of Florida may not have a frost line. On the other hand, some northern regions of the United States have frost lines deeper than 5 feet. The top of the footings must be below the frost line, or other provisions must be made to avoid damage to the footings and the house due to freezing action.

Footings may be formed by the trenches, when the soil is firm enough to maintain straight, neat sides. Footings may also be formed with form boards.

The concrete for footings must be the appropriate type for the job and region; must be mixed properly; must contain the type, correct amount, and appropriate size of aggregate; must be poured, finished, and cured under controlled and standard conditions; and may need to be tested under ASTM test procedures. Concrete will be discussed more fully later.

Footings may be poured separately from the foundation walls, or, if the foundation wall is to be concrete (as opposed to brick or concrete block), the footings and foundation wall may be poured as a single unit. One very typical residential method is to pour the footings first, forming a notch or "key" down the top center of the footing by impressing a shaped two-by-four into the wet concrete; after the footing has hardened, the foundation wall is poured, running down into the key portion, which helps keep the foundation wall from moving laterally. Many times, however, the footing is poured without the key and allowed to harden; the foundation wall is then poured on top of the level surface of the footing.

The size of footings is usually standard for the particular size of house, type of materials (frame, masonry, or combination), and building region. Generally speaking, the minimum thickness of the house footing is not less than 6 inches. The footing width depends on a number of factors, including soil bearing capacity, width of foundation wall, and whether or not reinforcement will be used. The footing is usually 4 inches or more wider than the foundation wall. Figure 3-3 indicates *typical* footing sizes for frame, masonry, and masonry-veneer houses of one and two stories, assuming typically loaded foundation walls with a soil bearing capacity of 2,000 pounds per square inch or more. These statistics will not, however, fit every house-building situation.

One-story houses can often be built without steel reinforcement. However, whether or not reinforcement is needed should be determined by a professional—usually the project architect or an engineer working with the project architect. The cost of professional design services in all areas of building will save the builder money in terms of both time and materials. When reinforcement is used in footings, it usually consists of two or more continuous lengths of steel bars, running near the bottom of the footing. (See Figure 3-4.)

FOOTING SIZES

	Frame		Masonry or Masonry Veneer	
Number of Stories	Minimum Thickness (inches)	Projection Each Side of Wall (inches)	Minimum Thickness (inches)	Projection Each Side of Wall (inches)
One story				
No basement	6	2	6	3
Basement	6	3	6	4
Two stories				
No basement	6	3	6	4
Basement	6	4	8	5

FIGURE 3-3 Footing sizes. (Courtesy of Federal Housing Administration.)

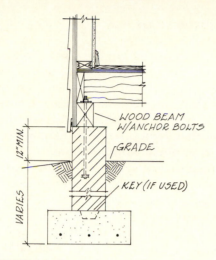

FIGURE 3-4 Typical residential foundation system.

The Foundation Wall

The foundation wall takes the load of the house and carries the load to the footings, which in turn distribute the load over the ground.

The height of the foundation wall may vary from 1 foot or so to 10 or 12 feet or even more, depending on the depth of frost line and on whether or not there is a crawl space or a basement under the first floor.

The foundation wall is built of concrete or, in some areas with mild climates, concrete block. Hollow brick and solid brick foundation walls are sometimes used.

The thickness of foundation walls depends mostly on their height and on the house load sup-

ported. Foundation walls are always thicker than the house wall that rests on them (except where the house exterior has a brick veneer, in which case the brick overhangs the outside face of the foundation wall by a maximum of 1 inch). The typical thickness for most residential construction is 8 to 12 inches. Figure 3-5 gives typical minimum recommended foundation wall thicknesses for frame, masonry, and masonry-veneer hours.

Concrete foundation walls usually require wood forms; plywood and two-by-fours are the typical materials used.

Steel reinforcement is often used in foundation walls, and, as with footings, this reinforcement should be specified by a professional—usually the project architect or an engineer working with the project architect.

The house must be secured to the foundation wall in some manner. This is usually accomplished using a *sill plate* and some type of *sill-plate anchorage*. The sill plate is the bottom two-by-four of the exterior wall of the house. The sill plate is usually anchored to the foundation wall in one of two ways:

1. For concrete foundation walls, ½-inch bolts are embedded a minimum of 6 inches in the foundation wall when the concrete is firm enough to hold the bolts in place but not so hard as to present difficulty in inserting the bolts. The bolts should be a maximum of 8 feet on center, and in no case should there be less than 2 bolts in a given sill plate; end bolts should not be farther than 12 inches

FOUNDATION WALLS

Foundation Wall Construction	Maximum Height of Unbalanced Fill (feet)[a]	Minimum Thickness (inches)	
		Frame	Masonry or Masonry Veneer
Hollow masonry	3	8	8
	5	8	8
	7	12	10
Solid masonry	3	6	8
	5	8	8
	7	10	8
Plain concrete	3	6[b]	8
	5	6[b]	8
	7	8	8

FIGURE 3-5 Foundation wall specifications. (Courtesy of Federal Housing Administration.)

[a]Height of finish grade above basement floor or inside grade.
[b]Provided forms are used both sides full height.

from the ends of any sill plate. The bolt spacing may need to be reduced for special conditions, such as in areas subject to hurricanes or earthquates; in such cases, the bolt spacing should be a maximum of 6 feet on center. When the foundation wall has cured, the sill plate may be drilled at the appropriate intervals, placed over the bolts, and secured.

2. For masonry or concrete block foundation walls, ½-inch bolts are embedded in the wall a minimum of 15 inches; the bolts will fit in the wall cavity and should be fully grouted. The spacing requirements are the same as for concrete foundation walls.

The purpose of sill plate anchorage is to tie the house to the foundation wall. In some exceptionally calm areas, sill-plate anchorage may be omitted (check with local authorities). However, sill-plate anchorage is always a requirement in earthquake-design areas and storm areas; it is also required for open structures such as porches and carports and for garages and other structures that are built on concrete slabs.

Wood-framed floors typically rest on the foundation walls and are secured to the foundation wall with sill-plate anchorage similar to that just described. Concrete-slab floors may rest on the foundation wall or on the earth within the foundation walls; where the floor slab rests on earth, the exterior wall of the house is usually anchored to the foundation wall; where the floor slab rests on the foundation wall, the exterior wall anchorage is to the slab. (See Figure 3-6.)

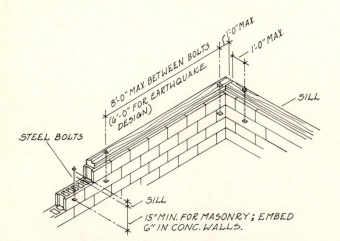

FIGURE 3-6 Typical sill-plate anchorage to masonry or concrete foundation walls. (If masonry, cores at anchor bolts are filled with concrete.)

Foundation Piers

Foundation piers are not likely to be used in areas where concrete floor slabs are typical. But piers may be used for wood-frame construction and for prefabricated modules, manufactured housing, mobile homes (which have metal floor framing), and other structures with framing similar to the old conventional wood framing.

Foundation piers usually support the house at regular intervals along the exterior wall and underneath the framing members of the house; the piers at the exterior wall are called *exterior piers* (see Figure 3-7) and the ones under the house are called *interior piers* (see Figure 3-8). Additional piers may be used as required to support additional house loads such as heating and air-conditioning equipment or some special equipment the owner may have.

Exterior piers are typically constructed of concrete, masonry, or concrete block (concrete block is sometimes an unacceptable foundation material in areas of extreme cold). When masonry, concrete block, or similar materials are used for exterior piers, these materials should be filled solid with concrete or grout; this is because the exterior piers are subject to wind loads (the lateral force or push of the wind on the house). Interior piers do not have to be solid unless for some reason they are subject to wind loads.

Piers that support wood-frame construction should be at least 12 inches above grade, to protect the wood from moisture. The maximum height of an exterior pier should be 3 times its least dimension; this height can be increased if the pier is reinforced. Interior piers of concrete or solid masonry should not be higher than 10 times their least dimension, unless they are reinforced. Interior piers of hollow masonry or concrete block should not be higher than 4 times the least dimension (all these heights for interiors assume no wind load).

The exact size and spacing of piers should be done by a professional. However, Figure 3-9 gives an idea of typical pier and pier footing sizes for residential construction.

In areas where earthquake design is required, special reinforcement will be necessary. Check with local authorities.

Concrete Grade Beams with Piers

In some areas, concrete grade beams with piers may be used. In appearance from the house exterior, the grade beam looks like a typical concrete

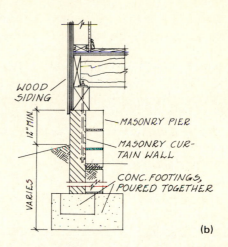

FIGURE 3-7 Examples of exterior foundation piers. (a) Masonry veneer, and (b) wood siding used in combination with a masonry curtain wall that screens a pier.

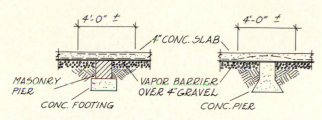

TYPICAL FLOOR SLAB PIERS. INSTALL EXTRA WIRE MESH OVER MASONRY PIER AS SHOWN. LIFT MESH UP OVER CONC. PIER AS SHOWN.

FIGURE 3-8 Examples of interior foundation piers made of masonry or concrete.

foundation wall. It is also about the same width as a typical foundation wall and serves the same purpose. But the grade beam, when used, may be as little as 6 inches below grade; at regular intervals, piers are placed under the grade beam to support it (the grade beam, in turn, supports the exterior wall of the house). The piers may be made with a posthole digger or power auger and are as little as 10 inches in diameter.

Grade beams should extend a minimum of 8 inches above grade when supporting wood-frame construction. The bottom of the grade beam should extend below the frost line, unless the area is not subject to frost or unless other engineering methods are used under the beam, such as replacing frost-susceptible soil with gravel.

The sizing of grade beams and piers, spacing of piers, reinforcement, and similar aspects of engineering design should be determined by a professional. Typically, the maximum spacing of piers is 8 feet on center; minimum pier diameter is 10 inches; and the bottom of the pier should be enlarged to cover about 2 square feet, for typical soils. Pier reinforcement is typically accomplished with one or more steel bars extending the full length of the pier and into the grade beam. Grade beams often require reinforcing, typically two to four steel bars running continuously through the beam. The typical minimum size of grade beam is about 8 inches wide by 14 inches deep (if wood framing is used, additional depth is needed to keep wood members up off the ground).

Figures 3-10 through 3-12 show typical uses of grade beams and piers.

FOUNDATION PIERS

Pier Material	Minimum Pier Size (inches)	Minimum Footing Size (inches)	Pier Spacing	
			Right Angles to Joists	Parallel to Joists
Solid or grouted masonry	8 X 12	16 X 24 X 8	8 feet on center	12 feet on center
Hollow masonry[a]	8 X 16	16 X 24 X 8		
Plain concrete	diameter or 10 X 10	20 X 20 X 8		

[a]Interior pier not subject to wind.

FIGURE 3-9 Foundation pier specifications. (Courtesy of Federal Housing Administration.)

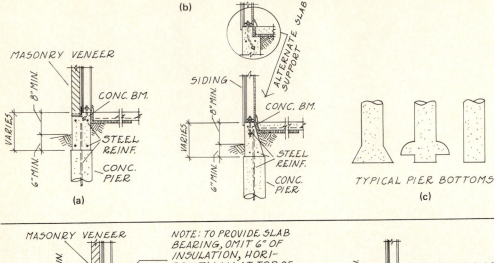

FIGURE 3-10 (a) and (b) Examples of concrete foundation walls supported by concrete piers. (c) Examples of some typical pier shapes.

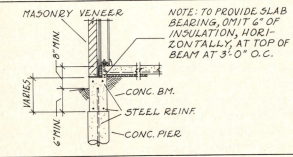

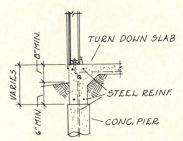

FIGURE 3-11 Grade beams and piers with structural slabs.

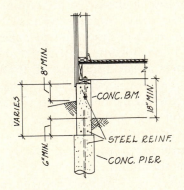

FIGURE 3-12 Example of grade beams and piers used to support a wood flooring system.

Sill-Plate Anchorage for Piers and Grade Beams

The purpose of sill-plate anchorage is to tie the house to the piers or the grade beams, which are the foundation in this case. The methods of anchorage are similar to those already discussed for anchoring houses to foundation walls. For masonry piers, embed ½-inch bolts a minimum of 15 inches in the pier; grout around the bolts. In concrete piers and grade beams, embed ½-inch bolts at least 6 inches. The bolts should be spaced a maximum of 8 feet on center in grade beams with not less than two bolts per sill length; bolts should not be closer than 12 inches to the end of sill pieces. When earthquake design is used, place the bolts a maximum of 6 feet on center in the grade beams. Power-driven steel studs may be used in place of the bolts for grade beams if earthquake design is not necessary; spacing of the studs should not exceed 4 feet on center with at least two studs per sill piece.

Anchorage is always required where earthquake design and concrete-slab construction are used. But in some areas, sill anchorage may sometimes be omitted; check with local authorities.

Pilasters

Pilasters are often seen in large commercial building and church interiors. They look like columns that have been pushed into the exterior walls with about a 2-inch remainder sticking out. In fact, they work very much like that. In foundation walls, pilasters are bonded into the wall, usually protruding from it about 2 inches. Used at regular intervals, pilasters increase the strength and rigidity of the foundation wall (they are often used in basement walls for this purpose). (See Figure 3-13.) Columns may also be used at the foundation wall to support girders. If made of wood, the base should be kept off the ground. Columns should not

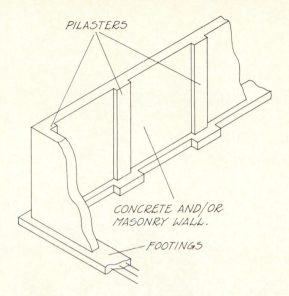

FIGURE 3-13　Example of pilasters used at periodic intervals to increase strength and stability.

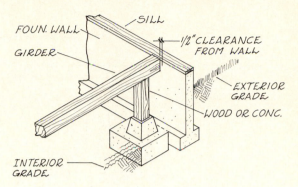

FIGURE 3-14　Column-supported floor girder.

be more than 12 inches from the foundation wall. (See Figure 3-14.)

Stepped Footings

In some cases, the building lot may slope severely. One solution is to excavate all or a portion of the area needed to locate the house on fairly level ground. But wholesale excavation like this is expensive and is usually avoided. Instead, the house itself is "stepped" along the slope, reducing the amount of excavation. Split-level houses are an example of this technique.

If the house steps along the slope, the footings must do the same. Care must be taken in the design and installation of stepped footings, or cracking damage to the foundation can easily occur. The following recommendations are typical, but professional design is required:

1. Horizontal distances along the footing between vertical steps should not be less than 2 feet, and the vertical step should not be greater than ¾ the horizontal distance between the steps.
2. The horizontal and vertical portion of the footing (which is assumed to be concrete) should be the same size and should be poured in one piece. Otherwise, the general considerations of footings pertaining to size, reinforcement, and so forth, are similar to other concrete footings with concrete foundation walls. (See Figure 3-15.)

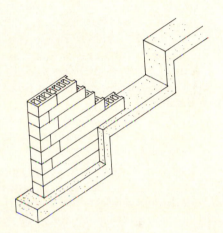

FIGURE 3-15　Typical stepped footing.

DAMPPROOFING AND WATERPROOFING THE FOUNDATION

The foundation itself should be protected from excessive moisture, but with most residential building sites, the basement or crawl spaces below finish grade are of greater concern. Water or excessive moisture should not be permitted to enter basements or crawl spaces. Under normal conditions, the following dampproofing will suffice for concrete foundation walls: Install one or more coats of bituminous dampproofing material (acceptable to local or applicable regulations) from the footing all the way up to finish grade. See also Appendix D-5.

Masonry foundation walls require additional precautions: Over a portland cement coating (parge) from footing to finish grade, apply one or more coats of bituminous dampproofing material; the portland cement coating should be at least ⅜ inch thick.

These recommendations provide minimum protection for basement walls. If the space within

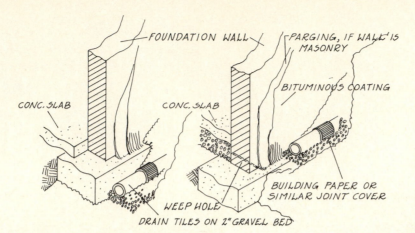

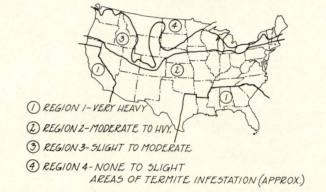

FIGURE 3-16 Foundation drains and dampproofing methods. Drain may be at footing (a) or below footing (b). Note weep hole.

FIGURE 3-17 Areas of termite infestation.

the foundation walls is below grade and is to be used as living space or is desired to be as dry as if it were habitable space (game rooms, utility rooms, storage, etc.), the following minimum recommendations are applicable:

1. Apply one coat of portland cement parging (masonry walls).

2. Then install a waterproof membrane all the way from the footing edge to finish grade. The membrane and its installation method should be satisfactory to local or applicable regulations; a typical material is polyvinyl chloride.

3. Foundation drains must be installed where they will pick up undesirable water and transport it away from the house (usually at the top or bottom of the rooting). These drains are typically concrete or tile. (See Figure 3-16.)

In some very dry regions, the dampproofing requirements for foundation walls without habitable space within are relaxed somewhat. But wherever below-grade space within foundation walls is to be made livable, these waterproofing recommendations are a minimum.

TERMITE AND DECAY PROTECTION

Termite Protection

Termites thrive in the damp southern regions of the United States and most of California. In colder areas, the danger of termite infestation lessens, and in extremely cold areas there is slight danger or none. (See Figure 3-17.)

The first step in controlling termites and their damage is simply to be sure that good construction practices are followed. These practices are usually required and spelled out by applicable authorities, and they are either outlined or implied in this text: Provide satisfactory site drainage; keep wood framing and siding an acceptable distance above the ground; provide adequate ventilation of the structure; flash properly; use vapor barriers and sheathing material as required; and so forth. Another early step is to be sure that stumps, brush, wood scraps, and other dead organic material that might attract termites are removed from the site.

Further protection is gained by blocking termite entry to the house at porches and entrances, planters and flowerboxes, garages and carports, and other susceptible points.

The most direct means of termite protection include covering masonry, concrete block, or similar foundation wall material with a cap of concrete; using metal shields to block termite movement to wood members; using a concrete foundation system (usually where the concrete floor slab and foundation wall are poured as one unit); using treated lumber; and using treated soil.

Capping the tops of masonry, concrete block, or similar unit foundation-wall construction with a minimum layer of 4 inches of concrete is sometimes an acceptable termite barrier; the cap also serves as a bearing surface for the house exterior walls.

Metal termite shields are made of aluminum, galvanized metal, copper, and similar materials. These metals look and are installed like flashing, except that their purpose is to keep out termites, not water. Termite shields are typically placed over the top of the foundation wall and either extend out from the foundation wall about 2 inches, for wood-framing protection, or into the slab (slab poured with the shield in place) if a concrete floor is used.

Concrete slabs, poured at the same time as the foundation wall so that the foundation wall and the floor slab are one piece, provide some termite protection.

Treated lumber offers termite protection. The lumber should be pressure-treated and should have identified (on the lumber) the name of the treating company, the type preservative used, and the amount of preservative contained, in pounds per cubic foot. There are many ways to treat lumber and many chemicals available for the treatment; the builder should be sure the wood passes authorized standards for its purpose and should be sure that the required wood members for the subject region are treated.

The soil may also be treated to help discourage termites and prevent damage. Various chemicals are acceptable, including DDT. The chemicals are typically mixed in an oil or water emulsion solution; the concentration of the chemical must meet applicable authority standards. These chemicals are applied uniformly under floor slabs, porches, crawl spaces, and other lateral constructions. Because these chemicals are often toxic to plants, animals, and human life, they usually are applied by specialists.

The information given here is necessarily general because requirements vary from region to region. Figure 3-18 shows typical uses of metal termite shields for both wood framing and concrete slabs.

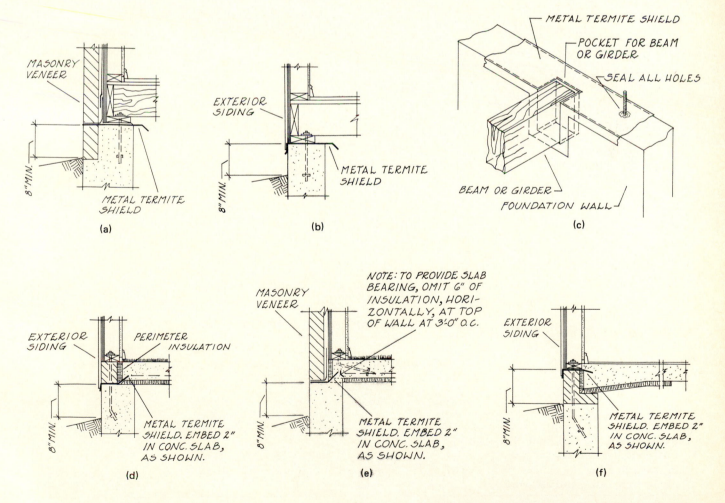

FIGURE 3-18 Metal termite shield installation methods.

Decay Protection

The need for decay protection, like termite protection, varies across the United States. Again, the southern states are areas of concern. Coastal areas also are subject to decay damage from high moisture and dampness. (See Figure 3-19.)

Generally, decay protection is provided by treating certain wood members to authority-recognized standards of pressure treating or by using decay-resistant woods such as California redwood, Tidewater red cypress, and western red

① REGION 1—NONE TO SLIGHT
② REGION 2—SLIGHT TO MODERATE
③ REGION 3—MODERATE TO HEAVY
   AREAS OF DECAY (APPROX.)

FIGURE 3-19 Areas of decay.

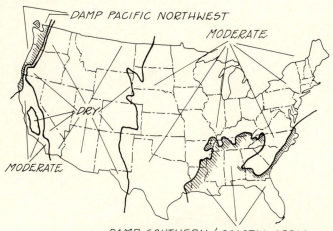

DAMP SOUTHERN & COASTAL AREAS
MOISTURE CONTENT AREAS (APPROXIMATE)

cedar. Some wood members will need to be protected in all areas of the country and some members will not, depending on moisture and climatic conditions for the particular region. Certain wood members that are directly exposed to excess moisture or weather—exterior door frames and fence posts, for example—will need protection in all areas; some joists, beams, columns, and other structures in more protected areas may not need special protection against decay. To be sure, check with local authorities.

Finish lumber should contain about the same amount of moisture as it will have in the house. The map in Figure 3-20 indicates relative moisture content by region; the table gives recommended moisture content as a percent.

WOOD FOUNDATION SYSTEMS

Wood foundation systems have reportedly been accepted by the Federal Housing Administration, the Veterans Administration, the Farmers Home Administration, the Basic Building Code, the Uniform Building Code, and the Standard Building

FIGURE 3-20 Finish lumber moisture content. (Table courtesy of Federal Housing Administration.)

FINISH LUMBER MOISTURE CONTENT

Use of Lumber	Dry Western Areas		Damp Southern and Coastal Areas		Remainder of US[a]	
	Average (%)	Individual Pieces (%)	Average (%)	Individual Pieces (%)	Average (%)	Individual Pieces (%)
Interior trim, woodwork, and softwood flooring	5–7	4–9	10–12	8–13	7–9	5–10
Hardwood flooring	6–7	5–8	10–11	9–12	7–8	6–9
Exterior trim, siding and millwork	8–10	7–12	11–13	9–14	11–13	9–14

[a]Local areas may require other moisture content limitations due to special climatic conditions.

Code. Thus it is likely that wood foundation systems will be seen more in the future.

It is not difficult to see how wood can function as a foundation from a structural point of view; wood can support two-, three-, and four-story buildings. There is no structural reason why wood cannot be used for foundations. The problem, of course, is waterproofing the wood; this is done with chemicals in a pressure-treating process. Caution must be taken to assure that wood foundation materials meet local or applicable requirements.

The wood foundation system is constructed like a stud wall sheathed with plywood. The foundation wall sits on a wide wooden plate, which takes the place of the normal concrete footing; the footing, in turn, rests on a bed of gravel.

There are certain advantages to wood foundations that deserve serious consideration. First, they are very simple and fast to install, even with unskilled labor. There may be materials cost savings. The foundation creates a crawl space that makes ductwork convenient, and there is additional space in the wall itself for wiring, plumbing, and so on. Wood is a better insulator than concrete to begin with, and when insulation is added to the wood foundation wall (insulation installed similarly to stud walls above), there may be energy savings to the owner. (See Figure 3-21.) For additional information about wood foundations, contact the American Plywood Association in Tacoma, Washington; The National Forest Products Association in Washington, D.C.; or the Koppers Company, Inc., in Pittsburgh, Pennsylvania.

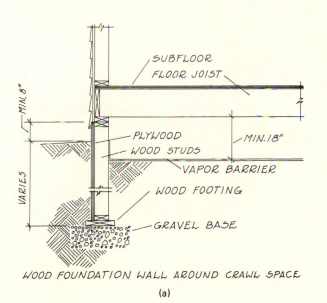

WOOD FOUNDATION WALL AROUND CRAWL SPACE

(a)

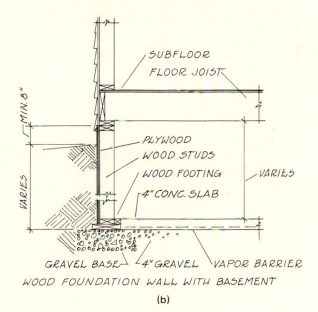

WOOD FOUNDATION WALL WITH BASEMENT

(b)

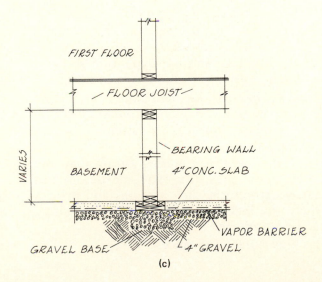

(c)

FIGURE 3-21 Wood foundation systems.

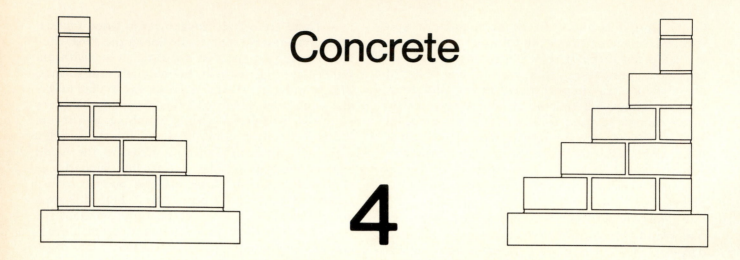

Concrete

4

PORTLAND CEMENT CONCRETE

Concrete is made up of aggregates held together by a paste. The aggregates are typically sand and stone and are inert. The paste is portland cement, and water and is the active ingredient. The ingredients vary; usually, concrete is about 60% to 80% aggregate and about 20% to 40% paste. Local building conditions and weather influence the makeup and placement of concrete, but in all cases the desirable end is to ensure that a firm bond is achieved between the aggregates and the paste. Properly designed and placed, concrete is perhaps our most durable building material; it has consequently been much used since the late nineteenth century.

There are five types of portland cement:

1. Used for general concrete work.
2. Liberates less heat during the curing process; thus may be used where temperatures are higher than favorable for type 1. Also has a moderate resistance to the action of sulfates.
3. Valuable where quick hardening is important for construction scheduling or other purposes. Will develop considerable strength within 24 hours.
4. Develops lower heat during curing than the other types. Thus, this type is often used in massive structures such as dams and bridges where the high heat generated during curing can be a problem.
5. Has high sulfate resistance.

In addition to these typical concrete types, there are special concrete preparations for specific purposes. White portland cement concrete, for example, may be made by using materials low in iron and manganese oxide, which produce the typical gray color of concrete; the resulting white concrete may be used as is, or color mixes or stains may be prepared. Waterproof concrete is also available.

In areas of extreme cold, an air-entrained agent may be added. The agent causes the concrete to be filled with microscopic air bubbles. These air bubbles permit water to freeze and expand without damaging the concrete.

Concrete Ingredients

Concrete is made of portland cement, coarse aggregates, fine aggregates, and water. The ideal mixture of concrete has a balance of these ingredients such that the aggregates fit together firmly and are held together by the chemical action of the bonding ingredient, portland cement. If there is an imbalance of any of the ingredients, the concrete will be inferior. A discussion of each ingredient follows.

Portland cement. Portland cement, in form, is a fine dry powder. It is composed of lime, silica, alumina, iron oxide, magnesia, and sulfur trioxide. When combined with water, the cement forms a paste that sets into a stonelike material; the set paste is fairly strong itself, but not strong enough for construction purposes. The desired quality of the paste is its ability to interact with the aggregates, forming concrete.

Coarse aggregates. Coarse aggregates are typically crushed rock, although natural gravels may be used where they are available. Slag from steel mills is sometimes used; so are crushed oyster shells. Basalt and limestone, however, are the preferred aggregates.

The size of aggregates varies. Even the definitions of size sometimes vary. In general, the smallest particles of coarse aggregate are those that will be held by a ¼-inch (no. 4) sieve. Pieces that go through are called sand or fine aggregate. The largest (most coarse) aggregate used in typical residential construction is about 1½ inches. The largest aggregate in a slab should be less than ⅓ the thickness of the slab; so, for a 4-inch driveway slab, the maximum size of aggregate would be a little over 1 inch. However, a retaining wall that is 18 inches thick could use aggregate largest than 1½ inches. Thus, the size of coarse aggregate varies with the particular job also. The aggregates must not be too close to the surface of the concrete or too close to reinforcing steel; typically, the largest aggregate should be less than ¾ of the spacing of reinforcing bars from each other and from concrete surfaces; otherwise, the aggregate will have a weakening effect on the concrete.

It is important in concrete work to maintain a variety of aggregate sizes. If the same size of aggregate is used, the relative volume of voids will be approximately equal whether the aggregate is large or small. But if the sizes are varied, the voids tend to be smaller, the smaller aggregate working between the larger pieces and the paste filling uniformly between all the aggregate, bonding the mass together.

Fine aggregates (sand). Fine aggregate, or sand, usually means a natural or crushed rock that will fall through a ¼-inch (no. 4) sieve. The smallest size is about 1/50 inch. Fine aggregate specifications vary, and, as with coarse aggregates, their selection is influenced by the particular job.

Ocean sand is not recommended. Sand particles with irregular shapes are preferred.

Water. Water starts the chemical process with the portland cement, which in turn becomes the bonding agent with the aggregates. This chemical process, called hydration, is rapid at first, then slows somewhat; generally, the process is assumed to be complete in about 28 days, but the concrete increases somewhat in strength thereafter. The concrete may be used before the 28 days in some cases. It is important to realize that the concrete's being hard does *not* mean that it has completed the curing process and is therefore ready for full use; usage is the key. A concrete walk, for example, might be used in a day or two; a concrete driveway, especially where heavy trucks will be used, requires more time.

The use of clean water is important. Water that contains soil or debris, even if it is minute, is adding an unwanted "aggregate" to the concrete that can weaken the chemical bonding process and result in inferior concrete. A simple rule to use is that water for concrete should be clean enough to drink.

Where the concrete will be exposed to sulfate compounds in the ground, special sulfate-resistant cements should be used.

Cleanliness of Ingredients

Concrete purchased from central mix plants may be assumed to be clean. If the contractor mixes the ingredients, care should be taken to avoid soil or debris in the cement, aggregate, and water. The bonding process between portland cement and the aggregates is critical. Oil, grease, and other impurities can create a film on aggregate, preventing bonding and resulting in unusable concrete.

Moisture

Aggregates may contain absorbed moisture, may have moisture on the surface, or may even present both conditions, depending on the type of aggregate, method of storage, weather, and other conditions. Contractors should be aware of the moisture state of the aggregates because the moisture affects the amount of water that will be used in the mix.

Storage and Handling of Aggregates

Coarse aggregates vary in size to minimize the size of voids. When aggregates are stored in loads over the size of a truckload, the aggregates tend to segregate according to size, leading to concentrations of large aggregate and concentrations of small aggregate; when a load of aggregate is removed from such a pile or bin, it will be not well blended. Aggregate is preferably stored in truckload-size piles.

Handling methods also influence the quality of aggregates. If bulldozers or trucks are allowed to roll over aggregates during handling, some of the aggregate will be crushed or broken, possibly resulting in an imbalance of aggregate size when the concrete is mixed. Aggregate dropped from considerable heights, as from elevated conveyors, may also break into smaller pieces. Cranes or similar equipment is the preferred method of handling aggregate to store it in separate, truckload-size batches.

Fine aggregates should be handled and stored similarly. Fine aggregates should also be protected from wind during handling and storage, as wind tends to blow off the finer particles.

Amount of Aggregates

The amount of aggregate versus cement may be classified as rich, lean, and harsh. Rich mixtures have high cement-to-aggregate ratios; lean, the reverse. Harsh mixtures have so much aggregate that they are difficult to work.

Air in Concrete

Air pockets in concrete reduce its strength, sometimes disastrously. Air pockets of visible size, called honeycomb, are typically caused by improper distribution of concrete; forms, reinforcement, and similar obstructions to distribution make this condition more likely. Poor mix proportion also helps create air pockets. Good job supervision is the best prevention for these potential problems.

Rotary concrete mixers pull excess air into concrete, but it usually escapes if proper placement methods are followed.

Air-entrained concrete, mentioned earlier, deliberately creates microscopic air bubbles in concrete. These air bubbles help prevent damage due to extreme cold. Even these microscopic bubbles weaken the concrete because they are voids, so stronger mixes are used to regain the strength lost.

The typical percentage of air in a good concrete mix ranges from about 3% to 9%. This percentage may be controlled by adjusting the size of the aggregates used and by adjusting the amount of entraining agent, where this agent is considered necessary.

Setting Time

Some hydration begins the moment the cement is mixed with the aggregates, the amount depending on the moisture in and on the aggregate.

The calculated time of hydration begins when water is introduced to the mixture. Hydration is a chemical reaction, and, as in most chemical reactions, high temperatures increase the process; thus, concrete tends to cure more slowly in cold weather than in hot. Other factors include the method of mixture (by hand or mechanical means), the amount of water in the mixture, the type aggregate, whether the concrete is above or below ground, and the shape of the concrete.

Concrete must be worked into place before it sets. Contractors must be aware of specific setting times and plan the method of pouring and the amount and type of labor and equipment needed accordingly. In piles and similar structures, setting may begin in less than an hour; it may do the same in a nonagitating truck; generally speaking, most structures begin to set in about 90 minutes, but careful study of exact local weather and building conditions should be made before pouring is begun.

Chemical additives are available that retard or accelerate the setting process. Under average building and weather conditions, these additives are not needed.

The Slump Test

The amount of water in the concrete mix is very important. Too much water may cause the aggregates to settle, leaving an excess of paste at the top and weakening the concrete. Too little water leaves dry areas within the mix that will not bond properly. Either condition makes the workability of the concrete poor; more important, an improper amount of water in the mix may make the resulting concrete unusable. The slump test is a quick, simple method of checking the consistency of concrete.

A testing cone is required. This cone is 12 inches high, with a truncated top 4 inches in diameter; the base is 8 inches in diameter. The cone is placed base down on a level board, concrete block, or similar hard surface. It is then packed to the top with the concrete to be tested. When the cone is lifted, the concrete will squash or slump down somewhat from its height when the cone was in place, causing it to bulge out at the sides. The slump rating is then obtained by measuring from the top of the concrete to the level surface and subtracting from 12, the height of the cone. Stiff mixes will slump an inch or so from the height when the concrete was in the cone; wetter mixes will slump about 4 inches; very fluid mixes will slump about 6 inches. How much the concrete slumps depends on the type, blend, and amount of aggregate in the mix, in addition to the amount of water. No aggregate should be removed for convenience in filling the test cone.

Stiff, low-slump mixes are typically high-strength mixtures; they shrink less and offer better water resistance than wetter mixtures. Stiff mixtures are more difficult to place, and care must be taken to avoid air pockets near forms and around steel reinforcement.

Wetter mixtures are easier to place, less likely to gather air pockets during placement, and fit well into smaller spaces such as house foundation-wall forms.

Proportions of Ingredients

The proportions of the several ingredients of concrete affect the quality of concrete and thus its usages. For large jobs, water, cement, and aggregates are expressed by weight. The same ingredients for small jobs are given as volumes. Air is always expressed as a percentage.

Water to cement. A quarter pound of water per pound of cement is technically enough to make the hydration process work; this 1:4, or 25%, ratio would result in the maximum paste strength for bonding the aggregates together. This ratio, however, would be too dry to be practicably workable. In fact, the stiffest practicable ratio is 40%, and this ratio is usually limited to concrete that will be exposed to sea water and sulfates. Typical usages in average climates and under normal conditions use a ratio of 53%; areas of extreme cold use a ratio of 49%. Dams and other structures where massive amounts of concrete will be used may have a ratio of 58% or higher. The preceding refers to non-air-entrained concrete. As mentioned earlier, air-entrained concrete is more workable than non-air-entrained concrete; thus, air-entrained concrete requires less water to be workable than non-air-entrained concrete.

Cement to aggregate. As mentioned earlier, aggregate makes up from 60% to 80% of the mix, by weight or by volume. The exact proportion of aggregate is usually determined in a laboratory. It may, however, be determined in the field. In the field, actual aggregate samples are used, typically in 20-pound batches.

Tables are used to determine test quantities for water-to-cement ratios and size of aggregate. The ingredients of the batch are mixed dry; then

noted quantities of water are added until the appropriate slump is achieved. At that point, the quantities are recorded. The resultant mixture usually contains as much aggregate as possible shy of a harsh mix.

Coarse aggregate to fine aggregate. The total aggregate used typically makes up 60% to 80% of the concrete ingredients; sand is 30% to 60% of the total aggregate figure. Generally speaking, the smaller the coarse aggregate, the more sand is used. Of course, there are size differences in the sand itself, and more sand can be used where there is considerable coarseness than in predominantly uniform, fine sand. Too much coarse aggregate relative to sand produces a harsh, soupy mix that, in the extreme, tends to segregate the ingredients; excessive sand raises costs because more cement is required, and the resultant mix tends to shrink.

Concrete Strength

The strength of concrete is determined by how much weight it can support per square inch without being crushed. Tests are made after 28 days of curing, unless the concrete is designed to be used prior to that time (early-setting concrete, for example).

The strength designed for depends, of course, on the usage; concrete for residential driveways, for example, does not have to have the strength of concrete used for expressways. Typically, concrete is rated between 1,000 to 12,000 pounds per square inch; pavements average somewhere between 4,000 and 7,000.

Concrete Manufacturing and Mixing

Portland cement is composed of lime, silica, alumina, iron oxide, magnesia, and sulfur trioxide. The raw materials consist of limestone, cement rock, oyster or marl shells, and shale, clay, sand, or iron ore.

These ingredients are crushed, then ground to a powder, mixed proportionately, then kiln-burned at temperatures approaching 3,000° F. The kiln burning chemically changes the combination of ingredients to a cement clinker. The clinker in turn is mixed with gypsum and ground again to a very fine powder (particles measure about 1 micrometer —one thousandth of a millimeter). This fine powder, called portland cement, may then be packaged in bags or barrels or shipped to distributors in bulk quantities; rail lines, barges, and trucks are used for shipment, depending on the most economic method for the amount of cement needed and its destination.

The other ingredients of concrete are typically available locally. The equipment and methods used to mix concrete locally vary widely, depending on the type of job. A dam, for example, with the huge masses of concrete required and the steady pouring schedule that such a job demands, justifies the expense of a stationary mixing plant at or near the job site; concrete transportation to the pouring area would be accomplished with an agitating truck or a dumper. Highway contractors use paving mixers; the concrete ingredients are placed in the paver from dry "batch" trucks, and the paver mixes the ingredients; then the paver moves along crawler treads, depositing the concrete on finished grade behind itself.

The most familiar mixer is the ready-mix truck. The mix can be done completely by the truck, the mixing starting as the truck starts for the job; or the mix may be accomplished at the central mix plant; or the mix may be partially begun before it is placed in the truck (called shrink mix). Ready-mix trucks are versatile in terms of speed, quantities deliverable, and access to the job.

Each job demands a relatively narrow range or combination of equipment, materials, methods, and labor that will most likely bring the job in on schedule and within the budget. Big contractors often must own or lease millions of dollars' worth of concrete equipment and employ hundreds of specialists to be competitive on the jobs they bid on. On the other hand, small residential contractors may be able economically to mix and place all their concrete work with a portable power mixer.

Joints in Concrete

There are two types of joints that the residential contractor is primarily concerned with: contraction and expansion. (See Figure 4-1.)

Contraction joints. Concrete shrinks enough in cold weather to crack. Contraction joints do not prevent cracking but rather are placed where cracks are anticipated. The typical contraction joint is made by sawing the set concrete: A narrow cut is made about one fourth the depth of the concrete; another, wider cut is then made an inch or so deep on top of the first cut, and a sealing material is installed. When the concrete cracks, it cracks along the cuts (if they were placed correctly), preventing unsightly cracks and cracks that might be more damaging.

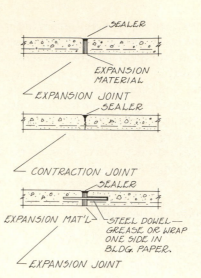

FIGURE 4-1 Typical residential expansion and contraction joints.

The sawing is usually done 4 to 6 hours after the concrete surface has been finished, when the concrete has set enough to be sawed but not so much that the blade will chip the edges badly or throw out aggregate.

Expansion joints. Concrete expands during hot weather and wet weather (when it absorbs water). When the concrete expands against another slab, building, or similar substantial object, buckling and cracking are likely to occur. To prevent this, expansion joints are used.

Expansion joints typically run straight down all the way through concrete slabs. The width of the joint varies and is usually filled with some compressible material with waterproof qualities; asphalt-impregnated fiberboard is a typical filler in residential construction. Some method of preventing slabs from getting out of vertical alignment due to the joint is sometimes necessary; steel dowels are one method. The dowel is set permanently in the slab on one side of the joint; on the other side, it is wrapped in building paper, oil-coated or otherwise made so that it does not bind with the concrete. When expansion takes place, the dowel slides horizontally but does not allow vertical movement.

Typical uses of expansion joints are where driveway slabs meet house foundation walls, between steps and the foundation wall, around columns and posts set in concrete slabs, and around the perimeter of basement slabs where they meet the foundation wall of the house. Expansion joints are often used in long drives, but in residential work, the dowels are often omitted.

Expansion joints can cause trouble if they are not sealed and maintained. Sealers keep out sand and other noncompressible materials that can cause the joint not to function. The joints must be kept sealed and free of debris.

The figures shown in this text relative to joints are for illustration purposes. Local building and weather conditions dictate the applicable construction standards used. Where no such standards exist, the joints should be designed by an engineer or architect.

Installing Concrete

Footings. Ready-mix drivers do not usually help with the pour. So, for each house footing pour, at least three workers will be required: one to hold and adjust the chute and two to work the concrete in the footing trench.

Pours are usually begun at a convenient building corner. The truck driver will move the truck away from the corner as required for a steady pour, adjusting the amount of pour as needed. The workers continue around the footing trench in this manner, the two in the trench working the concrete to the top of the footing forms (if forms are used) and leveling it as they go. Other footings adjacent to the foundation-wall footing—such as porch footings—may be poured at the same time.

Footing concrete is often poured a little wet because footings are thicker than slabs and more difficult to work in place, especially if steel reinforcement is used. Also, being a little wet allows more time to complete the footing pour before finishing the top of the footing, which is typically float-finished.

When the footing pour is complete and the surface gleam or shine has dulled, it may be floated. If the trench workers are good, they can level the top well enough for floating as they go; if they are not so experienced and the top is not level, it will have to be leveled prior to floating; a two-by-four screed is usually used.

Foundation walls. Concrete foundation walls are poured similarly to the footings. Forms are always used, and steel reinforcement is typical.

Basement slabs. If you have a basement, it should probably be poured first; otherwise, the ready-mix truck will have to drive over or around concrete walks, stoops, patios, and other such surfaces.

The proper subgrade should already have been established to accommodate the gravel fill and the appropriate thickness of the slab in order to get the floor at the elevation desired. Expansion material, typically asphalt-impregnated fiberboard, is installed around the edge of the foundation wall according to the construction detailing. Reinforcing wire is usually used for slabs other than walks, and this wire must be cut to fit around any protrusions through the slab, such as plumbing, drain lines, and water pipes. The construction detailing for these protrusions should also be provided: expansion materials around columns, drain grate housings, and so forth.

The pour is obviously easier before the wall framing is done, but it may be done after the framing is up by directing the chute between the stud framing at selected points.

It is possible to pour about 2 inches of concrete, then set the wire reinforcing over that, and pour the remainder of the slab thickness. But in practice, the wire is typically installed first, over pebbles or held up by metal rods and wire to the desired height (the wire mesh will be more than halfway into the depth of the slab). Walk boards or some other system must be provided to keep workers from tramping the wire too far down into the concrete.

The concrete should be poured as uniformly as possible, never allowing large heaps to accumulate or allowing the concrete to drop more than several feet; dropping too far tends to segregate the ingredients and thus weaken the concrete. Chute extensions are available that usually make it possible to avoid the above conditions; otherwise, some type of chute extension must be built or the concrete will have to be hauled in by wheelbarrow or other methods.

Enough workers must be employed—depending on the size of the slab—to work it in place, assuring no air pockets and assuring a uniform thickness. The concrete should be sloped to floor or other drains.

When the pour is complete, the surface is leveled, typically with a two-by-four screed. The finish used depends on the use the floor will receive: steel trowels give a dense, smooth finish, wood floats a rougher finish, and brooms a gritty, nonslip finish.

Contraction joints should be placed at least every 30 feet and at offsets. These joints are relatively quick and inexpensive to install compared to the unsightly cracking that may result if they are not used; generally speaking, if there is any doubt about whether a contraction joint should be used, it should be used.

Carport and garage slabs. Carport and garage slabs are constructed and poured similarly to basement floors. There must be expansion joints where the slab meets the foundation wall of the house and similar construction. Wire-mesh reinforcement is needed; so are contraction joints at least every 30 feet, preferably more, and where there are offsets. The finish is usually smoother than driveways and walks, depending on local preferences.

Concrete porch slabs, entrance stoops, and steps. Porch slabs, entrance stoops, and steps may be poured at the same time. If the porch is high enough to need handrails, provisions should be made in the formwork to accommodate them. If the porch or entrance slab has a foundation wall, it should have been installed when the house foundation wall was installed. Entrance stoops on grade should be laid over gravel and be a minimum of 4 inches thick; wire reinforcement may or may not be necessary, depending on the size of the slab and how much weight, if any, it must support other than normal foot traffic.

If the entrance slab or porch is high enough to need handrails, the steps will usually require reinforcement with wire mesh or steel rods.

Driveways. Driveways are typically 10 to 20 feet wide and 4 inches thick. They are reinforced with wire mesh in the same manner discussed for basement floors. Municipalities usually provide construction details for the entry apron, required slopes for drainage, crown (if any), and other details. Driveways are poured over gravel.

Expansion joints are required at the sidewalk, carport or garage slab, and anywhere else the slab butts against other construction.

After the forms and expansion joints are in place, the concrete may be poured; the concrete should be kept moist for about 3 days, to ensure proper curing.

The finish may vary somewhat with location; most typical is a level but somewhat rough surface.

Walks, sidewalks, patios, and accessories. Generally speaking, walks, sidewalks, patios, and accessories (such as concrete planter boxes) are left until last to spare them possible damage during construction. Municipalities provide construction detailing and specifications for public walks; they are generally 4 inches thick and are sometimes reinforced with wire mesh. Contraction joints are used approximately every 4 to 5 feet.

All walks are preferably laid over a bed of gravel, but this may sometimes be omitted. Finishes vary somewhat with locality, but overly smooth finishes are not desirable.

Patios usually do not need reinforcement; they should, however, be laid over gravel. Finishes vary greatly, often to gain aesthetic effects.

Concrete Forms

Forms are the supports or "molds" used to hold concrete in the desired shape until it sets enough to hold itself in place; the time forms are left in place varies with building conditions and with the type of structure being built.

Forms for structures like high-rise buildings and bridges are almost as complicated as the structures themselves and usually must be designed by engineering specialists to meet code approvals; these forms are significant cost items. Residential forms, however, and other low-rise construction forms are rarely this complicated and may be designed and built by the contractor. (See Figure 4-2.)

Forms may be built of wood, steel, or other materials suitable to hold the concrete in place; they may be built on the job or prefabricated. Where large numbers of the same item will be built, it is usually economical to build permanent forms or buy prefabricated ones of steel or other durable materials.

Forms also may become a part of the concrete itself. For example, where concrete slabs will have a brick or stone border, the border may sometimes be used as the form. Patios and terraces are often enhanced in appearance by using redwood grids that serve as forms as well as expansion joints and decorative elements.

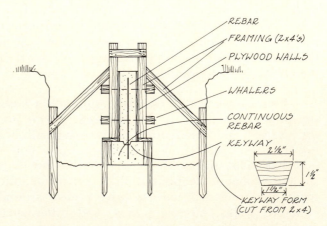

FIGURE 4-2 Typical footing and foundation-wall formwork.

Curing Concrete

The curing of concrete is a chemical process. Thus, there is more involved than simply letting the concrete set until it dries. In fact, keeping the concrete moist is esential to curing. The curing procedure varies with the weather, the structures themselves, and other building conditions. Curing procedures may sometimes require engineering consultation.

Generally speaking, residential concrete surfaces are kept moist during curing by spraying with a fine water mist. Burlap, laid over the concrete and kept moist, may also be used. Special paper is available from building supply houses that works similarly to burlap to keep the moisture in the concrete from evaporating before the chemical process of curing is complete.

Additives and sealers are sometimes used to aid curing. And in cold weather, it may be necessary to add heat to the concrete; steam may be used for this purpose.

The time needed for curing varies with weather, type of structure, and other building conditions. Most residential concrete structures will cure enough for limited use within a week or 10 days, but this should not be taken as a rule—sometimes it is less, sometimes more. Best curing temperatures generally range between 40° and 70° F.

Vibrating Concrete

Vibrating equipment is available to aid in consolidating concrete ingredients after a pour. This equipment should not be used to move concrete horizonally within the forms. If it is used too long or if the intensity of vibration is too great, the concrete may be damaged.

Tools and Equipment

Portable concrete mixers. Portable concrete mixers are available that will mix about 1½ to 9 cubic feet of concrete. The smallest units can be picked up and carried by one or two workers; the larger units have wheels, and some of them can be towed by automobile or pickup trick. Most of the mixers are available with a choice of electric or gasoline motors; they will mix mortar, stucco, plaster, feed, and liquids as well as concrete.

Because ready-mix is so easily available today, contractors even in small towns would find portable mixers too slow for driveways or similar slabs.

They are, however, practical for walks, patios, and similar usages, where the large size of the ready-mix truck might be inconvenient or where there is not enough work to justify the larger ready-mix quantities. Also, where the smaller jobs like walks and patios can be done without interfering with the scheduling of larger jobs (heavy trucks must not be allowed on walks, patios, etc.), the small jobs may be done simultaneously, saving time.

Portable mixers also are good for less experienced workers and homeowners, for jobs that can be broken down into segments—patios and terraces that are poured in grids, for example; this allows small areas of work to be done, then delayed, then started again, which is helpful when work time is fragmented.

Boots, shovels, hoes, gloves, and rakes. Prolonged exposure to concrete can burn the skin. Workers should wear calf-high or higher boots and gloves while working.

Shovels and hoes are the typical tools for residential concrete work; the ordinary garden variety will do, but builder-quality tools are usually cheaper in the long run.

A short-handled shovel with square edges, sometimes called a square-point, is the favored tool for distributing slab concrete. Where the forms are too narrow for a shovel, as in foundation wall forms, a two-by-four may be used. Hoes also are used for spreading; concrete hoes have a pair of large holes in the blade that makes spreading concrete easier. Special double-edged concrete rakes are available; one edge has teeth like a garden rake, for spreading. The other edge is serrated for leveling and preliminary floating.

Screeds. After concrete is placed, it is leveled with the top of the forms using some type of screed. Short spans—the tops of foundation footings, foundation walls, walks, and the like—can be leveled with scrap sections of two-by-fours. The screed is zigzagged across the top of the concrete at a slight angle, using the top of the forms as a guide. Such short spans may be leveled by one worker. Longer spans, such as double driveways and garage slabs, may require two-by-six or larger wooden members to avoid sagging; two workers are required for longer spans.

Where very precise leveling is required—floor slabs, for example—a metal screed is desirable. Magnesium screeds are available in typical sizes of 1 by 4, 2 by 4, and 2 by 5 inches; the lengths vary from about 6 to 24 feet. Magnesium screeds are light and rustproof, and concrete does not cling to them as readily as it does to wood. The initial cost of these screeds is obviously many times what a comparable wood member would cost, but the cost is recouped in time because the metal screeds last longer than wood.

Tampers. The purpose of tamping is to push the heavier aggregate far enough below the surface so that final finishing may be done. Where the appearance of the finish is of little or no concern, as on rough-finished garden walks and steps and similar surfaces, tamping may be done with rather primitive tools; short sections of four-by-fours with handles nailed on are typical where small areas are to be tamped. Larger, similarly made wooden tampers may be used for larger areas.

Where more precise finishes are desired, more sophisticated tampers such as the "jitterbug" or "rollerbug" are used. The jitterbug is essentially a frame (sizes vary, but 48 by 6 or 8 inches is typical) in which is mounted a metal grill with round or diamond-shaped holes. The handles are usually long, allowing the worker to stand while tamping. Tamping is done with short, quick up-and-down strokes. As the concrete is tamped, the finer ingredients of the concrete come up through the holes and the larger aggregate is pushed lower. It is this finer layer of concrete that will receive the final finish. In using any tool that moves the aggregate within concrete, care should be taken not to overdo the job; too much tamping will damage the concrete.

Rollerbug tampers do the same job as jitterbug tampers. But, as the name implies, rollerbugs use roller grills instead of the flat grills of the jitterbug. The rollers vary in size, but 39- to 77-inch widths with diameters of about 5 inches are common. Long handles allow workers to tamp the concrete without standing on it as they do with jitterbug tampers.

Floats. Tamping creates a shallow depth of fine materials at the top of concrete suitable for final finishing. When tamping is complete, floating is performed to smooth the surface of the tamped concrete. Quite a few types of floats can be used, depending on the size of the area to be floated and the final surface shape and finish desired.

"Bull floats" are typically made of 2- by 6- by 36-inch boards with long handles for smoothing large areas like driveways and garage and floor slabs; the edges of the board are chamfered slightly so that they do not dig into the concrete. Metal bull floats also are available. Wooden bull floats are often used for the final finish itself. Where a smoother, denser surface is desired, a metal bull float may be used.

Metal or wood bull floats are also often used as preliminary finish tools, the final finish being done with hand tools.

The typical hand float is wood, about 4½ by 12 inches long, with a wooden handle. Other sizes are available. Molded rubber floats also are available.

"Darbies" are essentially oversized floats. They are typically about 2 to 4 feet long and just under an inch thick; they are usually somewhat wider at the back, narrow at the front. The handles are similar to hand-float handles. Darbies are available in wood or magnesium. Darbies allow workers to reach farther than they could with ordinary hand floats.

Trowels are available in roughly the same sizes and shapes as hand floats. But trowels are always metal. In general, metal smooths concrete to a finer surface than does wood. For large areas, power trowels may be used; these trowels look something like four-bladed fans with a motor on top and with a long handle for pushing the blades around on the concrete surface.

Edgers and groovers. Edgers are tools similar in appearance to hand floats except they have a turned-down or flanged edge at right angles from the base piece. When run along the edge of a slab, the edger creates a smooth, rounded edge (sharp edges are undesirable because they are easily chipped and broken).

Groovers are also similar to hand floats but are equipped with a ridge down the middle of the base. When moved across the concrete surface, the ridge slices a neat cut into the surface. The depth of the ridge on groovers varies, depending on the purpose of the cut. Groovers may be designed to cut contraction or other functional joints in concrete. Or the cuts may be purely decorative.

Concrete saws. Concrete saws come with various power ratings—8-, 9-, and 14-horsepower gasoline engines are typical; saws in this power range can cut about 5 inches deep. The blades have no teeth; they are made of abrasive materials, some of which will cut either wet or cured concrete.

Push brooms. Push brooms with a variety of bristles are available, some with hose attachments for creating exposed-aggregate finishes. Hard-bristled brooms produce rougher finishes than soft-bristled ones.

Knee pads. Concrete finishing involves quite a lot of shuffling around on the knees. Pads protect against cuts and bruises and make the worker more efficient.

CONCRETE FINISHES

Exposed Aggregate

Exposed aggregate is perhaps the most familiar concrete finish. The aggregate that is exposed is usually added to the concrete when it is firm enough to support the aggregate but not yet hardened; the added aggregate is carefully pressed level to the surface of the slab. The aggregate is allowed to harden enough in the concrete to allow some of the surface cement to be brushed away without disturbing the aggregate; a broom and fine water spray (the water hose is sometimes built into the broom) are usually used.

A wide variety of effects can be achieved by varying the size, color, and shape of the aggregate selected for exposure.

Patterns

Almost any pattern desired is possible with concrete. The surface may be scored to create simulated flagstone, brick, cobblestone, and similar surfaces.

The contraction joints may be increased to form geometric or curvilinear patterns. This type of pattern is also functional, because it decreases the possibility of random cracks in the concrete due to contraction.

Redwood, western red cedar, and similar joint materials may be left flush with the concrete surface as decorative elements. More joints than are needed for functional purposes are often used for aesthetic effect.

Color

Concrete is typically a dull gray. Where large amounts of concrete are used and the color interferes with the overall color scheme, several methods of achieving color are used.

Integral-mix color formulas offer a fairly wide range of colors and are permanent. Some manufacturers will custom-mix colors to match or contrast with other colors in the project. White portland cement is used instead of the typical gray. Contractors may also mix their own colored concrete.

Dry-shake colors are sprinkled over the concrete as the surface is being final-finished. Dry shake requires considerable skill to be used effectively; if this skill is not available, integral-mix color is probably the better choice. The dry-mix method works as follows: The slab is first floated; then dry mix is shaken over the surface as evenly as possible. The surface is floated again. Another dry shake is added evenly over the surface. Then the final floating or troweling is done. Care must be taken in applying and finishing or the color will be splotchy. White portland cement is usually used. Individual concrete manufacturers' requirements may vary somewhat, and these requirements should always be followed.

Paints are another method of coloring concrete. Special portland cement paints are available. These paints may be applied with whitewash brushes or similar applicators. The application is usually made to wet concrete, and it is sometimes necessary to spray a fine mist over the concrete paint to make the color even.

CONCRETE REINFORCEMENT

Properly used, concrete is an excellent construction material, probably the most important material of the twentieth century. But the design or shape of concrete and the attendant reinforcement—typically steel—are critical to concrete performance.

The design of large concrete structural systems and their reinforcement—high-rise buildings, bridges, dams, and so forth—is usually done by structural engineers. Residential construction, especially light, one-story construction, has evolved many construction standards for concrete; still, where any question about design or reinforcement comes up, the cost of professional engineering design services, compared to the possible expensive consequences of poor design, are small indeed.

Typical concrete component standards and reinforcement in residential building are shown elsewhere in the text. Generally speaking, steel reinforcement is required in driveways, garage slabs, house floor slabs, foundation footings and walls, garden walls, and similar structures.

CONCRETE ESTIMATING

The prime responsibility of the contractor is to see that the job is done efficiently to the specifications. Thus, it is more important that the contractor have a knowledgeable overview of the construction tasks than it is to be physically capable of doing any or all of them.

Concrete work is more unforgiving of mistakes than, say, carpentry, where mistakes can be corrected with relative ease. The contractor must know when to use ready-mix and when to mix concrete on site.

Both ready-mix and on-site mixing require careful planning to see that the right amount of the right kind of concrete, the right equipment, and the right amount of qualified labor are all on the job when it is time to pour.

If ready-mix is used, less equipment is needed, and less-skilled workers may be used because they will not be responsible for mixing. Also, provisions for ingredient storage will not be necessary. How-

ever, the timing for ready-mix is more critical because once the ingredients of concrete are mixed, the time is limited until they must be poured. Thus, the job must be carefully planned in appropriately sized segments for the type of job, available labor, and delivery capability of local concrete suppliers.

If the contractor mixes on-site, dry storage must be provided for the ingredients. This can be difficult on some sites, maybe impossible. Bagged cement tends to absorb moisture and become lumpy. If the lumps cannot be loosened to their original consistency, the cement must be discarded. Aggregates must be kept clean and dry also and not dumped in large quantities, which makes them tend to segregate; they must also be protected from breakage.

Framing

5

The following chapter is directed primarily toward materials, terminology, and assembly. Appendix C contains span tables and other information helpful in house design.

WOOD FLOOR FRAMING

The foundation system, when complete, provides the base for the next component of the house structure: the floor framing system or floor deck. The floor deck consists of: the sill plate, which is where the deck rests on the foundation wall; the floor joists, which are the basic members of the floor framing system; beams and girders, which are sometimes necessary to support the joists (beams and girders are very similar in use to the beams we see supporting members at the second floor or higher levels); and, finally, the subfloor, which is a kind of preparatory layer between the rough framing members of the floor deck and the finished materials of the interior of the house, such as hardwood, tile, and carpet.

These are the essential components of the floor deck. However, other, more incidental items of the floor deck are usually necessary: trimmers, headers, tail joists, bearing strips, and joist hangers, among others. These items are discussed or illustrated in the text of this chapter.

The *sill plate* should be at least 2 inches thick. It must be wide enough so that the wall studs, when installed, will have full bearing (the entire surface of the wall studs will be supported, with no

overhang of the studs) and so that the floor joists will have at least 1½ inches of bearing surface. The sill itself should not overhang the top of the foundation wall. The sill must be straight and level on the foundation wall; it is sometimes necessary to set the sill in a bed of portland cement to assure that it is level and that uniform and complete contact is maintained with the top of the foundation wall. Sill anchorage has already been discussed. Where a concrete slab is used instead of a wood deck, the sill plate is simply the bottom member of the two-by-four stud wall.

Wood girders and beams may be solid members—four-by-fours, three-by-sixes, six-by-sixes, and the like; they may be built up of several solid members secured together; they may be laminated members; or they may be steel. In some cases, a girder may span from foundation wall to foundation wall, or the girder may have to be supported by piers or columns along its length. The length the girder must span, and the load it must carry influences the type and material of the girder and how many (if any) piers or columns it must have along its length.

Built-up girders, when three or more members are used, are often nailed together with 20d nails. Put two nails at each end of the girder, on both sides. If the girder is long enough so that the members must be spliced, use two nails at the ends of the splices, on both sides. Put a row of nails along the top and bottom of the girder at 32 inches on center, both sides; the nails should be staggered. A professional is often needed to design and size girders, but Figure 5-1 gives girder spans for some typical house building situations.

Where girders fit into concrete or masonry foundation walls, leave a ½-inch space at the ends and sides to allow air movement; otherwise, it will be necessary to use treated wood to protect against decay.

Depending on design needs, girders may span underneath the joist framing or be installed such that the joists butt against the girder sides.

If *wood columns* are used to support girders, the columns should bear on a base of concrete or solid masonry that rests on an adequate footing. If the column is in a basement, the column base should be at least 3 inches above the finished floor of the basement; in crawl spaces, the column base should be at least 8 inches above the grade. Column ends should be level to provide good bearing surface at the girder and base, and the column should not be spliced.

Columns, like girders and beams, often need to be designed by a professional. However, typical sizes are as follows: wood columns under first floor girders, not less than 6 by 6 inches; under lesser beams, trimmers, and the like, not less than 4 by 4 inches. Steel columns also may be used; they should be designed by a professional. If earthquake design is required, special precautions suit-able to the particular area must be taken to secure the column at the base and at the girder.

Wood floor joists are like the ribs of the floor deck, girders the backbone. Joists usually span the shorter distance between supporting members, which may be from foundation wall to foundation wall, from foundation wall to girder or beam, or from girder to girder. Joists are typically 2 inches thick and usually spaced at 16 inches on center, except where stronger subflooring, stronger finish flooring, or some other structural plus allows greater spacing. For design purposes, span is the clear span distance between the inner faces of support members (foundation walls, girders, etc.).

Floor joist framing for a given house may be:

1. Framed on top of girders or beams
2. Framed into the sides of wood girders or beams
3. Framed into the sides of steel beams
4. Framed on top of the wood sill on the exterior wall
5. Framed into masonry

Several of these conditions may sometimes be used on a single house; it is possible that all of them might be used on a single house, but not likely.

GIRDER SPANS

Width of Structure	Girder Size (Solid or Built-up)	Supporting Bearing Partition		Supporting Nonbearing Partition	Intermediate Girders (Other than Main Girder)
		Maximum Span			
		1-Story	1½- or 2-Story		
Up to 26 feet wide	4 × 6	—	—	5' 6"	7' 6"
	4 × 8	—	—	7' 6"	9' 6"
	6 × 8	7' 0"	6' 0"	9' 0"	12' 0"
	6 × 10	9' 0"	7' 6"	11' 6"	—
	6 × 10	10' 6"	9' 0"	12' 0"	—
26 feet to 32 feet	4 × 6	—	—	—	6' 6"
	4 × 8	—	—	7' 0"	8' 6"
	6 × 8	6' 6"	5' 6"	8' 6"	10' 6"
	6 × 10	8' 0"	7' 0"	10' 6"	13' 6"
	6 × 12	10' 0"	8' 0"	11' 6"	—

Note: These spans are based on an allowable fiber stress of 1,500 psi. These allowable stresses are average values taking into consideration upgrading for doubling of members in built-up beams. Where conditions vary from these assumptions, design girders in accordance with standard engineering practices.

FIGURE 5-1 Girder span specifications. (Courtesy of Federal Housing Administration.)

Builders should be familiar with all these techniques.

Joists over Girders or Beams

Joists over girders or beams may be joined to each other and secured to the supporting member in a number of ways; the following methods are typical:

1. Lap the joists over the supporting member, nailing the joists to each other and to the supporting member.

2. Butt the joist ends together over the center of the supporting member and tie the joists together with wood strips or metal straps; nail the joists to the supporting member.

3. Where a steel beam is the supporting member, a wood nailer may first be installed over the beam; the joists are then installed over the nailer. Or the joists may be laid directly on the beam with wood blocking between the joists. A continuous strip of wood blocking may be used instead of the blocking between the joists, if preferred.

The listed techniques are often used, but other methods may be acceptable. (See Figure 5-2.)

Joists Framed into the Sides of Wood Girders or Beams

In this condition, the joist ends butt the sides of the supporting member. Simply toenailing the joists to the supporting member would not provide a secure connection. Thus, steel angles, joists hangers, or wood bearing strips are the typical connectors. (See Figure 5-3.)

Joists Framed into the Sides of Steel Beams

In this condition, the joist ends butt the sides of the steel supporting member. Since the joists cannot be nailed to the steel beam, wooden blocking must be secured to the beam and the joist then nailed to the blocking. Similarly, the wood subflooring above must be provided with wood blocking. The joist must be trimmed to fit within the I of the steel beam and still maintain a minimum 1½-inch bearing surface. In addition to the wooden blocking (nailers), blocking must be provided between the joists to prevent lateral movement. (See Figure 5-4.)

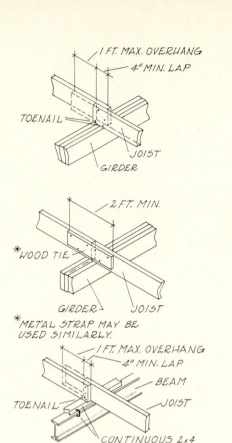

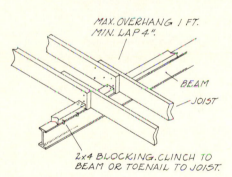

FIGURE 5-2 Typical methods for securing framing over girders and bearing partitions.

Joists Framed into Masonry

Where joists are framed into masonry (at masonry foundation wall or piers, etc.), the joists should have at least 3 inches of bearing surface on the masonry. There should be a ½-inch space between the joist end and sides and the masonry, unless the joist is treated wood. If the joists are 5 feet or more above grade, they must be secured to the wall; Figure 5-5 shows metal straps used to anchor joists at the sides and at the ends.

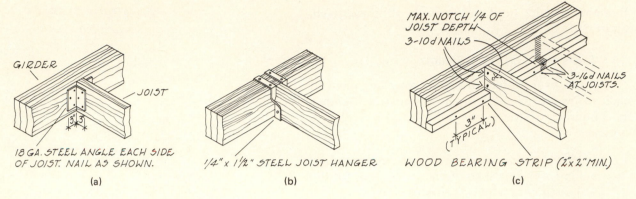

FIGURE 5-3   Three methods for securing joists to wood girders.

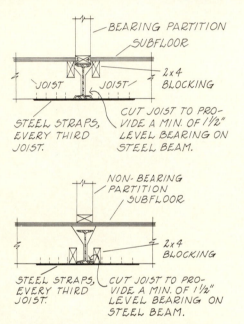

FIGURE 5-4   Two typical methods for framing joists into steel beams.

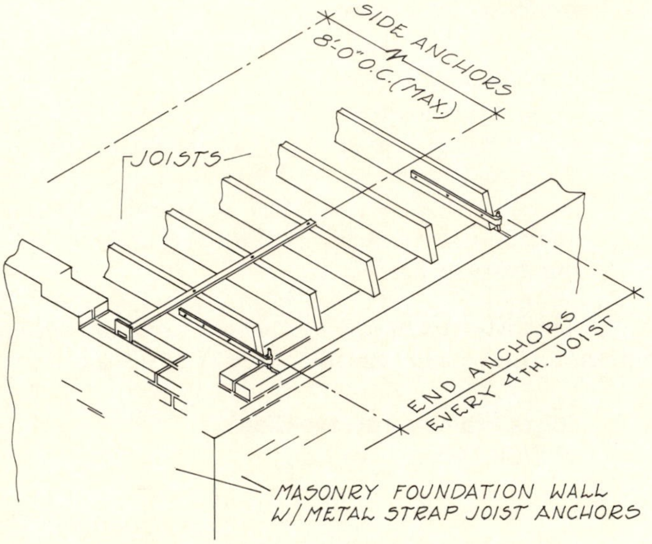

FIGURE 5-5   Method for tying framing members into masonry walls.

Joists on a Wood Sill over the Foundation Wall

Where the joists are secured to a wood sill, a 2-inch continuous header joist is usually provided; if the continuous header is not used, provide blocking between the joists at the ends with 2-inch members the same depth as the joists. The joists should have at least 1½ inches bearing on the sill. If the header is used, toenail the joists to the sill with two 10d or three 8d nails; if the header is not used, toenail the joists to the sill with three 10d nails or four 8d nails. The header should be secured to the sill with 8d nails at 16 inches on center; secure the header to floor joists with two 16d nails placed through the header and into the joist ends. The header does not have to be toenailed to the

sill if wall sheathing is nailed to the sill plate. (See Figure 5-6.)

Bridging is blocking installed between the joists to prevent the joists from twisting, prevent lateral movement, and increase the overall rigidity of the floor deck. Bridging is usually one of two types: solid or cross. Solid bridging utilizes members the same size as the joists, but to fit tightly between the joists and staggered, so that it may be toenailed or end-nailed to the joists. Cross bridging utilizes 1- by 3-inch boards angled between the joists; smaller metal cross bridging may be used if it is as strong as 1- by 3-inch wood members (see manufacturer's specifications). (See Figure 5-7.)

Headers and trimmers are framing members around an opening in the joists (for the chimney or stair opening, for example). The members that are

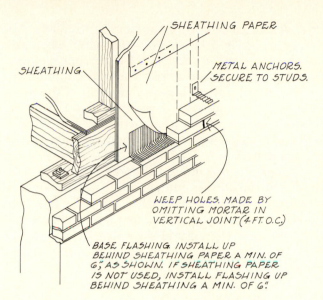

FIGURE 5-6   Typical frame wall with masonry veneer.

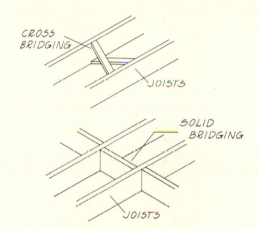

FIGURE 5-7   Examples of bridging between joists.

parallel to the regular joists are called trimmers, and the members perpendicular to the regular joists are called headers. Joists that butt the headers are called tail joists.

The following is a guideline for framed openings in the floor framing:

1. Headers 4 feet long or less may be a single member; headers over 4 feet long should be doubled.

2. Headers are secured to trimmers similarly to the way regular joists are secured to girders.

3. Headers over 4 feet long but not over 10 feet long should be double members except where some additional load is placed on them (such as appliances or a bearing wall), in which case the header should be designed as a beam with the particular load in mind. It is sometimes necessary to support headers with columns.

4. Trimmers are usually doubled members, unless the framed opening is near the end of the joists, in which case the trimmers sometimes may be single members.

5. Tail joists over 6 feet long should be connected to headers as if the header were a beam; tail joists under 6 feet may be nailed to the headers—use three 16d nails in ends and toe-nail with two 10d nails. (See Figures 5-8 and 5-9.)

Double joists are a frequently used method of supporting extra loads. Where a wall partition, for example, runs parallel with the regular floor joists,

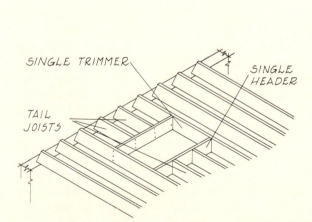

FIGURE 5-8   Framing small openings.

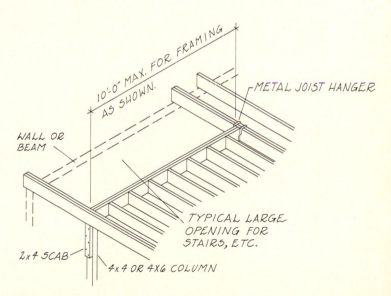

FIGURE 5-9   Framing large openings.

it is typical to install double joists under the partition; more joists may be added, as required.

If holes for wiring, pipes, and the like must be cut in the joists, the holes should not exceed 2 inches in diameter and should not be within 2 inches of the joist edge. Notches cut in the joists for wiring, pipes, and so on should be made near the end of the joist span and should not be more than ⅙ the depth of the joist; if notched on only the top side, and if the notch is not farther from the joist support than the depth of the joist itself, the notch may be as much as ⅓ the joist depth.

CONCRETE FLOOR SLABS

Concrete slabs are sometimes ground-supported (slab on grade) and sometimes structurally designed so that they are independent of the ground (structural slabs). If the ground is good—that is, stable and nonexpansive—the slab on grade is the cheapest, fastest way to install a floor slab. If the ground is not good or other construction conditions prevent the use of a ground-supported slab, the slab often must be reinforced.

In areas subject to serious frost heave, special precautions must be taken that add to the cost and reduce the desirability of ground-supported slabs. Local authorities usually have standard procedures and requirements for dealing with frost heave—typically involving removing soils with high clay content and replacing them with gravel or sand or soil that drains well. Extra steel reinforcement of the slab also may be required.

Ground-supported slabs may be laid over natural soil if the soil is of adequate bearing capacity and is otherwise acceptable for such use. If fill is used, it must be compacted to achieve the bearing capacity satisfactory to local or appropriate authorities (typically 2,000 pounds per square inch). Usually, the maximum fill for soil is around 8 inches but may be more. Sand and gravel is sometimes used as a fill component (it is also almost always used under the slab to help prevent moisture buildup); if so, the normal maximum depth is about 2 feet. Soil fill is typically installed in 4-inch layers, and the soil must be acceptable for the purpose; engineering inspection of both soil and installation of the soil as fill is often necessary. Soil on the building lot should be free of organic matter, debris, and other foreign matter; this is especially important where the soil is used as fill.

Structural slabs are typically reinforced with welded wire. The wire is laid in place so that it will be about an inch from the bottom of the slab; then the concrete is poured over the wire. Slabs also may be supported underneath by piers, in combination with the welded wire; it also is common for the structural slab to lap over the top of the foundation wall for support. (see Appendix D-6.) Steel reinforcing bars, when used, generally are a supplement to welded wire rather than the prime reinforcing material.

Generally speaking, fill under structural slabs need not meet the compaction requirements of compacted fill discussed earlier; some compaction, however, is necessary. The maximum amount of earth fill under structural slabs usually should not be over 3 feet; sand or gravel should not exceed 6 feet.

For both structural and ground-supported slabs, extra reinforcement usually is needed for interior bearing partitions or similar loads.

Vapor barriers are typically used under all concrete floor slabs. Asphalt-saturated felt, mopped with asphalt, is sometimes used, but a plastic membrane is easier to install and more common. (See Appendix D-7.) Usually, a layer of gravel 4 inches deep is laid, and the membrane is then placed over the gravel before the concrete is poured. Regardless of the vapor barrier used, it should be of one piece, or steps should be taken to assure that the joints do not leak.

Ground-supported slabs are normally about 4 inches thick; they may be as little as 3⅝ inches thick, but this is not recommended. Structural slabs should be at least 4 inches thick. Concrete is a relatively inexpensive material, and it is usually not wise to scrimp when the consequences of a too-thin slab could result in serious rework, which would be the contractor's responsibility. Ground-supported slabs are sometimes installed with two pours; where this is the case, the first pour is usually about 2 inches and the second about 3 inches. (See Figure 5-10.)

Sill-plate anchorage of frame walls is almost always required where concrete floor slabs are used, the only exceptions being areas of very mild climate where local authorities permit the omission of the anchorage. Anchorage may be either bolts or steel studs. If bolts are used, they should be ½ inch in diameter, embedded at least 6 inches in the concrete, and located not more than 8 feet on center; there must be at least 2 bolts per sill piece, and there should be a bolt not more than a foot from each end. If earthquake design is used, the bolts should not be more than 6 feet on center.

Steel studs (installed with a power tool), if

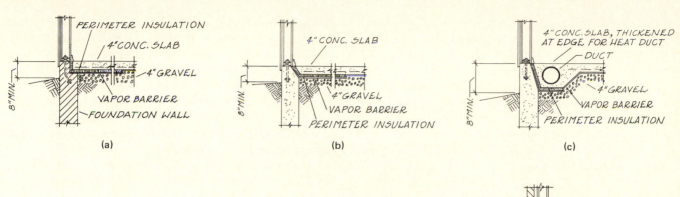

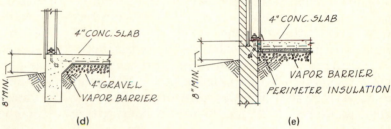

FIGURE 5-10 Examples of ground-supported slabs used with a variety of foundation-wall types.

used, should not be more than 4 feet on center; they should not be more than a foot from the ends, and there should be at least two studs per sill piece. If earthquake design is used, maximum spacing of steel studs should be 3 feet.

SUBFLOORING

Subflooring is installed to provide an interlayment for finish floor materials: hardwood floors, vinyl tile floors, ceramic tile, and so forth. The subfloor also strengthens the floor deck by reducing deflection and contributing to overall rigidity of the floor deck. The subflooring may be plywood, tongue-and-groove boards (T&G), or other suitable materials. Plywood is probably the most used material.

Plywood should be either of the exterior or structural-interior type. If any portion of the plywood is exposed to the weather, it must be exterior type. Allow a minimum of ½ inch between the plywood subfloor and concrete or masonry walls. Allow a 1/16-inch space between end joints and a 1/8-inch space between edge joints to prevent warping. Install plywood with outer plies perpendicular to the joists, and nail it to the joists with 8d common or 6d threaded nails at 6 inches on center for the edges; nail at 10 inches on center over intermediate joists. Install the plywood so that the end joints are staggered (end joints of adjacent plywood panels must not occur over the same joist).

Plywood may also be glued to the joists. Some authorities recommend gluing, claiming it increases floor-deck stiffness, reduces squeaks, and

allows smaller joists or longer spans. Generally, a ¼-inch bead of glue is laid over the joists, and the plywood panels are laid down (only a few at a time; otherwise the glue will begin to set on the joists) and nailed at 12 inches on center on the joists. Use a code-acceptable glue and follow the manufacturer's installation instructions.

Vinyl flooring (in tiles or sheets), carpeting, ceramic tile, wood block, and other floors may be laid directly over the plywood subfloor if the top ply of the subfloor is "C repaired" grade or better; this better grade simply presents a smoother surface for the upper materials to be secured to.

The thickness of the plywood subfloor depends on the load, the joist system underneath, the species of wood the plywood is made of, and similar factors. Typical plywood thicknesses are ½, 5/8, ¾, and 7/8 inches.

T&G boards hold together at the edges and make a smooth surface. The boards are usually installed diagonally so that finish wood-strip flooring may be laid without regard to joist direction (subfloor boards installed at right angles to the boards would not always make the best finish floor appearance). The boards are a minimum of ¾ inches thick and may be thicker, depending on the finish flooring and the joist system underneath. The boards usually should not be more than 8 inches wide. For nailing, use three 8d common or 7d threaded nails at each joist for 8-inch boards; for 4- and 6-inch boards, use two nails. As with plywood, allow ½ inch space between subfloor boards and concrete or masonry walls. The boards should be cut parallel to joists and centered on the joists; avoid more than two joints at adjacent boards. An additional underlayment may be necessary before vinyl, ceramic tile, wood block, and similar finish floors are installed.

Earthquake design may influence the thickness and installation of all subfloors; check with local authorities.

FRAMING METHODS

The three most commonly used methods of framing, in order of importance in residential construction, are platform frame construction, balloon frame construction, and plank-and-beam framing.

Platform frame construction is probably used more than any other framing method; perhaps this is because most residential construction for many years has been one- and two-story construction (single-family dwellings). A big advantage to platform construction is its simplicity and the speed with which it can be assembled. The platform frame is what the name implies: The floor deck is built up like a platform over the sill; subflooring is carried flush to the outside of the blocking or joist header. Then, typically, lengths of stud wall are assembled, using the platform as a work area, and the walls are simply lifted up and secured. The second floor is done the same way. Even when a balloon or plank-and-beam frame is used, builders often opt to use the platform system at the ground level (you will understand why in the following discussion of balloon framing and plank-and-beam framing).

If a second floor is built, another platform is built over the first-floor stud-wall cap plate, and the structure continues up as before. At the cap plate of the second floor, the roof structure is begun.

Balloon frame construction is more difficult to build than platform framing. At the ground level, both the first-floor joists and the wall studs rest on the sill, usually flush with the outside edge of the sill. The wall studs must be longer because they begin at the sill. They are continuous, and thus for two-story construction the wall studs can be as much as 18 feet long. This complicates the building assembly, and requires better lumber—which raises costs. Further, the subfloor must be cut to fit around each wall stud, rather than have the wall studs rest on the subflooring, as in platform framing. The second-floor joists present another condition: The wall studs must be notched to receive a ledger (often a one-by-six), which must be installed

to hold the ceiling joists, which are placed over the ledger. But balloon framing is still used, largely because the structure is more resistant to shrinkage differences between wood and masonry. This is important in three- and four-story residential structures (such as apartments and condos) where concrete fire walls are often required; balloon framing also reduces settlement differences between interior supports and exterior walls.

Plank-and-beam framing utilizes 2-inch planks laid over beams as much as 7 or 8 feet apart; the beams are supported by posts, either solid members or built-up members, usually not less than 4 by 4 inches. The floor deck, second floor, and second-floor ceiling (which may or may not be the roof) can all be framed in this manner, or the system can be mixed with platform or balloon framing to suit a particular house design. Plank-and-beam sounds more expensive than it is because the thick (2-inch) planks reduce the number of joists needed, and the planks make a handsome interior material (finish-quality lumber is usually desirable for floors and ceilings left natural). Labor costs may be reduced because fewer trades are involved and the framing is simple and fast to erect. Wiring, plumbing, ductwork, and the like may be troublesome with particular designs because these items are more difficult to conceal with plank-and-beam framing; a typical solution is to fur down around the items.

Figure 5-11 shows examples of platform, balloon, and plank-and-beam framing.

The Stud Wall

The two-by-four stud is the principal framing member in exterior walls and interior wall partitions. The combination of exterior walls and interior partitions must be strong enough to support the designed dead loads and live loads (including storm winds and earthquakes, in some areas).

Ordinarily studs are nominally two-by-fours, or they actually may measure 2 by 4 inches. The most common length is 8 feet, but two-by-fours are available in 10-foot, 12-foot, and longer lengths for high ceilings or where balloon framing is used. Studs are normally 16 inches on center, but 24-inch centers are allowed in some areas or in particular circumstances; 24-inch centers is the maximum spacing, however. The most common reason for 24-inch centers is probably to save money; 16-inch centers make better walls. The studs may be end-nailed to the sole plate *if* sheathing is used to tie the studs to the sill; end-nail with two 16d nails.

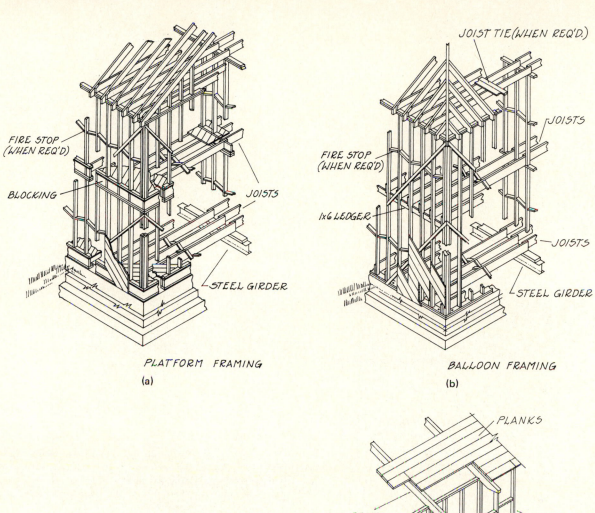

PLATFORM FRAMING
(a)

BALLOON FRAMING
(b)

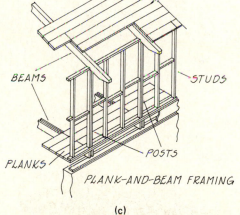

PLANK-AND-BEAM FRAMING
(c)

FIGURE 5-11   The three most common types of framing used today.

If the studs are toenailed to the sole plate, toenail with three 10d or four 8d nails. Nail similarly at the top. Studs should not be spliced. Studs may be notched ¼ their depth and drilled not more than 1¼ inches.

Corners

Exterior stud-wall corners and interior partition intersections may be constructed in several ways to provide a secure connection and to provide a nailing surface for the interior finish material. Corners should always have a minimum of three two-by-fours. The corner two-by-fours are nailed to each other with 16d nails a maximum of 24 inches on center; the corner members often use two-by-four filler blocks to increase the size of the members. Use at least three nails on the filler blocks, and always nail through the wide face of the two-by-fours. (The two-by-fours are usually arranged in one of the three methods shown in Figure 5-12(a).)

Interior framing is arranged with the finish materials in mind. The studs are arranged to provide backing for gypsum wallboard, paneling, or similar finish materials. (See Figure 5-12(b) and (c)).

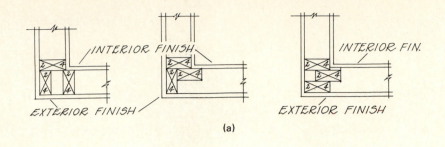

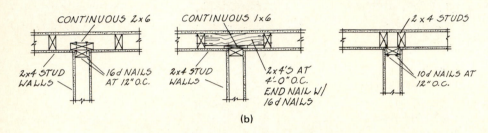

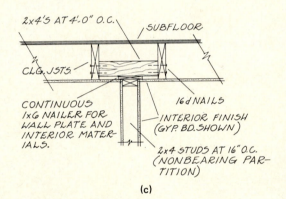

FIGURE 5-12 Typical stud wall corners: (a) exterior stud wall corners, plan view; (b) interior stud wall corners, plan view; (c) interior stud wall at ceiling, section view.

Wall Bracing

Exterior walls must be protected from possible racking. The bracing is typically accomplished by one of the following methods:

1. One-by-four or one-by-six boards let (notched) into the exterior stud wall. The boards may be let into the inner face of the stud wall or the outer face. They should be installed at about 45 degrees, near the corners of the wall, and should extend from the wall plate all the way to the top plate. Nail the brace to each stud and the plates with two 8d nails at each contact. There should be a brace board installed as described above at each corner, for each floor. (See Figure 5-13.)

2. Wood sheathing is a satisfactory bracing method when installed at about 45 degrees to the

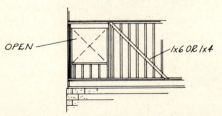

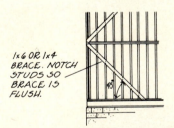

FIGURE 5-13 Corner bracing methods.

studs. Install the sheathing in opposite directions from each corner.

3. Plywood sheathing may serve as bracing. Install solid plywood sheets horizontally and provide two-by-four blocking members between every stud where the horizontal sheathing joints occur. Plywood is faster to install than boards or wood sheathing.

Other materials and methods may be used if they conform to approved engineering practices.

Framing Rough Openings

Doors and windows require a specifically sized framework to fit into. In general, the rough opening for a door is about 3 inches wider and higher than the door. Window rough-frame openings are generally an inch more than the height and width of the window. This extra space allows the doors and windows to be properly secured and positioned within the rough framing. However, the exact rough-frame requirements should be verified for the particular manufacturer of door and window that will be used *before* the rough frame opening is built. (See Figure 5-14.)

The rough framing for a door consists of a header, a jamb (vertical) stud under each corner (also called cripple studs), and a full-length stud nailed to each cripple stud.

Rough window openings are framed the same as door openings, except window openings need a rough sill; the rough sill is supported underneath with short cripple studs. Window openings over 36 inches wide must be double-framed; that is, they must have a full-length stud nailed to each cripple stud supporting the header. Window openings 36 inches wide and less do not have to be double-framed if the next full-length stud from the cripple stud is not more than 14 inches away, but the header must be extended over each cripple stud to the full-length stud. This sounds more complicated than it really is.

Nailing is about the same for rough window and door openings: Double-framed openings should have cripple studs nailed to full-length studs with 16d nails at 24 inches on center. The cripple studs may be toenailed to the wall plate (bottom) with two 10d nails or three 8d nails, or the cripple studs may be end-nailed to the wall plate with two 16d nails. The outer, full-length stud is nailed to the header with four 16d nails, toenailed to the wall plate with either two 10d or three 8d nails. The outer stud is nailed to the top plate like any other stud. For single-framed rough openings, toenail the cripple studs to the header with three 8d nails; nail to the top like other studs; toenail cripple studs to the wall plate with either three 10d or four 8d nails, or end-nail with two 16d nails.

Wall Sheathing

Wall sheathing provides racking resistance, sometimes provides a nailing base for finish exterior materials, and adds insulation value. Wall sheathing is usually one of four types: fiberboard, plywood, gypsum, or wood boards.

Fiberboard is typically made of wood pulp.

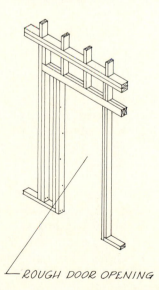
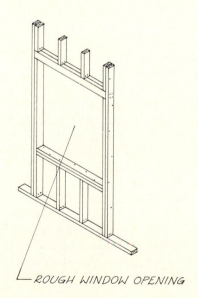

FIGURE 5-14 Rough door and window openings.

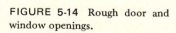
ROUGH DOOR OPENING

ROUGH WINDOW OPENING

It may be made of other materials, such as cornstalks or sugar-cane fibers. In any case, the fiberboard (also called hardboard) should be water-resistant but vapor-permeable and should meet applicable code and regulatory requirements.

Fiberboard is much used in residential construction. It is a relatively cheap material and is quick and easy to install. The minimum thickness for fiberboard sheathing is ½ inch. If wall bracing is used, ½-inch or $^{25}/_{32}$-inch fiberboard may be nailed to studs spaced 24 inches on center. But if wall bracing is omitted, studs must be spaced 16 inches on center for either fiberboard thickness.

If $^{25}/_{32}$-inch fiberboard is used over studs at 16 inches on center and let-in wall bracing is omitted, four-by-eight panels should be used, installed vertically. Vertical installation gives more resistance to wall racking. Nail the sheathing edges at 3 inches on center with 1¾-inch roofing nails; nail at 6 inches on center where the sheathing meets intermediate studs.

If let-in wall bracing is used, four-by-eight sheets may be installed either horizontally or vertically. Two-by-eight sheets should be installed horizontally. Use 1¾-inch roofing nails for $^{25}/_{32}$-inch fiberboard and 1½-inch roofing nails for ½-inch fiberboard. Nail the vertical edges at 4 inches on center and 8 inches on center where the sheathing meets intermediate studs.

Fiberboard is not a good nailing base, and additional nailing provisions must be made for some exterior materials—shingles, for example.

Plywood is an excellent sheathing material. It may be installed horizontally or vertically. More racking resistance is obtained if the panels are installed vertically. No additional bracing is required. The minimum thickness is $^5/_{16}$ inch. Studs should be spaced 16 inches on center if $^5/_{16}$-inch plywood is used; the studs can be 24 inches on center if $^3/_8$-inch plywood is used. Should they be needed for particular requirements, ½-inch, $^5/_8$-inch, and thicker panels are available. Nail with 6d nails at 6 inches on center along the edges; nail 12 inches on center where the sheathing meets intermediate studs. Siding should be nailed to the sheathing with threaded nails or other acceptable fasteners designed for use in low-density materials.

Gypsum sheathing has water-repelling qualities, if so marked; it also is fire-resistant. If the core of this sheathing has not been treated (against water and moisture absorption), the edges should be caulked or flashed with building paper (use 18-inch strips); tongue-and-groove edges may be used in place of flashing. The minimum thickness of gypsum panels is ½ inch; maximum stud spacing is 24 inches, and wall bracing is required. If studs are 16 inches on center and the sheathing is installed vertically, bracing may sometimes be omitted; check with local authorities. The sheathing panels are available in 8- and 9-foot lengths by 2- and 4-foot widths. Nail edges with 1½-inch roofing nails at 6 inches on center; nail at 8 inches on center on intermediate studs. Gypsum does not provide a good nailing base; additional backing must be installed for siding materials.

Spacing is usually required between all sheathing panels (and exterior siding panels as well). Plywood panels, for example, typically require a $^1/_8$-inch space between all panel edge joints and a $^1/_{16}$-inch space between panel end joints (the edges are the long side of the panel; the ends are the short side). In particularly wet or humid areas, these spacing requirements are sometimes doubled. Since the various sheathing and siding paneling materials vary somewhat in spacing requirements, you should check local installation practices and carefully study the manufacturer's installation instructions.

Obviously, there are advantages and disadvantages to each of the different sheathing types. And cost is not as simple to arrive at as it might first appear to be. This is why professional design services or consultation is important in house and building design. The selection of sheathing must be made with local weather, house design, initial cost of material, labor installation cost, and similar factors in mind.

Sheathing Paper

Sheathing paper is an asphalt-saturated material used to protect the wall-framing sheathing against moisture. The sheathing paper is installed over the framing similar to shingles, with a minimum of 4-inch laps; no. 15 is the usual weight of paper used. Some wall-framing sheathing materials, such as plywood, do not require sheathing paper; however, the paper is relatively cheap for the protection it provides the more expensive wall-framing sheathing.

When wall-framing sheathing is not required, sheathing paper must be used to protect against wind infiltration and to protect the wall framing from moisture. When wall-framing sheathing is not used, install no. 30 sheathing as discussed here.

Exterior Wall Layout
and Installation (Platform Framing)

In platform framing, the outside face of the exterior stud wall is normally flush with the outside face of the deck, which is complete before exterior stud walls are begun.

The exterior stud-wall sole plate is first laid out on the deck with chalk lines, showing both sides, with indications for studs, cripples (the short studs under the headers of framed openings for doors and windows), intersecting interior walls, and so forth.

Next, the exterior stud wall is built. If the deck is a concrete slab, or, otherwise, if anchor bolts are to be used to hold the wall plate in place, the location of these bolts must be marked on the wall plate and holes drilled so that the wall plate can fit over the bolts. Next, the exterior wall is lifted up, secured, and temporarily braced. The exterior wall at this time consists of the wall plate, the studs with any rough openings, the top plate, and the cap plate.

As erection proceeds, frequent checks should be made to assure that the walls are plumb and in alignment. When the exterior walls are in place, they should be rechecked and adjusted, if necessary. When the walls are accurately plumbed and aligned, install the corner bracing.

Interior Partitions

The decision about whether to build the roof or the interior partitions first depends largely on the house design and on how important it is to get the house under roof. Small, one-story houses can sometimes have the roof installed with only the exterior walls in place. Some one-story houses and most two-story houses often require that the bearing wall or walls be installed before the roof is started. If weather is not a problem, many builders install all the walls before starting the roof.

When the interior partitions are begun, it is typical to begin with the bearing walls. Before the interior partitioning begins, remove all scraps and debris from the floor deck and sweep the deck clean; the partitions must be aligned carefully, and they are laid out with chalk lines, which require a clean surface. The dimensions given on the working-drawing floor plan are from edge of stud to edge of stud. Chalk-line the bearing-wall plates (sometimes called *wall plate*, sometimes just *sole*, and some-

times *wall sole plate*) onto the subfloor with two parallel lines that represent the edges of the wall plate; then mark the location of studs, intersecting partitions, door location, and so forth. Do not bother to chalk more than you can construct in a day or so; otherwise the chalk lines will get smeared or lost and will have to be done again.

Before you begin installing the partitions, check to be sure the walls represented by the chalk lines are aligned as they should be and that the right angles are true. You can use a 3-4-5 right triangle to check the angles; build the triangle with one-by-fours or one-by-sizes and make it 6 feet by 8 feet by 10 feet for greater accuracy of measurement. A smaller triangle or steel square may be used to check short wall sections.

The bathroom wall is usually the only interior partition that is thicker than the other walls; this is because the bathroom wall must house the plumbing lines. The needed thickness varies somewhat, depending on whether the pipes are copper, plastic, or iron; copper and plastic usually require a little less space. The wall may be made thicker by using two-by-sixes for studs, or two-by-four studs may be staggered to make the wall thicker (the wall plate and top plate would be two-by-sixes or whatever the thickness of the built-up wall happened to be).

Mark the stud locations accurately, making sure they are at right angles to the wall plate. When locating doors or other spaces, note what the space is, somewhere near the center of the space. A felt-tip pen with a wide point is good for this job, or you may use a carpenter's crayon or other convenient and easily visible marker.

After marking the stud locations on the floor deck, count the studs you need for the first partition you intend to erect and precut the studs; then lay the studs on the deck, perpendicular to the wall plate (which, at this time, is a pair of chalk lines). The studs will be at 16 or 24 inches on center along the wall plate. Check the wall section in the working drawings and compute the exact length the studs must be; this height will depend on the desired finish ceiling height, which is usually a minimum of 8 feet, and on the type of ceiling material, which is often Sheetrock but may be other materials, as desired. These calculations are simple arithmetic and will be obvious from the wall section.

Next, lay the two-by-four wall plate near the chalk lines that now represent it and lay the top plate at the other end (top) of the studs. The wall plate and top plate should be longer lengths of two-by-four to reduce the number of joints; joints in the top plate should fall over studs. Joints in the cap plate (installed later) should be at least 32 inches away from splices in the top plate. You already have the stud locations marked on the floor deck, along the chalk-lined wall plate; put the wall-plate studs close to the chalk lines and, using the marks as a guide, mark the stud locations on the wall-plate material. Do the same thing for the top plate.

Next, assemble the wall. Nail through the wall plate and top plate into the ends of the studs with two 16d nails for each end. When all the studs are nailed in place, install the cap plate, using 16d nails at 16 inches on center.

Next, lift the wall up, position it carefully over the chalk lines, and nail it to the subfloor with 16d nails at 16 inches on center.

Temporary bracing with two-by-fours may be necessary. But if you carefully choose the wall sections you build, bracing can be very minimal. Usually, you build interior bearing partitions first (although you may start with any wall that seems convenient, if getting the roof installed as quickly as possible is not the main objective), and these partitions typically butt the exterior wall at some point. So, in a typical situation, you would start with a bearing partition, installing the portion that butts the exterior wall first, and move on from there. To cut down on bracing for a long stretch of straight wall, you may branch off the main wall, as you go, with lesser partitions; these will serve to stabilize the main wall and reduce bracing.

It is unlikely that your working drawings will accurately dimension the location of plumbing lines, stack vents, and so forth. However, the toilet, sinks, bathtub, and other large fixtures are located on the plans fairly accurately. Study these locations and avoid installing studs opposite the center line of these fixtures. Instead, offset the studs about a foot away from the toilet center line. Offset them about 8 inches from the tub centerline of kitchen sinks and at least 6 inches from bathroom sinks. Leaving spaces in the studs in this manner allows room for vent stacks, waste lines, supply lines, and so forth, and avoids relocating studs.

TYPES OF ROOFS AND THEIR USES

Five types of roofs are commonly used in residential construction: the gable roof, hip roof, gambrel roof, flat roof, and shed roof. The sawtooth roof is occasionally used in residential construction but is used more often in commercial and industrial building. (See Figure 5-15.)

The *gable roof* is perhaps the most typical residential roof. In this style, rafters extend from the cap plates of opposite exterior walls upward, where they meet at the ridge board. The ends of the structure formed are called gables, from which this roof system gets its name.

Gable roofs are versatile and widely used throughout the United States. The distance spanned between exterior walls can be increased by increasing the size and pitch of the rafters; the pitch itself may be varied considerably for varying weather conditions (usually steeper for snow country).

Hip roofs are like gable roofs except that the ends (where the gables would ordinarily be) are sloped up to the ridge, as well as the length portion of the house. Thus, the "hip" eliminates the gable, cutting down on expense and maintenance (gables often are wood and have to be painted; if the gables are brick or other masonry, initial cost is higher than a hip roof, where roofing is the only material needed).

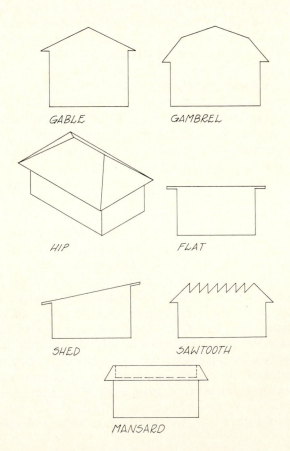

FIGURE 5-15 Commonly used residential roof.

The *gambrel roof* is a pitched roof similar to the gable roof except that somewhere near the center of the rafters the rafters bulge outward. This bulge is formed by the addition of vertical support members under the rafters, near the centers; this creates two sloped portions to the roof, one slope from the exterior-wall cap plates to the interior vertical members mentioned and another slope from the interior vertical members to the ridge board.

The effect of the bulge in the roof is to create a wider attic space with adequate headroom. And this extra space is economical because the roof also serves as an exterior wall. Dormers often are used with gambrel roofs, and the ends of the roof may be gabled or hipped (the hip, however, cuts down on the habitable length of the upstairs because it reduces headroom).

The *flat roof* uses the ceiling joists as the roof framing (these joists usually must be heavier than ceiling joists for pitched roofs). As the name implies, the roofs are flat or nearly flat; flat roofs often are sloped slightly so that water drains off. The flat roof may be used where the house covers a relatively large surface area—too large to span with normal-sized roof rafters.

The *shed roof* is like half of a gable roof. Rafters slope up from the cap plate of an exterior wall to an opposite exterior wall, which must be higher to receive the upward-sloping rafters. Shed roofs may be used for relatively narrow sections of houses, where a gable or hip roof would look peculiar; thus, shed roofs are often used in modular construction, where the sections manufactured may be only 12 or 14 feet wide. Also, shed roofs often are used against higher building sections; for example, a one-story kitchen with a shed roof may butt against a two-story section containing family room, living room, and bedrooms.

Sawtooth roofs, like flat roofs, are used where relatively large surface areas must be roofed that would be too large to be conveniently covered with a pitched roof. Also, because of the shape of the sawtooth roof (as the name implies, it is a series of small pitched-roof sections resembling a saw edge), glass or windows may be inserted to let in natural light.

Roof Construction Terms

The following terms are basic to roof design and construction.

Roof *slopes* are ratios. These ratios measure the amount of vertical rise for every horizontal foot: a slope of 3 in 12, 4 in 12, 5 in 12, 6 in 12 means the roof will rise, respectively, 3, 4, 5, or 6 inches for every horizontal foot. These vertical units of rise—3, 4, 5, and 6—are typical in residential and other light construction. Slope is frequently used to indicate rafter angles from the horizontal; if slopes are not given, pitch will be.

Roof *pitch* is the ratio between the vertical rise and the span. If a roof has a rise of 4 feet and a span of 24 feet, the pitch is ⅙. (See Figure 5-16.)

Span is the horizontal distance from cap plate to cap plate.

Cap plates are the top members on the wall framework; the rafters are secured to the cap plates, ceiling joists, and ridgepole. Rafters are usually notched to make a better contact at the cap plate.

A *bird's-mouth* is a cut or notch made in a structural member to make it fit better with a

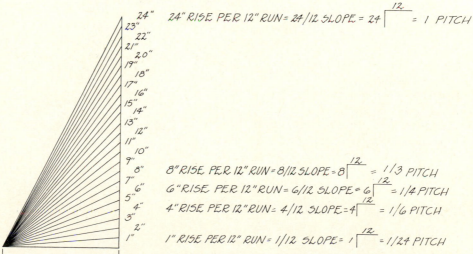

FIGURE 5-16 Chart showing the relationship between roof slope and roof pitch.

cross member; a rafter notch, mentioned in the preceding paragraph, is called a bird's-mouth.

Run is the horizontal distance from the cap plate to the ridgepole (run is usually half the span). The *unit run* is always considered to be 1 foot.

Rise is the vertical distance from the cap plate to the ridgepole.

The *ridgepole* is the uppermost member of the roof, to which the rafters are nailed.

Rafter tails are the lengths of rafter that extend beyond the cap plate, away from the house.

Common rafters extend at right angles from the exterior walls to the ridgepole.

Hip rafters extend from the exterior wall framing at the corner (45-degree angle to the walls) to the *ends* of the ridgepole.

Valley rafters extend from inside corners formed by exterior walls (such as where porches tie into the main body of the house) to *a* ridge (the ridge is not necessarily the main ridge of the house but may be some lower ridge, such as the ridge of a porch, for example, which may be lower than the ridge of the house itself).

Jack rafters are typically located between ridges and valley rafters, between hip rafters and cap plates, or—less frequently—between valley rafters and hip rafters. Thus, because these rafters are located within angles—formed, for example, by hip rafters to cap plates (45 degrees)—they are shorter than common rafters.

Purlins are bracing members tied perpendicularly to rafters.

The *ceiling frame* ties the exterior walls together, forming a rigid base for the roof frame, and provides a strong surface for ceiling finish materials such as Sheetrock or plaster. The top side of the ceiling frame provides a place for insulation, wiring, plumbing, mechanical equipment, and similar uses and may be floored with planks or plywood or left unfinished.

The *cornice* is the portion of the roof that overhangs the exterior walls. The component parts of the typical cornice are the rafter tails, the fascia board, the shingle strip (drip), the soffit (if the rafter tails are enclosed), the frieze and molding, and the blocking necessary to build the cornice.

The *carpenter's framing square* is an essential tool in framing work. The "square" is shaped like an L. The body, or long part of the L, is 2 inches wide by 24 inches long. The tongue, or shorter part of the L, is 1½ inches wide by 16 inches long. The two parts meet at a right angle. The framing square has various increments marked on its face for computing roof pitches and other information.

Ladders refer to the framing over gables.

Stub joists are short members placed perpendicular to the regular ceiling joists. These members are used in place of the first ceiling joist under a hip roof; otherwise, the first joist would hit the jack rafters. (See Figure 5-17.)

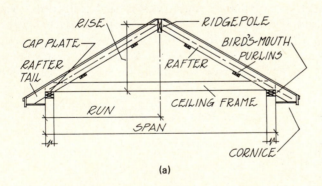

(a)

FIGURE 5-17 Typical residential roof system. (a) Components and assembly. (b) Typical framing members of hip roofs (right, and gable roofs (left and front). (c) Ceiling and stub joist layout.

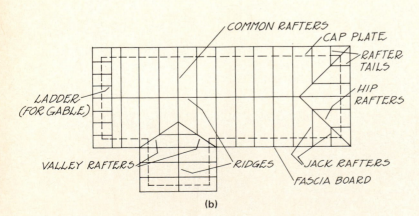

(b)

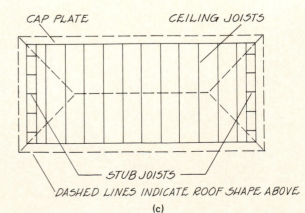

(c)

Trusses are pre-engineered components that normally consist of the ceiling framing and rafters, all in one package. They are strong and quick to install, and they free the area beneath for any size room arrangement. (See Figure 5-18.)

Roof sheathing is a cover secured to the roof

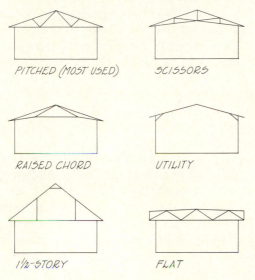

PITCHED (MOST USED) SCISSORS

RAISED CHORD UTILITY

1½-STORY FLAT

FREQUENTLY USED LIGHT TRUSSES
(TYPICAL SPANS 20' TO 32', SOMETIMES MORE)

FIGURE 5-18 Frequently used light trusses.

rafters. Plywood is a typical sheathing material, although other materials are sometimes used. Roof sheathing ties the rafters together, providing structural rigidity; distributes loads to the framing members; offers insulation and vapor-barrier qualities; and acts as a base for roofing materials.

A *cricket*, also called a *saddle*, is a wooden structure built at the high side of a chimney to divert water around the chimney and thus prevent water buildup behind the chimney. (See Figure 5-19.)

BUILDING THE ROOF FRAME

If roof trusses are used, the ceiling framing is usually omitted, since the bottom chord of the truss performs this function. Trusses will be discussed later.

If roof trusses are not used, ceiling framing is installed before the roof framing. This is not to imply that ceiling framing and roof framing are independent of each other; in fact, ceiling and roof framing are designed each with the other in mind.

Typically, the ceiling frame is made up of members called *joists*, which are two-by-sixes, two-by-eights, two-by-tens, two-by-twelves, or some combination of these members. The function of ceiling framing joists is to tie the exterior walls

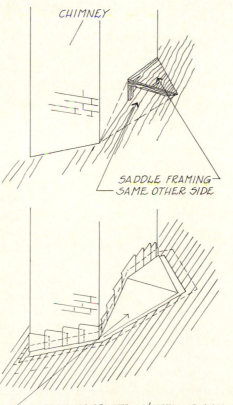

CHIMNEY

SADDLE FRAMING
SAME OTHER SIDE

SADDLE COMPLETE W/METAL FLASH-
ING INSTALLED.

(a)

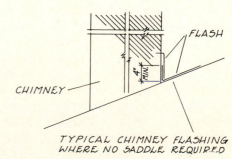

FLASH

CHIMNEY

4" MIN.

TYPICAL CHIMNEY FLASHING
WHERE NO SADDLE REQUIRED

(b)

FIGURE 5-19 (a) Proper saddle installations on steep roofs. (b) Proper flashing installation on shallow roof slopes that require no saddle.

together, in coordination with the rafters; that is, joists usually extend across the narrow dimensions of the house or building, as do the rafters. When the ceiling joists span the same direction as the rafters, they offer resistance to the outward thrust of the rafters, which push out on the walls; the joists counteract this thrust. This is the typical way of arranging ceiling joists and rafters, although there may be exceptions to the rule or partial exceptions, brought about by room partition arrangements below the ceiling joists or other conditions.

Ceiling framing joist sizes are usually indicated above directional arrows on the floor plan of the working drawings. The most typical spacing is 16 inches on center, but wider spacing is sometimes used, depending on the span, load (if any other than ceiling materials on the joists), and species of wood. The depth of the joists is determined by similar circumstances.

Before the ceiling joists are installed, they are cut to fit the appropriate span, including a trim cut at the end, such that the joist does not protrude above the later-installed rafters, interfering with the application of roof sheathing.

The ceiling joists are secured at the cap plates and where they rest on wall partitions. There is considerable variation in nailing techniques in securing the joists, and structural aids such as metal straps and joist anchors are often used. Later, when the rafters are installed, they are secured to the ceiling joists.

When the ceiling framing is complete, the roof frame itself may be built.

The Roof Frame

The roof framing members are cut to fit and placed around the perimeter of the house or building or otherwise laid close to where they will be installed.

After the members are prepared, the ridgepole is lifted onto the ceiling frame. Light scaffolding can be quickly built to hold the ridgepole in nailing position if it is too high to be held comfortably by hand or if there are not enough workers to hold it. The ridgepole is secured by rafters at each end (long ridgepoles will be done in several sections), and several rafters—enough to keep the ridge board from sagging—are secured between the ends. Checks are made to assure that the ridgepole is level and properly in place; the remaining rafters are then filled in and secured. Usually, common rafters are installed first, then hip rafters, valley rafters, and

so forth. Gaps may be left where special framing conditions are needed—such as around dormers; these special conditions may then be framed later, after a substantial amount of the typical framing is in place, for stability.

With a number of common rafters in place and secured, hip and valley rafters may be installed. As with common rafters, hip rafters and valley rafters are nailed to both the ridgepole and the cap plate. Then jack rafters are installed.

If the roof has gables, the ridgepole is usually allowed to extend beyond the exterior walls the desired overhang distance (this distance may be the same as the cornice overhang, but it can be more or less than the cornice overhang); the ladder is then constructed.

Roof bracing may be used to increase allowable rafter spans, reduce roof thrust at exterior or other walls, protect against wind, or all these things. The typical *wind beam* or *collar beam* is a 2-inch member secured from rafter to rafter a short distance below the ridgepole. Roof bracing is usually accomplished by using two-by-four purlins secured to two-by-four braces (the braces extend from the rafters at the purlins to a bearing partition or beam below).

The roof framing system is usually described in the working drawings by one or more sections taken through the house.

Where *roof trusses* are used, the ceiling framing may be reduced or eliminated. In the truss, the members comparable to rafters are called the *top chords* and the member comparable to the ceiling joist is called the *bottom chord*. Braces called *webs* are used to secure the top and bottom chords. Trusses have advantages and disadvantages over conventional roof framing.

On the advantage side, trusses speed up construction where the roof is relatively simple and repetitive in nature. Trusses may be built off the job site, thus reducing the amount of work time lost to bad weather. Trusses are easier to install at the cap plates than conventional rafter systems. Perhaps the greatest plus of trusses is that, because they are not usually dependent on bearing walls or other supports along the bottom chord, they allow more design freedom in arranging room partitions below. Also, less-skilled labor is required to install trusses than to install a conventional roof.

But trusses are not always the answer to roof framing. Trusses must be built in specially constructed jigs to assure that each truss is the same. Trusses must be carefully stored in a vertical position to avoid warping. If the roof is complicated, it may be more time-consuming to design and place trusses than to have a conventional roof installed by skilled carpenters. The same type of truss system, used in a large number of houses, can result in a look of sameness that may detract from the appearance of the houses, when viewed together.

The contractor must think through the advantages of conventional roofs versus trussed roofs for the particular job rather than assume that one system or the other is automatically the answer.

The following guidelines are helpful but should not be taken as rules for roof framing:

1. Rafters should be cut to provide a full and even bearing surface on the cap plate and should be toenailed with 10d nails.

2. In coastal and other areas subject to high winds, metal straps are recommended; straps are nailed to ceiling joist or rafters (or both, should the weather warrant such precautions) and nailed to the exterior wall studs; use 6d nails. Special anchors also are available for this purpose.

3. If the rafters do *not* extend in the same direction as the ceiling joists, provisions should be made to counteract roof thrust; the typical solution is to use cross members, secured to the rafters near the ridgepole.

4. Rafters and ceiling joists framed in the same direction also should be provided with cross members near the ridgepole, in addition to being nailed together where they meet at the cap plate.

5. Rafters should be as nearly opposite each other at the ridgepole as possible.

6. The ridgepole should be at least deep enough to provide full bearing for the rafters; that is, the rafters should touch the ridgepole over the full end area where they meet it; rafters are toenailed to the ridgepole with 8d nails.

7. Collar or wind beams should be placed a maximum of 4 feet on center.

8. Hip and valley rafters should be:

    At least 2 inches thick

    At least the depth of the end cut of the rafters that join them

    Supported at the ridge with braces to some solid bearing (such as a wall partition or beam below)

    Toenailed to the cap plate with 10d nails (jack rafters should be toenailed to the hip and valley rafters with 10d nails)

9. Where jack rafters frame onto an adjacent sheathed roof, the valley rafter may be omitted.

The Roof Cornice

The roof cornice, or eaves of a house, is the portion of the roof that overhangs the exterior walls. The visible components of the cornice include the shingle strip, fascia, soffit, frieze, and any trim members that may be used.

The *shingle strip*, sometimes called the *drip*, is a small strip of wood that supports the shingles at the cornice edge, allowing the shingles to project beyond the fascia board enough to aid drainage.

The *fascia* is the outside face of the cornice and is often a one-by-six. The fascia board is usually attached to blocking members at the rafter ends, sometimes called the *rough fascia*.

Underneath the rafter projections (rafter tails) is the *soffit*, which forms the bottom of the cornice. If the rafter tails are left exposed, there is no soffit. The soffit may be horizontal, in which case there must be added blocking members for support. These blocking members, called *lookouts*, are attached horizontally to the rafters and secured at the exterior wall; the soffit is then nailed to the lookouts. The soffit does not have to be horizontal; it can be attached directly to the bottom of the rafter tails, assuming whatever slope the rafters have.

The *frieze* is used where the soffit meets the exterior wall; the frieze is decorative, forming a transition between the cornice and the exterior wall. There may be trim at the corner formed between the soffit and the frieze, such as a quarter-round member, or some decorative molding, or the soffit may simply butt the frieze.

Cornices, when they project a foot or so beyond the exterior wall, function to protect the exterior walls from the elements. They may project only a few inches or they may project several feet; in any case, they should complement the general architectural style of the building.

In addition to these functions, the cornice, by way of the soffit, provides a means of ventilating the space between the top-floor ceiling and the roofs of buildings. This ventilation is required for both pitched-roof buildings and flat-roof buildings; if ventilation is not provided for these spaces, condensation and the resulting deterioration of materials are likely to occur.

Ventilation is often accomplished by drilling holes periodically along the soffit, such that air is allowed to circulate in the space above the top-floor ceiling; these holes should be screened. Another typical method is the use of a continuous screened vent along the soffit. In addition to the soffit vents, gable vents (often prefabricated) are usually used.

The vented area should be about 1 square foot for every 150 square feet of top-floor ceiling area. Thus, a 3,000-square-foot house would need about 20 square feet of vent area. But this is only a rule of thumb because varying roof structures require unique treatment and local weather conditions influence design. Check with local authorities.

Ventilation should be designed with possible additions in mind. In attics, for example, the gable vents should be placed above the level of any new ceilings that may later be installed in the attic for room or storage additions.

Building the Roof Cornice

There are many ways to construct cornices; the following is perhaps one of the more typical ways that works. First, it should be mentioned that window head heights are commonly 6 feet 8 inches; there are other heights, but this one is common if there is no particular design reason for the height to be otherwise (this assumes an 8-foot finished ceiling height, which also is typical in residential construction).

Assuming a 6-foot 8-inch window head height, it is clear that the rafter tails of a pitched roof can extend down a relatively short length before they cause a horizontal soffit to interfere with the window. And even if the soffit fits directly to the bottom of the rafter tails, it is not normally desirable for the fascia of the cornice to fall below the window head. So the window head height influences the length that the rafter tails can be, to put the soffit in the desired position relative to the window head. With a horizontal soffit, the rafter tails would normally be terminated such that there would be enough space above the window head to allow for an appropriately sized frieze board (the frieze board is usually the same size as or somewhat smaller than the fascia board of the cornice). To give a very practical example: Assuming an 8-foot finished ceiling with a 6-foot 8-inch window head height and a roof pitch of $8/12$, one could use a one-by-six fascia and frieze and a one-by-two drip (shingle strip), and the distance from the finished exterior wall (brick) to the outside face of the fascia board would be about 12 inches; this is simply one typical cornice arrangement among many.

With the foregoing in mind, consider how the cornice would actually be constructed. The plans should show the roof pitch and give the desired distance from the finished exterior wall to the outside face of the fascia; the contractor knows where to terminate the rafter tails, which can be done before the rafters are set in place or afterward. If the tails are cut first, less cutting is required on the scaffolds. But the rafters will vary slightly in length after placement, so some adjustment of the fascia board will have to be made to correct for these variations. If the rafters are not trimmed before placement, a chalk line may be stretched along them, and then they may be trimmed more accurately. Either way is acceptable; it is a matter of personal preference which is selected.

After the rafter tails are trimmed, the lookouts may be installed. The lookouts are usually two-by-fours; they are nailed to the rafter tails and at the exterior wall framing (a two-by-four or similar member is usually installed as a ledger at the wall framing to facilitate nailing). If a rough fascia is needed, it may now be installed; contractors try to avoid cornice details that call for a rough fascia because this is an extra expense, both in labor and materials. The soffit is usually installed after the rough fascia and before the finish fascia.

When the soffit is nailed or screwed in place (use rustproof nails or screws), the fascia may be installed. The drip, frieze, and any trim are then installed.

The gable soffit may be constructed in a variety of ways, but, staying with the pitched roof example above, the process would be as follows: The cornice would be extended the amount of the gable overhang, forming a kind of box, often made of 1-inch or similar stock. The soffit at the gable ladder, nailed either to the bottom of the ladder or to some blocking at the ladder, would simply butt the cornice. Many examples of this condition may be seen by driving through any subdivision in the country.

Cornices, as described here, require time and a degree of carpentry skill to construct. In time, some of the materials will have to be replaced, even if they are maintained well. Therefore, the use of prefabricated metal and synthetic cornices has increased recently. A number of manufacturers produce prefabricated cornices. Most are similar: A metal channel installed at the exterior wall frame holds the soffit at the wall; at the rough fascia (a rough fascia will often be required to accommodate the finish metal or synthetic fascia, which is usually not a structural member, as in traditional construction), the finish fascia is installed and drops down enough to receive the soffit. In turn, the finish fascia fits into a prefabricated drip. This is a typical example of a manufactured cornice; the installation of a particular cornice should be done according to the manufacturer's instructions.

The prefabricated cornice eliminates the need for lookouts, reduces the amount of skill needed for construction, and requires less homeowner maintenance once installed. However, wood cornices often have a higher-quality appearance than do many prefabricated cornices, which sometimes have too much of a "manufactured" look. As always, the contractor or builder must strike a balance between construction costs and marketability.

INSTALLING ROOF SHEATHING

Plywood Sheathing

Plywood sheathing is one of the most frequently used sheathing materials. It is strong, offers sound and weather insulation benefits, is quickly installed, and is relatively inexpensive for its high quality.

The following is presented as a general guideline for plywood sheathing.

1. Plywood should be structural-interior type, exterior type, or a type equal in quality.

These types of plywood conform to recognized lumber or inspecting authority standards, if so marked on the plywood; the contractor should verify exactly what quality of lumber is being purchased.

2. The plywood sheets (typically 8 by 12 feet) should be installed with the outer plies perpendicular to the joists or rafters; this usually means simply that the plywood will be installed lengthwise over the rafters, since the outer plies (that is, the outer layers of wood) run the length of the sheet.

3. Plywood sheets should be staggered so that the end joints in adjacent sheets occur over different rafters; this prevents long, undesirable joints (in the direction from cornice to roof ridge).

4. Where the ends of the sheets bear on the rafters, nail at 6 inches on center, using 6d common or 5d threaded nails for $5/16$- and $3/8$-inch plywood, 8d common or 7d threaded nails for other sheet thicknesses. Where the sheet covers intermediate rafters, nail at 12 inches on center, using the same size nails.

5. Where staples are used instead of nails, consult local authorities for standards.

6. The thickness of plywood depends on factors such as joist (in the case of flat roofs) or rafter spacing, type of plywood used, and the kind of load the roof is designed to withstand. (See Figure 5-20.)

PLYWOOD ROOF SHEATHING

Plywood Species	Plywood Thickness (inches)	Asphalt or Wood Shingles or Shakes		Built-up Roofing		Slate, Clay Tile, or Asbestos-Cement Shingles
		Edges Blocked	Unblocked	Edges Blocked	Unblocked	Edges Unblocked
Douglas fir	$5/16$	16	16	16	—	—
	$3/8$	24	20	24	16	—
	$1/2$	32	24	32	20	16
	$5/8$	42	28	42	24	24
	$3/4$	48	32	48	28	32
Western softwoods	$3/8$	16	16	16	—	—
	$1/2$	24	20	24	16	16
	$5/8$	32	24	32	20	20
	$3/4$	42	28	42	24	28

Maximum Rafter or Joist Spacing (inches on center)

FIGURE 5-20 Plywood roof sheathing specifications. (Courtesy of Federal Housing Administration.)

Wood Board Sheathing

1. Lumber should meet all applicable authority standards, including grade, moisture content, and dimensions. Typically, the minimum thickness is ¾ inch; resawn boards $^{11}/_{16}$ inch thick may be used where the roof slope is greater than 3 in 12 and the rafter spacing is not over 16 inches on center. The maximum width should not be over 12 inches.

2. Rafter spacing should not exceed 24 inches on center.

3. The boards should be tongue-and-groove, shiplapped, or square-edge with ends cut parallel to and centered over the rafters or joists. As with plywood sheathing, the end joints must be staggered; not more than two end joints should touch at the rafter or joist. Tongue-and-groove boards may have joints between rafters and joists, but adjacent boards should not both have joints between supports; also, where there is a joint between supports, the board should be secured to at least two supports.

4. Wood-board sheathing should be laid without spaces between, except where wood shingles will be the roofing material; if wood shingles will be used, some spacing between boards may be tolerated, the exact spacing depending on the shingle size.

5. Nails should be 8d common or 7d threaded, two nails in 3- to 8-inch boards and three nails in 10- and 12-inch boards, installed at every rafter or joist.

6. If the roof boards are installed diagonally over rafters 24 inches on center, tongue-and-groove boards should be used to increase strength.

7. All loose knots and knotholes over an inch in diameter should be repaired if asphalt, fiberglass, or built-up roofing is to be used. Tin is the typical material used for such repairs.

Various other types of roof sheathing are available (fiberboard, planks, synthetic materials), but these materials usually are used for custom houses and require engineering or architectural consultation. Plywood and boards are commonly used for roof sheathing throughout the country.

The Roof Covering

6

Once the roof framing is in place, the roof covering may be installed. The basic parts of this covering are the sheathing, underlayment (if called for), flashing, and the roofing material itself, which in this text is either some type of asphaltic shingle, wood shingles, or wood shakes.

ROOFING TERMS

The following terms are common in the roofing business.

Blind nailing refers to nailing in some area that will later be covered by shingles.

A *blind valley* is a flat valley that usually is not visible from the ground.

Bond lines are the alignment of *cutouts* on two- or three-tab shingles. These cutouts, or slots, form *tabs* in the exposed portion of the shingle, giving the viewer the impression that the roof is made of smaller shingles than is actually the case; a *two-tab shingle* has two tabs, a *three-tab shingle* has three.

Bull is a common name for plastic cement used in roofing. The cement is applied with a *bull paddle*, typically a narrow wood member or shingle.

The *butt* is the bottom edge of a shingle.

Capping in or *drying in* is the application of roofing felt over a roof deck.

Courses or *runs* of shingles are similar to brick courses. But shingle courses may be considered diagonally and horizontally as well as vertically.

A *drift down* is a shingle course that angles down at one of its ends.

Double-coursed shingles have two layers of shingles per course; this technique is usually used on side walls where wood shingles are employed.

Exposure refers to the amount of the shingle that is left visible, exposed to weather.

Face nailing is nailing the exposed portion of shingles.

The *factory edge* is the edge as it comes from the factory; shingles must sometimes be trimmed in the field.

A *fishmouth* is a rise or buckle in a shingle or area of shingles.

Freezebacks or *ice dams* are icy areas that creep back into the shingles, due to freezing and thawing action.

Gauge refers to the measurement of shingle surface.

Granules are the crushed particles on the shingle surface.

A *hip pad* is a protective cushion worn by roofers.

A *membrane* is an asphalt-saturated cloth used in combination with plastic cement to waterproof joints and edges.

The *roof pattern* is the design formed by the shingles.

Pin nails are small nails usually used on the ridge; typically, they are galvanized 3d or 4d nails.

The *rake* is the name given the cornice or eaves at the gables of a house.

A *shadow course* of shingles is a double course of shingles, laid to create a shadow line.

A *square* is a unit of roof measure: One square

is an area that measures 10 by 10 feet, or 100 square feet. Shingles for the job are figured in squares.

The *starter course* is the roofing material at the cornice edge of the roof. The starter course may be a certain width of roll roofing, over which the shingles are then started. Or the starter course may be shingles; thus, the roofing at the cornice is somewhat thicker than the rest of the roof.

The *tail* is the top (thin) end of a wood shingle or shake.

A *tie-in* is where shingles from separate roof areas or angles are joined.

Underlay is roofing felt, laid over the roof sheathing.

TYPES OF SHINGLES

Asphalt Shingles

Asphalt shingles are the dominant roofing material in residential construction. This material is initially inexpensive compared to wood, metals, and other roofing materials. It also is quick to install and lasts well for the expense—as much as 15 years, even longer in mild climates. When asphalt shingles become too worn to be serviceable, they may be roofed over with new shingles. Opinion is somewhat divided about whether the old shingles should be removed before applying new ones. Some authorities hold that the old shingles should be removed so that the deck may be repaired, if need be. Others contend that the old shingles provide insulation and should be left in place.

It seems reasonable to suggest that the decking be inspected from the attic side, when possible, and any needed repairs be made from underneath. The appearance of the roofing materials often affords clues as to the condition of the deck underneath, should it not be possible to inspect from the attic side. If the worn roofing is fairly smooth and level, chances are the deck is in acceptable condition to receive another layer of shingles; this is a matter of judgment on the contractor's part.

Depending on the capability of the deck, the roof structural system, and applicable regulations, some roofs can hold several layers of roofing. It is not recommended here, however, that more than three layers be applied, even if local codes permit. Most roof decks and structural systems are not designed for such loads, and although many structures would hold the extra shingles without danger of collapsing, the rafters would often be forced

into permanent deflection, causing the roof to sag noticeably. This condition can often be seen in old homes where as many as five or even more layers of roofing have been installed over the years.

Asphalt shingles get their name from the base mat of the shingle, which is saturated with asphalt. This base mat may be made of organic material (cellulose fibers) or of inorganic material (fiberglass or other synthetic fibers). Many contractors recommend the fiberglass mat, claiming it holds up better than the organic ones. The base mat is covered with mineral granules, which give the shingles their color and protect them from the elements.

Asphalt shingles are available in a variety of shapes. The most popular shape probably is the three-tab. Other shapes include the two-tab, the three-tab hexagon, the T-lock, and the Dutch lap shingle (this one is a single, rectangular shingle without slots).

Wood Shingles and Wood Shakes

Wood shingles are relatively smooth on both sides because they are sawed. Shakes are rough on at least one side because they are split. Both these shingles perform the same function, but the shake forms shadows and has a more rustic appearance. Both types of wood shingles take on a weathered appearance in time.

Wood shingles are more expensive to buy and install than asphalt shingles and may result in higher insurance costs to the owner; however, shingles may be bought impregnated with fire-retardant chemicals, and fire-resistant underlayments may be used to reduce the chance of fire; these precautions, of course, raise the cost. The main reason for using wood shakes or shingles is their inherent beauty.

Depending on local availability, the most used woods for shingles are western red cedar, redwood, and cypress. Each of these wood species is available in a variety of grade qualities. Typical lengths are 16, 18, and 24 inches in widths of 3 to 9 inches (some shingles come in widths greater than 9 inches, but if they do, they should be split before installing; 9 inches is the recommended maximum width).

ASPHALT SHINGLE INSTALLATION

The following discussion covers the basic installation procedure for asphalt shingles, focusing on the most popular type shingle, the three-tab, and on the T-lock.

When roofing an existing house, inspect the deck sheathing and structural system carefully; this can be done from the attic in most houses. The new roofing will not be any smoother or more level than the surface it is installed over. All needed repairs should be noted and pointed out to the owner; any repair work done is an added cost, and what is to be done and what it costs should be spelled out completely in the contract.

Before installing new shingles, on new or old roofs, carefully sweep the roof clean. New roofs normally require a felt underlay of 15-pound asphalt felt. The felt is installed from the bottom edge of the roof, working up. The nails for the felt should be used sparingly—only enough to keep the felt in place until the shingles are installed; meanwhile, use the shingle bundles to help hold the felt in place. The bulk of the shingles are stored toward the roof ridge, since work begins at the bottom edge. Carefully placed, the bundles will be where they are needed as work proceeds without undue materials handling.

If roofing over existing shingles, omit the felt underlay. As with new roofs, start at the edge. If a metal edging is used, do not install it over existing shingles as this will cause condensation and, later, rotting of wood members. The old shingles at the edges must be trimmed back for metal edgings.

Gutters can be a problem to roofs. If they are stopped up, the overflow can back up into the shingles at the edge and result in leaks. The owner should be notified if this condition exists and proper repair work (another extra cost) initiated. Roofers are paid to do roofing, but as a practical matter, they must be aware of needed repairs that, unattended, may reflect poorly on the roofing job.

The actual installation of shingles is fairly simple, once basic procedures are mastered. First, consider the job individually: Three layers of shingles is the maximum recommended load. If the roof already has three layers, one should be removed before new shingles are installed; local building codes should be observed. Some roofers learn how to do one roof pattern and use it for all jobs. However, this approach proves more difficult than planning each job individually. For example, the 5-inch pattern (each shingle is offset 5 inches left or right of the shingle below) is best on hip roofs, and the 6-inch pattern (each shingle offset 6 inches left or right of the shingle below) is best on gable roofs—"best" meaning the shingles look good and are easy to install. Tie-ins also should be considered; 6-inch patterns are easiest to tie in.

First, inspect the roof deck. In new construction, there are usually no serious problems; knotholes, however, may still be present, and these must be repaired if they are an inch in size or larger.

Where side walls must be shingled—at dormers, for example—the 6-inch pattern is again the better choice, because using the 5-inch pattern would require that each shingle be cut.

Stairstepping with a 6-inch pattern is a much used roofing method. The following discussion is for new roofs. To install using this method, first measure in 30 inches from a point ¾ inch out from the rake; the ¾ inch allows for the shingle overhang. Snap a chalk line from the roof edge to the ridge, parallel to the rake. Then snap another chalk line 6 inches beyond this line (in a direction toward the roof, away from the rake). This gives you two vertical lines, 6 inches apart; the line nearest the rake is 29¼ inches from the rake edge, the second line 35¼ inches from the rake edge. These lines are your guide for the starter course.

The starter course is begun with one shingle, located at the bottom corner of the roof (where the cornice meets the rake). The starter course runs along the cornice edge of the roof; it is an extra course, so the starter shingles are trimmed off 3 inches on the high side to prevent excess thickness when the regular roofing shingles are laid. Also, the starter course is laid upside down (with the slots up, toward the ridge). Lay the first starter shingle such that there is a ¾-inch overhang both at the roof cornice and at the rake edge. If the vertical chalk lines were positioned correctly, as described in the preceding paragraph, and if the first starter-course shingle is positioned with the ¾-inch overhang at rake and cornice, then the vertical edge of the starter shingle away from the rake should be just on the second chalk line. If this is not the case, recheck the chalk lines and the positioning of the starter shingle. When the starter-course shingle is properly placed, nail it (use four nails: one near each edge at the bond line and one just down from each of the two slots at the bond line); the nails should be driven firmly and straight down, flush with the shingle, not pounded down below the shingle surface.

Lay the next starter shingle adjacent to the first and nail it as you did the first one. The vertical edge of this second shingle should also be on the second chalk line. Continue the starter course in this manner until it is complete.

With the starter course installed, you are ready to install the first course of the regular shingles. To start the first shingle of the regular shingle coursing, lay the shingle (slots down) over the starter shingle such that one vertical edge is on the first chalk line and the other vertical edge overhangs the rake; this shingle should overhang the cornice edge by ¾ inch. There will be three nails in this shingle: one nail about an inch in from the chalk line and one slightly over the two slots; the nails will be near the bond line. There will be several inches of overhang at the rake—do not trim yet. Place the second shingle of the regular coursing adjacent to the first one and nail with four nails: one nail about an inch from the vertical edges and one nail just over each of the two slots, as with the first regular shingle. One vertical edge of this second shingle will be on the first chalk line.

The third shingle of the regular coursing may now be installed; this shingle is placed over the first shingle, 5 inches above the first shingle, with a ¾-inch overhang at the rake and with the remaining vertical edge over the second chalk line. Thus, the third shingle placed will overlap, horizontally, the shingle beneath by 6 inches (the distance between the first and second chalk lines, which are the guides for shingle positioning).

To help keep the horizontal lines of the shingle coursing parallel, you now begin *gauging*, or measuring, the shingles; this usually is done with the roofing hatchet. The hatchet used for a 5-inch exposure has a tick mark or a drilled hole near the blade edge; the distance between this tick mark or hole to the hatchet head is 5 inches. Place the mark or hole at the lower horizontal edge of the shingle, and the top of the head should just touch the horizontal edge of the shingle above; this measurement should be taken near the ends of the shingle.

The fourth shingle is installed above the third shingle with a vertical edge on the first chalk line and with several inches overhang at the rake. Gauge the shingle before nailing; nail as described previously.

Next, some shingle cutting will be necessary. Cut two of the three-tab shingles, making two two-tab shingles and two one-tab shingles. Install one two-tab shingle above the second regular shingle installed, along the second chalk line; install the other two-tab shingle above, on the first chalk line. When the two-tab shingles are installed, install the fifth and sixth shingles as usual, stepping them up the roof at the rake edge. The one-tab shingles are then installed, using two nails per shingle and placing the shingles such that they fill in the stepping of the previous shingles. The seventh and eighth shingles are then installed. When two more one-tab shingles are installed at the seventh and eighth shin-

gle, the first "stairstep" from the cornice will be complete; you then return to the bottom (cornice) edge and start another stairstep, and so forth. (See Figure 6-1.) Rake trimming is discussed later.

Diagonal or stairstep installation is best used for lengthy slopes. Short slopes, such as the roofs over small porches and similar areas, are best done in successive vertical rows from cornice to ridge; this is called the *straight-up* method.

The following text discusses the use of a 6-inch pattern installed by the straight-up method on a new roof.

The early part of this method begins as the stairstep method discussed above: Allow for a ¾-inch overhang. From a point ¾ inch out from the rake, measure in 30 and 36 inches and snap vertical chalk lines as discussed before. Install the starter course with the slots up and with 3 inches trimmed off the tabs.

The first starter-course shingle is installed with a vertical edge on the second (36-inch) chalk line. Therefore, the first whole shingle used in the regular shingle coursing will be installed with a vertical edge at the first (30-inch) chalk line. The second whole shingle will be installed alongside the first whole shingle. Moving up, the next shingle will have a vertical edge on the second chalk line; then install a shingle alongside this one. Continue the two rows of shingles up to the top in this manner; having two rows at the rake will make it easier to keep the shingle courses horizontal because the two rows serve as a guide for the remaining shingles. As a further aid in keeping the courses horizontal, snap several horizontal chalk lines across the roof; these chalk lines will be on the felt underlay, snapped such that the chalk lines represent the top of the shingles.

As you install successive single vertical rows, do not take them all the way up; take them only up to the next horizontal chalk line. When you reach a chalk line with a vertical row, the tops of the shingles should touch the line; the chalk lines, of course, correspond to the first two vertical rows of shingles you installed. Thus, you move vertical rows up to the first chalk line until the roof is finished under that line, then move vertical rows up to the second chalk line until the roof is finished under that line, and so on. (See Figure 6-2.)

Two roofers can use the straight-up method easily and quickly. One roofer starts a vertical row, and the second roofer follows with the succeeding row. When the first roofer reaches the horizontal chalk line, he returns to the cornice and starts the next vertical row. When the second roofer reaches

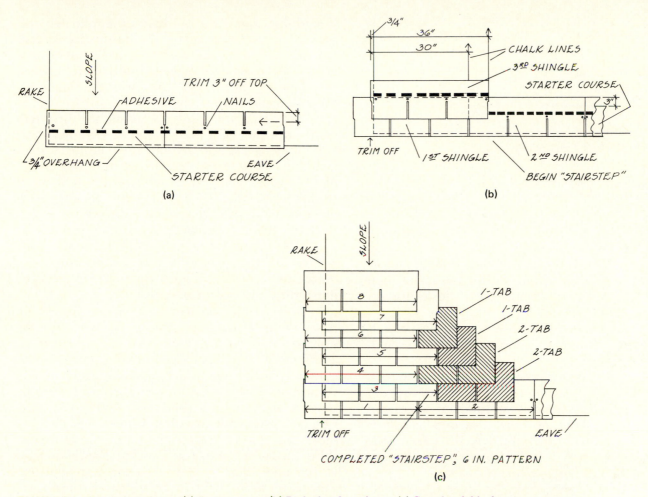

FIGURE 6-1 Shingle installation. (a) Starter course. (b) Beginning the stairstep. (c) Completed 6-inch stairstep pattern.

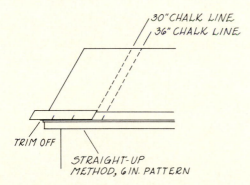

FIGURE 6-2 Straight-up method of shingle installation.

the chalk line, he returns to the cornice and begins the next vertical row, and so forth.

In shingling with the straight-up method, using a 6-inch pattern the shingles in each course are offset 6 inches from each other. When nailing, the shingles that offset to the right use only three nails, omitting a nail at the right tab; this is done

to allow the shingles that offset to the left to be placed under the right offset shingles; they are nailed at that time.

To install a 5-inch pattern with the stairstep method, first place the starter course as discussed earlier; you do not, however, need to snap vertical chalk lines. After the starter course is installed, place the first whole shingle such that it offsets the first starter shingle by 5 inches.

Place the second shingle directly above the first shingle, offsetting 5 inches to the left with a 5-inch exposure. Nail just above the bond lines.

Now, cut two two-tab shingles and two one-tab shingles. The third and fourth shingle laid will be two-tab shingles and the fifth and sixth shingles will be one-tab shingles. Together, the first through sixth shingles will complete the beginning run in the stairstep.

Next, go back down to the cornice and begin the next stairstep. Continue in this manner until you reach the roof top, then trim the rake edge. (See Figure 6-3.)

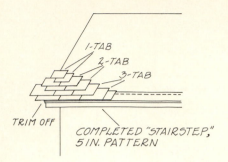

FIGURE 6-3 Completed 5-inch stairstep pattern at the rake edge.

T-lock shingles form a basket-weave pattern. These shingles are particularly good for reroofing where existing wood or asphalt shingles might otherwise have to be removed for the installation of three-tab or other type shingles. T-locks also are resistant to strong winds.

T-lock shingles of different brands vary. But, as with other asphalt shingles, installation procedures are very similar, regardless of brand.

After performing any patching work or repairs that may be necessary on the existing roof, trim back the existing shingles at the roof edges and install a metal edge or drip; this drip usually is about 4 inches wide and laps over the edge of the roof approximately 1 inch. Obviously, it is easier to install the drip on new construction.

The starter course may be made from the T-lock shingles by cutting off the stem portion of the T. This is best done while the shingles are in the bundle, cutting them all at once. The installation of the starter course is about the same as for other type shingles. Roll roofing may be used as an alternate to cut shingles for the starter course, but roll roofing also involves cutting. If, for example, a 20-inch-wide T-lock is used, the roll roofing would need to be cut into 40-inch sections to ensure that the joints of the roll-roofing starter did not fall at the corresponding lock portions of the T-lock shingles, thus causing a leak. The roll roofing has to be cut because temperature changes cause expansion and contraction of the material, which could damage the shingles; the cutting might be omitted where temperatures are mild and constant.

The starter course should overlap the roof edges by ¾ inch. A horizontal chalk line representing the high side of the starter course will help keep the course straight.

After the starter course is installed, the regular shingle coursing is begun. The first regular course is of cut shingles, made by cutting off the stem of the T but leaving the slots, which are needed to secure the second course. Begin by installing the

first shingle so that it overhangs the rake and cornice by ¾ inch, the same as the starter course. Nail two more shingles adjacent to this first shingle; a gap of approximately ¼ inch is often left between the shingles, but check the particular manufacturer's instructions.

Now you are ready to install the first two whole shingles. The shingle nearest the rake will lack reaching the rake by half the width of a shingle; this is because the shingles at the rake are all half shingles; this will be discussed next. Install the first two whole shingles by first locking them together and checking their alignment before nailing; when they are properly in place, nail them.

Next, install the half shingle at the rake. T-locks often come with marks that locate the center for ease in cutting. Cut the shingle straight down the middle, vertically.

Now install two more whole shingles above. After these shingles are installed, another row of shingles may be brought up from the bottom. Horizontal chalk lines may be used at about every fourth course to help keep the coursing straight, and a vertical chalk line may be used to keep the half shingles at the rake straight. The stairstep installation is often used in installing T-locks; one installer locks the shingles together, another nails, and a third carries the shingles. Usually, T-lock shingles at the rake edge must be cemented under each shingle to keep them in place; check the manufacturer's instructions. (See Figure 6-4.)

It should be mentioned that some roofers pre-

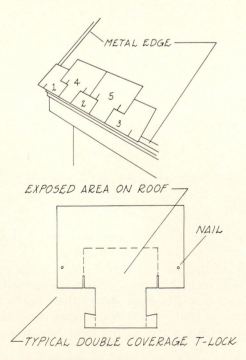

FIGURE 6-4 The T-lock shingle.

fer to install whole shingles at the cornice and rake edges, then come back and trim later. This is not a good practice if there is a strong wind that could pull the shingles loose, and it is not good to allow the shingles to droop over the edge longer than necessary. Installing the shingles whole at the edges does, however, make for very straight edges because a chalk line can be snapped and the shingles then cut with a hook knife.

The Rake Edge

At the bottom edges of the roof, the shingles are simply laid flush with the starter course, so no trimming is necessary. At the rake edge, the shingles may be allowed to overhang until the rake is complete, when they will be trimmed with snips or a knife. The rake shingles may alternatively be precut to the exact size needed for the desired overhang, typically ¾ inch. There are advantages and disadvantages to both methods.

Shingles may be precut by removing the bundle wrapper and knifing down through several shingles at a time; this is fast and convenient. But when precut shingles are used, more attention is required during installation, because each shingle must be installed perfectly if the rake edge is to be smooth. Often, precut shingles form a less-than-straight rake edge. Nevertheless, many roofers prefer precut shingles for the rake edge because they can be installed rapidly.

Some roofers use a starter strip at the rake. The starter strip is installed at the desired overhang; the regular rake shingles may then be installed with the starter strip acting as a guide. The regular shingles may be precut and installed flush with the starter strip, or whole rake shingles may be installed, overhanging the starter strip; the whole shingles are then trimmed with the starter strip as a guide. This is, however, a time-consuming method.

If whole shingles are used at the rake edge, there will be some overhang. When the rake edge is complete, a chalk line can be snapped at the desired overhang (usually ¾ inch), and the shingles can then be trimmed, using a knife or snips. This method eliminates the time needed to precut shingles, and it produces a very straight rake edge. However, the rake shingles should not be allowed to overhang during strong winds, since they may be damaged, and they should not be allowed to overhang longer than necessary, especially during hot weather. For a right-handed roofer, the right rake will be fairly easy to trim; the left side, however, will be more difficult.

Ridges

Installing shingles at the ridges of hip and gabled roofs is one of the simplest jobs in roofing, but it must be done correctly.

Start hip ridges from the bottom. The first shingle installed, at the bottom corner, must be mitered to fit. The remaining shingles up the hip ridge usually are cut as follows, using the three-tab shingle as an example: First, get a bundle of shingles in a comfortable working position; a workbench or the tailgate of a pickup will do. Holding the shingle with the slots away from you, trim to the slots. (See Figure 6-5.) This will give you three tapered shingles. Typical residential houses with hip roofs will require several bundles of cut shingles for the hip ridges; to reduce the amount of legwork, cut at least two bundles before starting to install the shingles. Precut ridge shingles also are available, as an alternative to cutting the shingles, but you will have to calculate the number of shingles you need. If you cut the shingles, each bundle contains about 27 shingles (check the wrapper for the exact number), and you get three ridge shingles from each three-tab, or 81 ridge shingles per bundle.

Before you begin installing the new ridge shingles, remove the old ones. If this is not done, the hump will be too high. On new construction, be sure that the slope shingles are trimmed back so that they do not quite reach the ridge peak; otherwise the ridge will be too high, have bumps in it, or both.

Install the first shingle at the bottom corner. Now move to the top of the hip ridge and place another shingle, but do not nail it; instead, mark

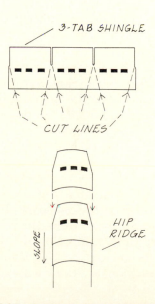

FIGURE 6-5 Cutting three-tab ridge shingles.

where the edge falls; then snap a chalk line from that edge to the corresponding edge of the bottom shingle. This chalk line will help you keep the shingles in a straight line. You need a chalk line only for one edge.

Start installing the shingles, moving up from the bottom. The shingles will have the same exposure as the slope shingles—usually 5 inches for three-tab shingles. The shingles are easier to install if you prebend them. Do this by holding the shingle at the edge and bending it "horseshoe fashion" so that it will more readily fit over the ridge. In cold weather, care is required in bending the shingles, or they will break. If the shingles break repeatedly, you will have to wait until warmer weather to install the ridge shingles or find some way to warm them before installing.

Lay each shingle with the edge at the chalk line and nail that side down (nail above the adhesive to ensure that the nails will be covered; this also makes a smoother ridge); then lap the shingle over and nail the other side. Only two nails are needed per shingle, typically 1¼ inches for new construction and 1½ inches for old roofs. Move up the hip ridge in this manner until you reach the juncture at the top.

The juncture at the top, where the hip ridges meet the peak, is folded in a particular way. Therefore, do not install the last shingle until you have worked both hip ridges within one shingle of the peak. When you get both ridges of the hip roof to the peak, install the shingles as follows: Lap one of the hip-ridge shingles over the other; then install the first shingle of the peak, nailing it to the two underneath, typically with a galvanized 4d or 5d nail; this nail is exposed, and exposed nails are used only for special conditions such as this. Use a minimum number of nails at ridge junctions, and be sure each junction is smooth and free of humps. (See Figure 6-6.)

Continue shingling the ridge at the peak as you did the hip ridge. However, the ridge at the peak of hip roofs and the ridges of gabled roofs do not require tapered shingles. Simply cut the shingles straight through at the slots.

If you are installing ridge shingles over a dormer, porch, or similar structure where the ridge of that structure meets the sloped shingles below the peak, lap the shingles in the same manner as if the ridge were sloping downward, like a hip ridge. This is because runoff from the roof slope drains water, sometimes rapidly, onto the lower ridges. If the shingles are lapped the wrong way, that is, opposing the drainage, they will scoop the water

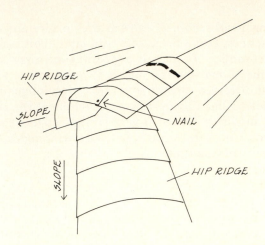

FIGURE 6-6 Proper installation of shingles at hip ridge junctions.

under the shingles rather than shed it. It is a simple matter always to consider the direction in which the roof runoff drains and lap the shingles accordingly. (See Figure 6-7.)

Ridges for gabled roofs are shingled the same way as hip ridges, except the shingles do not have to be tapered and the termination of the shingles at the gable end is different. The following is one accepted way to install shingles on gabled roofs: Start shingling at one end of the ridge and work along the ridge until you are within 2 feet of the second end, then stop. Now, shingle back until you intersect the first run; cement the intersection thoroughly, then cover the intersection with a cap shingle, securing the cap shingle with 4 nails. These nails will be exposed, as the single nail was exposed at the hip ridge juncture with the peak ridge.

If you are installing asphalt shingles over

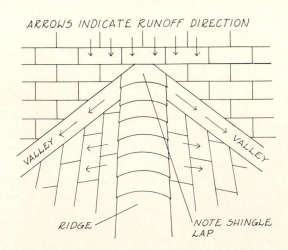

FIGURE 6-7 Ridge shingles lapped to promote proper drainage.

existing wood shingles, it is often best to remove the metal ridge (a metal ridge is typical on wood-shingle roofs). The metal ridge has a hump in it that may be difficult to cover smoothly with asphalt shingles. However, as you begin removing the metal ridge, you may notice that many of the wood shingles are becoming loose or breaking up. To avoid undue repair work and extra cost to the owner, you may therefore decide to leave the metal ridge in place.

If you leave the metal ridge, cut the new asphalt shingles so that they follow the same angle as the hump in the metal ridge, but do not overlap it. By avoiding an overlap at the metal ridge, you will cut down on the height of the ridge when the asphalt shingles are installed, making for a smoother, neater job. Otherwise, install the new asphalt shingles at the ridge as described previously. (See Figure 6-8.)

FLASHING

If the shingle surfaces of roofs were always uninterrupted, shingle installation and flashing would indeed be simple. In fact, on all but the smallest houses, just the reverse is true: Shingle surfaces are interrupted by dormers, porches, chimneys, valleys, and similar components. These "interruptions" cause the shingle surface to change directions—sometimes horizontally, sometimes vertically. For example, roof shingles may turn up to form the side walls of dormers or some other wall; the shingles of the main roof slope may have to change directions horizontally for the intersecting slopes of porches or hip sections. Thus, to prevent leaks, some type of flashing is needed around dormers, chimneys, plumbing vent stacks, ventilator pipes, within valleys, and so forth. And this flashing must be installed with the shingles in mind; that is, the shingles must be related properly to the flashing, and where the shingles change directions, proper tie-ins between shingles must be provided if the roof is to prevent leaks and present a neat appearance. On a new roof, flashing and shingle tie-ins are well standardized. For reroofing work, unfortunately, compromises must often be made, and there is no adequate way to generalize the many roof conditions that may need repairing, except to say that, wherever possible, the standards of new construction should be duplicated.

Valley Flashing

Four types of valleys are commonly used: the full lace valley, the half lace valley, the W valley, and the smooth valley. Each valley has certain functional and appearance qualities; no single valley is appropriate for all roofs. Perhaps the most important rule of thumb is simply to choose the valley type that most effectively deals with the amount and direction of the water runoff. Study the path the water takes on its way off the roof.

The full lace valley. In a full lace valley, the shingles are "laced" together up the valley. Some roofers may use this valley too much because it is fast and simple and requires no trimming. Where the two slopes that make the valley are the same slope, each shingle is slipped under the adjacent shingle on the opposite slope; the number of shingles used will be equal for both slopes. Where the slopes are not equal, two and occasionally three shingles of the lower slope are worked under the steeper shingles (remember the direction and amount of water; the steeper roof will shed water toward the lower slope, so the lower slope is worked under the steeper slope).

The following pointers may be useful.

1. When you begin shingling up one slope, leave the shingles at the valley loose so that the second slope shingles may be worked under.

2. Only whole shingles should be used over the valleys. There should be no joints in the valley,

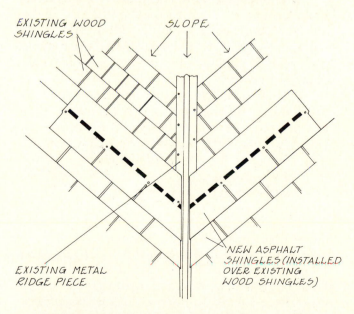

EXISTING WOOD SHINGLES

SLOPE

EXISTING METAL RIDGE PIECE

NEW ASPHALT SHINGLES (INSTALLED OVER EXISTING WOOD SHINGLES)

FIGURE 6-8 New asphalt shingles installed over wood shingles, retaining metal ridge.

and joints near the valley should be sealed with plastic cement.

3. Install heavy roofing paper or metal flashing under full lace valleys to assure no leaks.

As noted, the full lace valley is quick and simple to install. However, it has some disadvantages: First, it often looks sloppy. If a leak develops (and leaks may develop without the metal or roofing-paper flashing), it is difficult to find and repair. Also, T-lock shingles cannot be used for full lace valleys. (See Figure 6-9.)

The half lace valley. The half lace valley is made by lapping one slope entirely over the valley and up the other slope about 12 inches; when the second slope is shingles, it is first lapped over the valley center, then trimmed back from the valley center about 2½ inches. Thus, there are no open edges, and the chance of a leak is not great. The half lace valley is a much used method. It is not to be used on lock shingle roofs, but it may be used on strip shingle roofs and any roof using three-tab shingles. The half lace does not require a layer of heavy roofing paper or metal flashing underneath, as does the full lace, but the extra precaution is still a good idea if the job can assume the extra cost.

Joints and nails should be kept a minimum of 10 inches from the valley center. Therefore, some measuring will be necessary as you approach the valley with each course. Simply measure, when you are within half a dozen shingles or so from the valley, and determine where the shingle joint at the valley is going to occur. Then, add a one-tab shingle to ensure that no joint or nail will be required within 10 inches of the valley center; this check is simple and fast to do. You are simply determining where the joint will be rather than leaving it to chance.

The roof slope that will shed less water (generally the lower slope) is usually lapped over the valley. The steeper slope is then installed and trimmed; this directs the water from the steeper slope over the trimmed edge to the gentler slope, rather than the reverse, which would direct the water under the edge. If any water does get under the trimmed edge, the surface it touches will be shingled, preventing leaks. (See Figure 6-10.)

The W valley. The W valley gets its name from the metal form, shaped somewhat like a shallow letter *W*, that is placed in the valley. The metal form has a 1-inch hump in the middle and may or

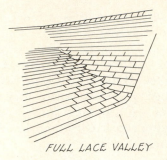

FIGURE 6-9   Full lace valley.

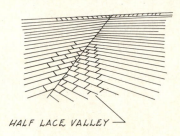

FIGURE 6-10   Half lace valley.

may not have water guards at the edges; the water guards are a good precaution on new construction but are undesirable for reroofing because the raised edges are difficult to cover smoothly with new shingles. When using the metal form on new roofs, metal clips, located approximately 12 inches on center, are used to hold the form in place. When necessary, it is possible to nail the metal itself at the edges.

The shingles should be trimmed back from the hump, or center of the form, about 2 to 3 inches from the center line on both sides, making the exposed trough 4 to 6 inches wide. Do not nail more than 2 inches from the edge of the form toward the center.

This type valley is often used in commercial construction and may be used for most roofs. It is difficult to use where the valley changes direction, however; if the valley changes direction, the metal form must be cut to fit the change in direction, then soldered back together. Obviously, this can be time-consuming. (See Figure 6-11.)

The smooth valley. The smooth valley is widely used in residential construction. It is made using 90-pound roll roofing or a sheet of galvanized metal. If metal is used, it should be wide enough to allow 10 inches of coverage on each side of the center line. The 90-pound roll roofing is generally available in 34-foot-long rolls, 3 feet wide; it is usually available in the same colors as the shingles.

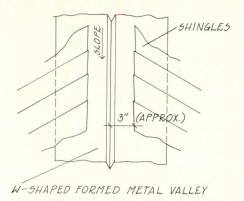

FIGURE 6-11   W-shaped formed metal valley.

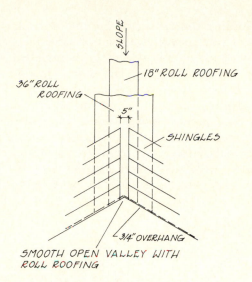

FIGURE 6-12   Smooth valley.

To install a 90-pound roll-roofing smooth valley, cut a length of roll roofing equal to the length of the valley plus a foot or so. Then cut this length so that it is 18 inches wide.

Lay the 18-inch strip granule side down in the valley and press it to fit the shape of the valley. It is important that the valley material be smooth because humps detour water, sometimes under shingles, and foot traffic (for later ordinary maintenance, such as the removal of leaves) on the humps can cause breaks in the material. When the roofing is smooth, secure one edge with nails at about 1 foot on center. Then nail the other side similarly; nail about 1 inch from the edge. Trim the roofing flush with the eaves line.

With gloves or a trowel or other suitable tool, lay plastic cement the full length of the valley; the cement strip should be at least 9 inches wide, with a uniform thickness of ⅛ inch or more. The cement is not a precaution—it is necessary to bond the two valley coverings together; otherwise, the top layer of material is unsupported and thus is more subject to hail, rain, and foot-traffic damage.

Next, install a 36-inch-wide strip of roll roofing, which is the final valley surface. This strip is installed with the granule side up. Secure this strip the same way you did the first one.

When the shingles are installed, they should be trimmed about 2½ inches from the center line of the valley, making the valley trough 5 inches wide. (See Figure 6-12.)

Two types of metal valleys are used: roll metal valleys and formed metal valleys. Roll metal valleys must be pressed to fit the valley; formed metal valleys are preformed for a more exact fit. The preformed valleys are preferred.

Roll metal valleys are made as follows: First, cut the length of metal to the length of the valley plus a foot or so. If there is another valley on the

other side of the peak, the metal will have to be longer still because the metal should be lapped over the ridge and then over the other ridge flashing by about 1½ to 2 inches.

The roll metal will be 20 inches wide, providing 10 inches of coverage on each side of the valley center line. Press the roll metal into the shape of the valley. This is not an easy job because the roll metal is 28- or 30-guage galvanized sheet metal. A smooth fit is important, however, because humps detour water in smooth valleys. Also, the roll metal should be installed with care because wrinkling and stomping the material into shape can break the galvanized coating, causing eventual rusting of the material.

When the roll metal is smoothly in place, nail one side, then the other. The shingles should be installed and trimmed so that the valley trough is about 5 inches wide.

Formed metal valleys are preferred over roll metal valleys because the preformed valleys are easier and safer to install and less likely to leak because of damage during installation.

Install as follows: First, chalk a line down the center of the valley. The preformed metal valley material typically comes in 18-inch-wide sheets, 8 to 10 feet long. Most residential valleys can be done with two sheets. Place the first sheet with the center over the chalk line, starting at the bottom. Let the material overhang the roof edge somewhat. Nail the corners; the metal can be trimmed flush with the eaves line later. With the bottom corners nailed, move up, nailing one side 1 inch from the edge, the nails at 8 to 10 inches on center.

Holding the center of the metal in place, nail the other side, as you did the first edge.

The second piece of metal is installed with a 3-inch lap over the first. The nailing installation is the same as for the first piece.

Smooth valleys are much used, but they should not be used where one of the slopes drains more water into the valley than the opposite slope. Smooth valleys should not be used where for some reason the valley slope is not uniform. And smooth valleys are not recommended where the valley curves.

Dubbing Shingles

Shingles at the valley are trimmed to present a straight, neat line at each side of the valley trough. When this is done, the overlapped shingles are left with pointed edges in the direction of the roof top. As water drains down the valley, it will hit the pointed shingles, and part of the water will run along the top of the shingle, sometimes finding its way to an opening in the shingle underlayment, resulting in a leak.

To prevent this condition and the accompanying leaks, the overlapped shingle points are cut off, in a line parallel to the shingle slots. This process is called *dubbing* the shingles. When the shingles are dubbed, water that gets under the edge of the shingles that overlap the valley material is turned back to the valley trough. (See Figure 6-13.)

Wall Flashing

Roofs are often interrupted by vertical walls: dormers, second-story portions of the house, and so forth. Also, roofs often butt against walls where the exterior wall of a house jogs. In any case, wherever a roof meets a vertical wall, there is the danger of water draining down the wall and into the joint where the roof meets the wall.

Where an asphalt-shingled roof meets a masonry wall, metal flashing is usually stepped up the masonry wall, inserted about an inch into the masonry joints. Under this flashing, in turn, metal shingles are stepped up the roof at the masonry wall. Last, the asphalt shingles are tied in. (See Figure 6-14 bottom.)

Where an asphalt-shingled roof meets a wall covered with siding, the flashing is somewhat simpler. Before the carpenter installs the siding, metal flashing is secured to the wall surface; this flashing may be individual pieces, stepped up the wall, or the flashing may be one or two long lengths of flashing. In either case, the flashing turns up the

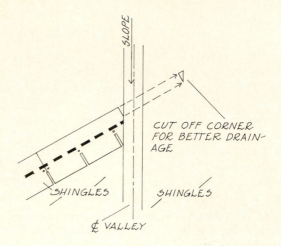

FIGURE 6-13 Dubbing valley shingles.

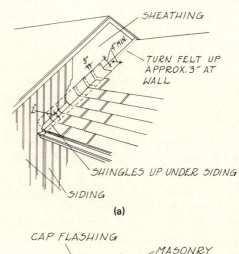

(a)

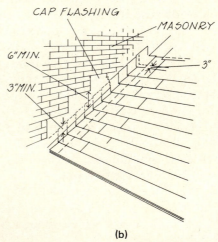

(b)

FIGURE 6-14 Flashing installation: (a) with asphalt shingles against siding; (b) with metal flashing against masonry.

wall several inches and turns several inches onto the roof that will later be shingled. The wall siding then laps over the flashing where it turns up, and the roof shingles lap over the portion of the flashing that laps onto the roof deck. (See Figure 6-14 top.)

Chimney Flashing

Water runs down the roof slope and splashes against the chimney where the roof butts the chimney. The valley created by the intersection of the roof slope and the chimney is a potential leak area. The wider the chimney, the longer the water takes to drain away and the greater the potential for a leak. If the chimney is 30 inches wide or wider, a saddle or cricket should be used. The saddle or cricket is simply a built-up area behind the chimney, a kind of "miniroof," that sheds the water away from behind the chimney. The saddle is built by blocking up from the roof deck at the chimney center and sloping down each side from the center; this raised portion is then covered with flashing material. (See Figures 5-19 and 6-15.)

The flashing material on the cricket and up the sides of the chimney is usually metal. Metal flashing is stepped around the chimney, inserted in the joints about 1 inch. The saddle flashing fits up under the chimney flashing and extends onto the roof deck several inches, where it is overlapped by asphalt shingles.

Where the chimney is less than 30 inches wide, the water buildup behind it is usually not enough to justify building a saddle. Instead, metal flashing is inserted into the chimney, similar to that described earlier. Another metal flashing element is then inserted under the chimney flashing and extended onto the roof deck, where it is overlapped by asphalt shingles. (See Figure 5-19.)

Eaves Flashing

The metal drip edge is sometimes omitted from roofs with slopes greater than 4 in 12. For roofs with slopes of 4 in 12 or less, the metal drip should be used. The metal drip extends under the underlayment a minimum of 1½ inches; secure the drip to the roof decking with rustproof threaded nails at 4 inches on center. The nails should penetrate the roof sheathing; if the underside of the sheathing is exposed, the nails may be driven through the sheathing and into the fascia board.

In cold areas where ice buildup at the roof edges may occur, install a course of 90-pound roll roofing along the eaves, over the metal drip (at the rake, the drip should overlap the underlayment). The roll roofing should extend from the eaves to a point up the roof approximately 1 foot beyond the interior wall surface. The roll roofing overhangs the drip at the eaves about ¼ to ⅜ inch.

For low-pitched roofs, a double course of roll roofing at the eaves may be necessary. The double course is installed similarly to the single course described earlier and is cemented together with plastic cement.

Stack and Vent Flashing

Stacks—pipes that exit through the roof—must be flashed properly to avoid roof leaks. The flashing may be fabricated by the builder or roofer but is usually a manufactured assembly consisting of a base piece and a cap. The base piece is bedded in cement and nailed to the roof deck; it fits around the pipe and extends up to where it is covered by the cap, which also fits around the pipe. The cap may have a clamp that secures it to the base piece, if the pipe stands fairly high above the roof; if the pipe is relatively short, the cap can fit over the top edge of the pipe and down around the base piece. (See Figure 6-16.)

Typically, the low side of the base piece laps the shingles and the high side fits under the shin-

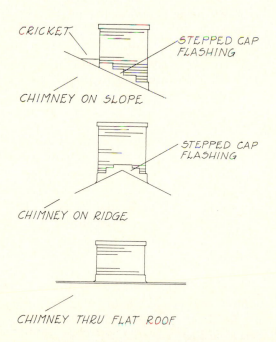

FIGURE 6-15 Chimney flashing.

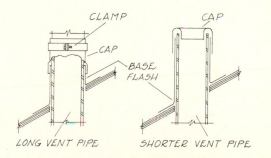

FIGURE 6-16 Stack flashing.

gles. This facilitates drainage. However, it is possible to shingle over the whole base piece.

Roof ventilators also pierce the roof. As with stack flashing, vent assemblies usually come with their own flashing components, and the base piece usually fits under the shingles on the high side and over the shingles on the low side. As with stacks, it is usually possible to shingle all around vents. (See Figure 6-17.)

When flashing around stacks and vents using manufactured assemblies, follow the manufacturer's instructions and local authority requirements.

ROLL ROOFING

Roll roofing is seldom seen nowadays, except occasionally on utility buildings. The following steps are for a typical installation:

1. Install a metal drip at the roof edge. The drip should extend up the roof a minimum of 3 inches, secured with rustproof threaded nails at 4 inches on center.

2. Cut a 19-inch-wide strip from the roll and install it as the starter course.

3. Install the first full course, using galvanized 11- or 12-gauge nails near the top at 12 inches on center.

4. Continue coursing up the roof, lapping each course 19 inches (roll roofing is 36 inches wide, so there will be 17 inches of exposure). Ends of the material should be lapped 6 inches.

5. Starting at the top, lift up the courses one by one and apply cold-application adhesive made especially for this purpose. The entire lapped area under each course should be uniformly and generously coated with adhesive, but take care not to apply so much near the bottom edges that the adhesive comes out onto the roof. The end-lapped areas should be coated

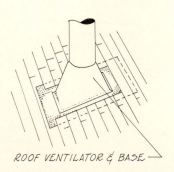

ROOF VENTILATOR & BASE

FIGURE 6-17 Ventilators flashing.

with adhesive and nailed down. Apply pressure with a broom to help bond the adhesive.

6. The ridge and hip areas are installed using 12- by 36-inch pieces cut from the roll roofing; lap and cement just as you did for the regular coursing. The ridge and hip pieces are started from the bottom, and the same general rules as used for ridge and hip shingles apply.

7. Inspect the entire roof to be sure that there are no bulges and that the adhesive has bonded uniformly. It may be necessary to roll the roof with a light roller. Where bulges persist, apply more adhesive and put additional weight (such as concrete blocks) on the area until the cement bonds; additional nailing may be necessary, but avoid excessive nailing, as leaks may result.

Roll roofing cannot be used where a truly flat roof is required because roll roofing as described here requires a minimum slope of 1 in 12. However, for houses and buildings with mansard or parapet roofs, roll roofing (with proper construction detailing) can sometimes be used in place of built-up roofing. Roll roofing does not present a handsome appearance where it is exposed.

Roll roofing makes a very dry roof, because it has so few joints and there is so much overlap. It is also quick and easy to install.

THE BUILT-UP ROOF

Whole books have been written about built-up roofs. The built-up roof is used a great deal in commercial construction, where flat roofs are typical. Flat roofs are not typical in residential construction, however, so only the highlights of built-up roofing will be presented here.

Follow these instructions for installation:

1. Install one layer of unsaturated building paper over the entire roof deck; the minimum headlap should be 2 inches. The roof deck may be any of a number of materials: concrete, steel, or wood, to name a few. In residential construction, the deck is usually plank or plywood.

2. Install one layer of no. 15 roofing felt with a headlap of 19 inches. Start at the bottom of the slope ("flat" roofs are actually sloped somewhat) and work up, nailing the bottom of the sheets to the deck at 18 inches on center. (See Figure 6-18.)

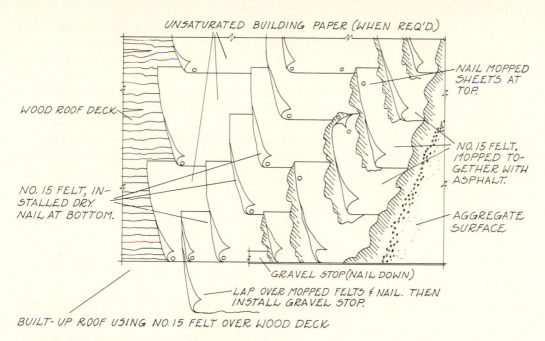

FIGURE 6-18 Built-up roofing over wood deck using no. 15 felt, dry sheets.

There is an alternative method: Install one layer of no. 30 or no. 45 roofing felt, starting at the bottom of the slope, head-lapping each sheet 2 inches. Nail at top and bottom of the sheets, 18 inches on center (or only the bottom of the sheets, 9 inches on center, through the laps). (See Figure 6-19.)

The nails should be corrosion-resistant, roofing type, at least 12 gauge, with ⅜-inch heads; there should be standard tincaps (tincaps function similarly to washers, providing more surface contact with the roofing materials and thus preventing tearing of the materials), a minimum of 32 gauge. Or similar large-headed nails may be used. The nails should be about 1 inch long, or just long enough to penetrate the wood deck. If the underside of the deck is exposed to view,

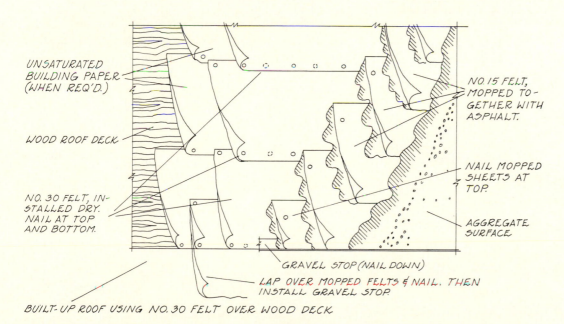

FIGURE 6-19 Built-up roofing over wood deck using no. 30 felt, dry sheets.

¾-inch threaded nails may be used at 12 inches on center with 1¼-inch roofing nails at 8 inches on center along framing members (such as two-by-eights and two-by-tens), which they will not penetrate. If the sheathing is plywood or a similar low-density material that the nails might pull out of, use screw-thread roofing nails or similar. Special nailing requirements may be necessary; check with local authorities.

3. Install a metal gravel stop, if required. Gravel stops are used to terminate the roof-surfacing materials at the roof edge, unless the roof has a parapet or similar edge. With parapet and similar walls, the roof surface is terminated by butting the vertical wall. (See Figures 6-20 and 6-25.)

4. Install a second layer of no. 15 felt (over the first layer of no. 15 felt or over the no. 30 or no. 45 felt, if you chose the alternate mentioned in step 2). As you install each course, mop underneath with bitumen, about 23 pounds per roof square. The entire surface under the second felt coat should be thoroughly mopped with bitumen and should be broomed or rolled to ensure tight bonding; nowhere should dry felt touch dry felt.

5. Next, flood the entire roof surface with a uniform coat of roofing asphalt (see Figure 6-21 for amount of asphalt).

6. Spread aggregate while the asphalt is still hot (see Figure 6-21 for amount of aggregate).

These instructions are typical for installing a built-up roof on a wood deck. A mineral-surface cap sheet is sometimes an acceptable alternative to spreading aggregate over the final coat. (See Figures 6-21 and 6-22.)

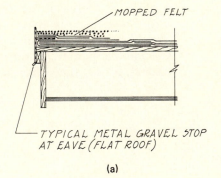

TYPICAL METAL GRAVEL STOP
AT EAVE (FLAT ROOF)

(a)

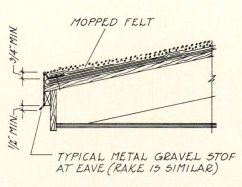

TYPICAL METAL GRAVEL STOP
AT EAVE (RAKE IS SIMILAR)

(b)

FIGURE 6-20   Metal gravel stops.

BUILT-UP ROOFING

Roof Type	Minimum and Maximum Roof Slope	Bitumen			Surfacing
		Average Mopping Coats	Minimum Flood Coat	Minimum Total Bitumen	
Asphalt (Type I) and aggregate	½ in 12[a] to 1 in 12	25 lbs. asphalt[b]	60 lbs. asphalt[b]	110 lbs. asphalt[b]	400 lbs. gravel or crushed rock or 300 lbs. slag on level roof; 300 lbs. gravel or crushed rock or 225 lbs. slag at 3 in 12 roof slope; proportional weight for intermediate roof slopes
Asphalt (Type II) and aggregate	1 in 12 to 3 in 12	25 lbs. asphalt[c]	50 lbs. asphalt[c]	100 lbs. asphalt[c]	
Coal-tar pitch (Type I) and aggregate	0 in 12 to 1 in 12	25 lbs. pitch	75 lbs. pitch	125 lbs. pitch	
	1 in 12 to 2 in 12	25 lbs. pitch	65 lbs. pitch	115 lbs. pitch	
Asphalt (Type II) and cap sheet	2 in 12 to 6 in 12	25 lbs. asphalt[c]	—	75 lbs. asphalt[c]	19-in. double-coverage mineral-surfaced cap sheet

[a]Roof slope may be reduced to 0 in 12 if asphalt complies with "Specifications for Asphalt for Low Slope Roofs," Chief of Engineers, Department of Army, Guide Specifications CE220.12.

[b]Use Type I 160° to 180° asphalt.

[c]Use Type II 180° to 200° step-roof asphalt.

FIGURE 6-21 Built-up roofing specifications. (Courtesy of Federal Housing Administration.)

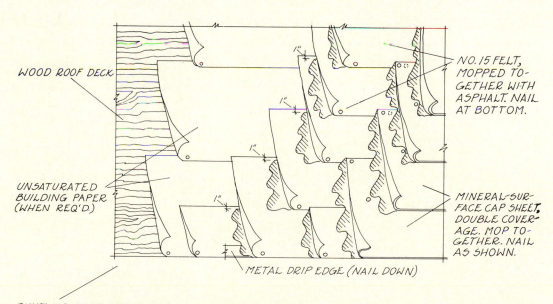

WOOD ROOF DECK

UNSATURATED BUILDING PAPER (WHEN REQ'D)

NO. 15 FELT, MOPPED TOGETHER WITH ASPHALT. NAIL AT BOTTOM.

MINERAL-SURFACE CAP SHEET, DOUBLE COVERAGE. MOP TOGETHER. NAIL AS SHOWN.

METAL DRIP EDGE (NAIL DOWN)

BUILT-UP ROOF USING DOUBLE-COVERAGE MINERAL-SURFACE CAP SHEET OVER WOOD DECK

FIGURE 6-22 Built-up roof using double-coverage mineral-surface cap sheet over wood deck.

Flashing a Built-up Roof

As with all roofs, stacks, vents, and other roof interruptions are potential sources of leaks and must be flashed properly.

Where the roof surface does not terminate by butting a vertical wall, a metal gravel stop is required at the eaves and the rakes. As a minimum, the gravel stop should extend onto the deck 2½ inches and down ¾ inch below the fascia trim (this trim piece is usually a one-by-two); the gravel stop should be at least ¾ inch above the roof surface, to hold the gravel in place and keep the asphalt surfacing from running down the fascia in hot weather.

For slopes less than 1 in 12, it is recommended that the portion of the gravel stop on the roof deck be covered with two strips of felt (one 6 inches wide and one 9 inches wide) set in a uniform layer of plastic roofing compound; the two strips are then mopped over. For slopes more than 1 in 12, the gravel stop may be applied over the dry felt and the 6- and 9-inch strips may be omitted.

It may sometimes be necessary to use a gravel-stop joint cover plate to prevent leaks (check local regulations). Figure 6-23 describes such a cover plate.

If a mineral-surface cap sheet is used, install metal edge flashing over the starter felt at the eaves and over the first mopped felt at the rakes. (See Figure 6-24.)

Typical nailing for metal edging and gravel stops is 4 inches on center, penetrating the roof sheathing, except where the nails would be exposed; if the nails would be exposed, nail down into fascia or roof framing members.

Where a built-up roof will butt a vertical wall, a cant strip should be installed before any roofing materials are installed over the decking. The cant

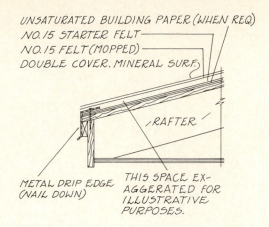

FIGURE 6-24 Metal edge flashing for mineral-surfaced cap sheet (section at eaves; rake is similar).

strip is usually about 4 inches high and 4 inches wide and forms a triangular shape in the corner between the roof deck and the vertical wall; the cant strip serves as blocking on which to turn up the roofing felt. Flashing is secured to the wall in some manner; it covers the cant strip and is turned under the asphalt flood coat and gravel. Figure 6-25 shows some typical flashing conditions where built-up roofs meet vertical walls.

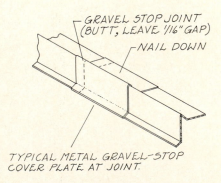

FIGURE 6-23 Typical metal gravel-stop cover plate at joint.

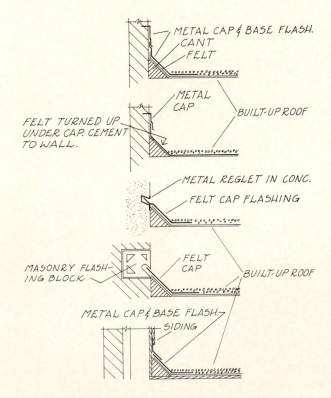

FIGURE 6-25 Flashing installation where built-up roof meets a wall or other vertical surface.

Flashing around protrusions through the roof is usually done with manufactured assemblies. Using the flashing around a vent pipe as an example, the installation procedure would be as follows: First, hot asphalt is spread around the pipe where the flange will fit. Spread enough asphalt to coat the area under the flange uniformly.

Next, set the flange in place around the pipe and nail it to the roof deck. To further ensure a watertight fit at the flange, two or more collar pieces of no. 15 felt may be cut to fit around the flange; each of these collars should be set in asphalt.

In most residential work, the next step is to flood the area around the flange with hot asphalt and then spread the gravel, completing the installation. However, the area around the pipe may be further built up with felt or similar material, sloping the built-up area down from the pipe to ensure positive drainage.

WOOD SHINGLE
AND WOOD SHAKE INSTALLATION

Because of the way they are made, wood shingles differ somewhat from wood shakes. Wood shingles are sawed and thus are relatively smooth on both sides. Shakes are split and have at least one rough-textured side, giving them a more rustic appearance than shingles. There are other considerations involved with wood shake and wood shingle roofs. (See Appendix D-4.)

Roof slope is an important and early consideration in installing wood shingle or wood shake roofs. The minimum recommended slope for both shingles and shakes is 4 in 12; porches may be as little as 3 in 12, however.

Exposure is influenced by slope. Generally speaking, the exposure of the shingles or shakes is reduced as the slope decreases: For example, a 16-inch shingle with a 5 in 12 slope or greater could use a 5-inch exposure; the same sized shingle on a 4 in 12 slope should not have an exposure greater than 4½ inches. Follow the manufacturer's instructions.

The *quality* of wood used for shingles and shakes is also an early consideration. Moisture content affects quality; generally speaking, lumber used for shingles, shakes, siding, and other finished uses should be kiln-dried or seasoned to the moisture content that it will assume in use. Of course, the moisture content that is tolerable will vary with the geographic region, since rainfall and other factors influencing moisture content are different in different parts of the country. The average allowable moisture content in Arizona, for example, would be 8% to 10%, while in Florida, it would be 11% to 13%. Check with local authorities for the region in question.

The *fire rating* of shingle, shake, and siding lumber may also be considered part of its quality. Shingles and shakes are available in Underwriters' Laboratories (UL) class B ratings and UL class C ratings; to gain a UL class C rating, the lumber is treated with a fire-retardant chemical. Special underlayments that make the lumber more fire-resistant are available.

In general, lumber quality is established by the association recognized as expert in the particular species. The West Coast Lumber Inspection Bureau, for example, publishes standard grading rules for Douglas fir, western hemlock, western red cedar, white fir, and Sitka spruce.

Sheathing for the roof may be either solid or spaced; there are advantages and disadvantages to both. Spaced sheathing is usually obtained by installing one-by-fours or one-by-sixes at the exposure centers; for example, one-by-fours across the roof at 5 inches on center for a shingle exposure of 5 inches. The maximum exposure is usually 10 inches; in any case, spaced sheathing should not exceed 10 inches on center. Spaced sheathing costs less because it takes less wood to do the job. Also, in areas where high moisture is a problem, spaced sheathing allows the shingles or shakes to dry out quicker; one of the chief reasons for wear to any wood is excessive moisture.

Solid sheathing, typically ½-inch exterior plywood, offers insulation advantages over spaced sheathing. In areas of extreme cold or where wind-blown snow is typical, solid sheathing is recommended. In addition to plywood, 1-inch boards may be used for solid sheathing.

Underlayment

Underlayment is best avoided under wood shingles. This is because the felt encourages condensation under the shingles, which increases wear. However, underlayment may be necessary in areas where windblown snow is encountered, in areas of extreme cold, or where air infiltration may be a problem.

Eaves flashing is required for wood shingle roofs where the outside design temperature is 1° F or colder. Where the roof slope is 4 in 12 or greater, apply a double layer of underlay at the eaves, extending up the roof to a point 24 inches inside the interior face of the exterior wall; for slopes less than 4 in 12, seal the double underlay with plastic cement (about 2 gallons per 100 square feet of underlay).

Underlayment is used with wood shakes more often than not. In any case, wood shakes require a 36-inch-wide starter strip (typically 30-pound felt), placed along the eaves line; a double and sometimes a triple starter course of shakes is then installed before the regular coursing begins.

The underlayment for wood shakes at the regular courses is actually an "interlayment," because of the way it is installed: An 18-inch layer of felt is placed over each course, lapping the shake course by double the weather exposure. For instance, if you are using a 24-inch shake with a 10-inch weather exposure, the distance from the shake butt to the bottom of the felt interlayment would be twice 10, or 20 inches. The ends of the interlayment should be lapped at least 4 inches. It is not necessary to nail the interlayment at the bottom, but it should be nailed at the top with ¾- to ⅞-inch roofing nails at approximately 12 inches on center.

Nailing

Nails should be corrosion-resistant shingle type. The length of the nails depends on the particular job and on whether the job is new construction or a reroofing (which would, of course, require longer nails). In any case, the nails should be long enough to penetrate the sheathing, solid or spaced, by at least 1 inch. If plywood sheathing is used, the nails should be threaded because plywood is less dense than planks and thus does not hold nails as well.

INSTALLING WOOD SHINGLES

To install wood shingles over one-by-four spaced sheathing, proceed as follows:

1. Install the one-by-four (or one-by-six) sheathing, spacing it equal to the shingle exposure—4½ inches on center for a 16-inch shingle with a roof slope of 4 in 12 to 5 in 12, for example. (See Figure 6-26 for recommended shingle exposures for given slopes and shingle lengths.)
2. Install the double starter course. The shingles should extend about 1 inch beyond the roof sheathing at both the eaves and rake edges. Lay the shingles ¼ inch apart and, for the

WOOD SHINGLES

Shingle Length (inches)	Maximum Exposure (inches)		
	Slope Less than 4 in 12	Slope 4 in 12 to 5 in 12	Slope 5 in 12 and Over
16	3¾	4½	5
18	4¼	5	5½
24	5¾	6¾	7½

FIGURE 6-26 Wood shingle specifications.

second layer of the starter course and the courses thereafter, sidelap the shingles 1½ inches. Alternate-course joints must not align or there may be leaks.

3. Install the remaining courses, moving up the roof, nailing the shingles ¾ inch from the edges and 1 to 1½ inches above the butt line of the following course. Thus only two nails are required per shingle, but they must be placed carefully, as noted. If the nails are placed closer to the edge, the shingle may split; if placed too far from the edge, the edge of the shingle will curl up. If placed too close to the shingle exposure, nails can cause leaks.

4. On hips and ridges, install a narrow strip of 30-pound felt before installing the ridge or hip units; this felt is good, cheap insurance against a leak. Install the hip and ridge units (preferably prefabricated units) at the same exposure as the slope shingles. The hip and ridge units are very visible and thus should be as straight as possible; use a chalkline down one side of the hip or ridge to align the units. Longer nails are required for the ridge and hip units because these units overlap the slope shingles.

5. Valleys should first receive an 18-inch strip of 30-pound felt (centered in the valley). Then install a 26-gauge (or heavier) galvanized metal form, either preformed or W-shaped. The metal should extend 10 inches on each side of the center line of the valley; very steep slopes (such as 12 in 12) may extend somewhat less. If two or more pieces of valley metal are used, headlap the pieces 6 inches or more at the joints. Shingle up to about 2½ to 3 inches from the valley center; this will give you a trough of 5 to 6 inches. The shingles will meet the valley at an angle. The points are trimmed off the shingle at the valley to improve drainage. (See Figure 6-27.)

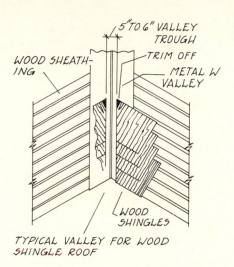

FIGURE 6-27 Metal W valley for wood shingle roof.

It is fastest to precut the angled shingles on the ground, then lift them all to the roof for installation.

INSTALLING WOOD SHAKES

The general installation procedures for installing wood shakes is about the same as for wood shingles.

1. Install the sheathing, spacing it equal to the shake exposure. (See Figure 6-28 for recommended shake exposures for given slopes and shake lengths.)

2. Install a 36-inch strip of 30-pound felt, flush at the eaves. Use only enough nails to keep the felt in place; the additional nails on the starter courses will aid in holding the felt securely.

WOOD SHAKES (HAND-SPLIT)

Type	Length (inches)	Maximum Exposure (inches)
Hand-split and resawn shakes	18	8
	24	10
	32	13
Taper-split shakes	24	10
Straight-split shakes	18	5½
	24	7½

FIGURE 6-28 Specifications for hand-split wood shakes

3. Next, install the first starter course. This course should overhang 1 inch at the eaves (and rakes). To keep the shakes in a straight line, some roofers install a shake with a 1-inch overhang at each corner of the roof, then stretch a string between these shakes to use as a guide for the intermediate shakes; in this method, it may be helpful to add another shake or two between the corner shakes, to help keep the string in place. Some roofers find the string troublesome, especially when the weather is windy, and so use an alternate method: They tack on a temporary wood member along the eaves; this gives a rigid guide for keeping the first starter course straight. Leave ¼-inch gaps between the shakes. This same alignment method may be used for wooden shingles as well.

The second starter course should be placed directly over the first, keeping the edges flush at the eaves. The joints, however, should not be allowed to align on alternate courses. A third starter course is sometimes laid; if so, it is laid directly over the first two, keeping alternate joints staggered.

4. The interlayment is now begun. The interlayment is 18-inch strips of 30-pound felt. Locate the first interlayment so that the bottom edge is twice the exposure distance from the butt of the starter shakes. For example, if the exposure will be 4½ inches, let the bottom edge of the interlayment line up 9 inches up from the starter shakes. Remember, it is only necessary to nail the interlayments at the top.

Chalk lines may be used effectively to keep both the interlayments and the shakes on a straight horizontal line.

5. Install the second course of shakes at the proper exposure and nail them in place, using only two nails, as discussed previously.

6. Next, install another 18-inch interlayment, then a course of shakes, then another interlayment, and so forth, working up the roof slope. Many roofers prefer to install all the interlayment first, then install the shakes. This is the fastest method, because you are then working first with only interlayment, then with only shakes, which tends to make the work go more efficiently. Also, if the interlayment is done first, the roofer may then do as many as five courses at a time, cutting down on movement time.

The shingle and shake installation described

here uses perhaps the simplest pattern to install. Other patterns may be used for special design effects.

FLASHING FOR WOOD SHINGLE AND SHAKE ROOFS

Flashing for eaves and valleys was described earlier. Flashing for side walls, chimneys, and roof protrusions such as pipes and vents are handled much the same as discussed for asphalt shingle roofs.

Chimneys

Chimney flashing is usually galvanized sheet metal or copper; copper is recommended. The exact installation of chimney flashing will vary somewhat, depending on chimney size and where the chimney occurs on the roof—at the ridge, through a roof slope, at a gable, or in some other location.

However, chimney flashing is usually made up of three components: the base flashing or apron, the side flashing, and the cap flashing. It is best if the cap flashing, which is stepped into the mortar joints, is installed when the chimney is built; if not, the mortar must be gouged out of the brick mortar joints deep enough for the flashing to be installed and remortared.

The base flashing fits underneath the shingles, around part or all the base of the chimney, extending approximately 4 inches under the shingles or shakes and 4 inches up the sides of the chimney.

The base flashing is covered and waterproofed by the side flashing, which in turn is covered and protected by the cap flashing.

Side Walls

Side-wall flashing is usually galvanized sheet metal or copper; copper is preferred. Side-wall flashing is made up of two components: the base flashing and the cap flashing.

The base flashing is secured to the roof under the shingles or shakes, extending under the shingles or shakes several inches and up the sidewall several inches.

The base flashing is covered and protected by the cap flashing, which is secured to the side wall under the siding, if there is siding, or into the

masonry, if the side wall is masonry. As with the chimney, the side-wall flashing may be installed early by the mason (preferred) or later by gouging out the mortar deep enough to accept the flashing and then remortaring.

Where the roof is sloped (as opposed to flat roofs) and the sidewall is masonry, the cap flashing must be stepped into the masonry to gain the desired angle; again, it is easiest if this component of the flashing is installed by the mason when the side wall is laid.

Pipes, Vents, and Other Protrusions

Manufactured flashing for pipes, vents, and other roof protrusions is the fastest method. The exact components vary somewhat from manufacturer to manufacturer, but not much; typically, there is flashing at the base and the side, and there may be some type of cap. Pipe flashing, for example, usually consists of a flange (base piece), a barrel (side flashing), and possibly a cap piece, depending on the length of the pipe protrusion above the roof.

Flashing materials, in order of preference, are copper, lead, galvanized sheet metal, and rubber.

In general, all manufactured flashing should be installed according to the manufacturer's instructions. If there is going to be a leak, it usually is related to the base piece of flashing. Some roofers feel that manufactured base pieces are too small. To overcome this problem, a simple precaution may be taken: Install two layers of 30-pound felt around and over the base flashing. (See Figure 6-29.)

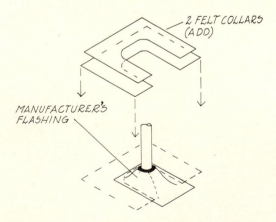

FIGURE 6-29 Augmenting manufacturer's flashing around vents.

REROOFING WITH WOOD SHINGLES AND SHAKES

Placing new shingles or shakes over an old roof will increase the insulation values and result in a more substantial roof, if certain steps are taken:

1. The roof should be carefully inspected before another layer of roofing is installed.
2. If the house is in a wet climate, the shingles may be too moist to overlay with new shingles —excess moisture is probably the most important deterrent in laying new shingles over old ones. If there is substantial deterioration of the existing roof because of moisture, tear off the old shingles before installing new ones.
3. Remove all vent, pipe, and other flashing and install new flashing before laying new shingles.
4. Trim back shingles at the eaves and rakes and install a new starter course. Some roofers do this step by installing a one-by-four along the rake and eaves line.
5. Remove hip and ridge units. The old hip and ridge units are sometimes replaced with thin, beveled siding, placing the beveled side down.
6. Replace any curled, split, or otherwise damaged shingles with new shingles.
7. Apply new shingles or shakes using the instructions given earlier.

This procedure assumes that the roof is structurally sound: no rafter deflection, rotted or otherwise damaged structural members, or other serious defects. With all reroofing jobs, regardless of the materials, the finished roof will be no smoother than the surface it is installed over.

PREFABRICATED SHINGLES AND SHAKES

Factory-produced shingles and shakes are available in large panels. Obviously, installing a panel with a dozen or more shingles on it is faster than installing the shingles one at a time. However, the same techniques as previously described for shingle and shake installation and flashing still apply. The roofer should be aware of these techniques, even if using factory-produced panels. Panel sizes and specifications vary; the roofer must be certain that the factory specifications meet the job specifications and that the factory installation instructions are acceptable.

The panels are fast to install, but not quite as fast as they might first appear, because the panels must be cut and placed around pipes and other roof protrusions, which must be flashed as previously discussed. Whenever panels must be cut, some time is lost. Nevertheless, factory-produced panels do offer speed, especially on simple roofs or large building projects.

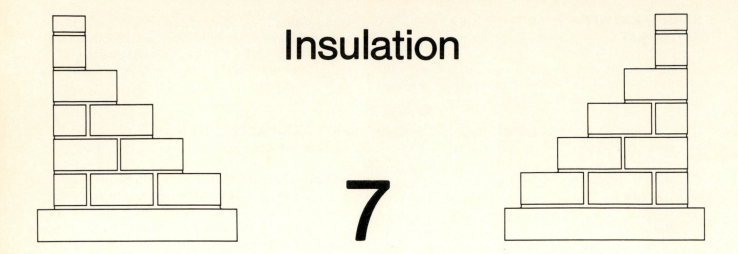

Insulation

7

The actual installation of insulation is one of the easiest and fastest jobs in building. Local requirements for insulation and methods of installation are fairly standard locally, and local U.S. Department of Housing and Urban Development (HUD) offices, building departments, and utility companies are familiar with them and often have data sheets available. All the needed materials are available locally.

What the builder needs is a practical understanding of how insulation works, so that different building situations can be handled correctly; although insulation installation is simple, it must be done correctly, and there is more involved than simply the insulating material itself.

The need for insulation is brought about by the basic fact that heat moves. In the winter, heat tries to get outside to the cold air. In the summer, heat tries to get inside to the cool, air-conditioned air. Insulation is material that reduces that heat movement.

The problem of insulating a house is complicated by another basic fact: When warm air and cold air are separated by some material, condensation sometimes forms on the material. A typical example of this condition can often be seen on windows, when the warm, moist air of the inside (the air is best when moist—it keeps you warmer and minimizes sore throats, dry noses, and the like) cools rapidly when it touches the cold glass, and the individual particles that make up the inside air collect together, becoming condensation. On windows, the problem of condensation can usually

be solved by turning on the fan of the central heating and cooling system. This will keep air moving by the window so that it does not have time to condense. This brings up an important requirement of insulating a house: *ventilation*.

Soffit vents and gable vents are typical methods of house ventilation used to help prevent condensation on building materials that in turn can cause mildew and eventual decay of materials. Condensation has often been responsible for serious damage in major commercial buildings as well as houses. A rule of thumb for gable vents is 1 square foot of ventilation area for each 300 square feet of ceiling area, assuming a ceiling vapor barrier is used; use 1 square foot of free vent area for each 150 square feet of ceiling area when no vapor barrier is used.

One more aspect of house insulation remains to be considered: the inside air itself. The inside air is moisture-laden. And it will pass through many of the materials of construction: Sheetrock, plaster, and wood wall paneling, to name a few. If this moisture-laden interior air is allowed to pass through the interior materials, it can condense when it reaches the colder materials closer to the outside. To prevent this from happening, a *vapor barrier* is placed at or near the warm face of the wall. For example, a vapor barrier is typically installed between Sheetrock and the wall studs (actually, most Sheetrock has the vapor barrier bonded to the back side, for ease of installation). Thus, the moisture-laden, warm air, or vapor, is prevented from reaching the colder surface where it can condense. Vapor barriers are generally needed at walls and ceilings and under concrete slabs.

HEAT MOVEMENT AND MATERIALS

Although builders do not normally compute insulation requirements, it will be helpful to know something about heat movement and the ability of various materials to resist that movement. An understanding of several definitions is necessary:

1. Heat is often measured in *British thermal units* (BTUs). A British thermal unit is the amount of heat it takes to raise the temperature of 1 pound of water 1° F.

2. The ability of your walls, ceilings, and floors to hold heat (BTUs) inside must be measured somehow, to determine what capacity your heating and cooling unit must have. Heat will pass, or leave, through Sheetrock, wood paneling, plaster, and so forth, at different speeds; thus, we must know the *U values* or *overall coefficient of heat transmission:* The overall coefficient of heat transmission, U, is the number of BTUs that will pass through a 12-inch square of the wall, ceiling, or floor in 1 hour per Fahrenheit degree of temperature difference between the air on the warm side (of the wall, ceiling, or floor—each should be computed) and the air on the cooler side.

   To figure the heat transfer in a given square foot of wall, ceiling, or floor, all the materials in that square foot must be considered. A typical wood-frame wall, for example, may be Sheetrock on the inside, nailed to two-by-four studs. On the outside, the studs may be sheathed with plywood, fiberboard, or some other material, or they may be unsheathed. Finally, there is the exterior siding, which may be wood or some synthetic material or brick veneer—the heat-transmission values of all these materials, plus the air space within the studs, must be considered in determining the overall coefficient of heat transmission for the wall. The same must be done for the walls and ceiling.

3. In computing values for individual materials—Sheetrock, wood paneling, plaster, and others—another definition is necessary: The *K value* is the number of BTUs per hour, per square foot, per Fahrenheit degree of difference, per inch of thickness for any one particular material. Thus, U values for walls, ceilings, and floors are the sum of the K values for the various materials in them.

4. Finally, there is the *R value* of materials, which is the resistance to heat flow of a particular material.

The engineering computations for U, K, and R values are beyond the scope of this text and of little concern to builders beyond certain very practical considerations. *The builder must try to obtain an adequate amount of insulation value for the least cost.* The typical use of the information given will be in determining whether or not to use, say, exterior sheathing (assuming there is a choice for the particular structure in the particular locality), and if so, what type of sheathing, what thickness, and so on. Otherwise, the builder is largely concerned with the R value of the various kinds of synthetic insulation in their typical forms: batt or roll insulation and loose insulation (these are for living spaces; stiff insulation for floor slabs, for example, is shown in Figure 7-4.)

Every building material resists heat flow, to some extent—some materials resist heat flow significantly; some do not. If the builder can use a material economically that has a significant insulation value over one that does not, the better insulator should be used.

In choosing synthetic insulation, first determine what R value is required, given the particular house wall, floor, and ceiling sections (which should have been designed with their insulating qualities in mind). The highest R values will be needed in the ceilings, then the walls or floor, depending on the area of the country (in some northern areas of extreme cold, the frost line is several feet deep, and the floors may need more insulation than the walls). (See Figure 7-1.)

Typically, batts are used in the walls and under the floors. Loose insulation may be used above ceilings. (See Figure 7-2.) Different manufacturers vary their products for different installation requirements; install per the manufacturer's instructions, but verify that those instructions meet local codes and applicable requirements. Batt insulation typically comes in 16- and 24-inch widths; R values to 38 are available, and the 38 rating will suffice for very cold areas.

Remember, where there is a considerable difference between inside and outside temperatures, vapor barriers must be used. Most batt insulation is available with a factory-installed vapor-barrier foil on one side. The insulation is installed with the vapor barrier facing the warm side of the wall, to

R VALUE	BATT OR BLANKET INSUL.	LOOSE
19	5½"-6½"	6½"-8¾"
22	6½"	7"-9½"
30	9"	10"-11"
38	12"	13"-17"

ZONE A – R19 FOR ALL ENERGY TYPES
ZONE B – R19 FOR GAS, OIL, AND ELECTRIC
     HEAT PUMP; R22 FOR ELECTRIC
     RESISTANCE HEAT
ZONE C – R22 FOR GAS, OIL, AND ELECTRIC
     HEAT PUMP; R30 FOR ELECTRIC
     RESISTANCE HEAT
ZONE D – R30 FOR ALL ENERGY TYPES
ZONE E – R30 FOR GAS AND OIL; R38 FOR
     ALL ELECTRIC HEAT
ZONE F – R38 FOR ALL ENERGY TYPES

FIGURE 7-1 Insulation recommendations by region.

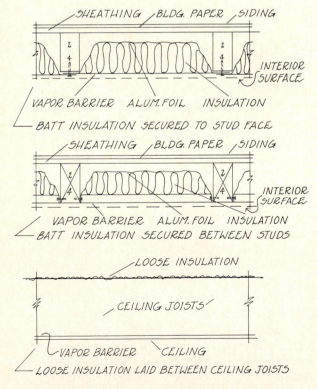

FIGURE 7-2 Installation of batt insulation in walls and loose insulation in attic.

prevent the warm, moisture-laden air of living or work spaces from forming condensation on contact with colder outside materials.

When batt insulation is used above the ceiling, the insulation flaps are usually nailed between the joists, so that ceiling materials (such as Sheetrock or paneling) can make tight contact with the ceiling joists. When insulation is installed between joists or studs in the manner just described, it leaves the edges of joists and studs exposed; in some areas of the country, and unless the interior wall material comes with a vapor barrier on one side, these exposed joists and studs must be covered with a vapor barrier. Vapor barriers may always be installed separately, but builders usually find the vapor-barrier–backed materials more convenient for installation purposes; otherwise, typical vapor barriers are made of waxed paper, polyethylene, copper foil, and aluminum foil. Vapor-barrier materials are readily available in building materials supply stores.

Insulation is usually installed over obstructions such as light fixtures and electrical boxes. But if the light fixtures are recessed, insulation is usually kept about 3 inches away from the light (follow the code and the manufacturer's instructions).

In attic spaces above a ceiling, either batt or loose insulation works well. Install the batts with the vapor barrier face down, toward the warm space below. If the attic is unfloored, the insulation is installed between the joists; if it is floored, the insulation can be placed over the floor. Start at the eaves and work toward the center of the house. The eaves often need blocking at the cap plate to keep the insulation perimeter intact and straight and to prevent interference with soffit ventilation. (See Figure 7-3.)

The insulation is ordinarily installed over

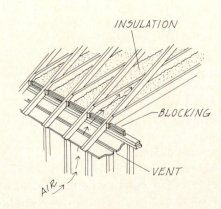

FIGURE 7-3 Blocking at the eaves cap plate.

wiring, unless doing so compresses the insulation; insulation value is diminished when the material is compressed. Workers should take appropriate safety precautions when installing insulation, wearing gloves, protective clothing, and masks.

In basements, crawl spaces, or similar unheated spaces beneath heated spaces, install insulation so that the vapor barrier faces up, toward the heated space above. The insulation should be lapped over the sill plate at the foundation walls. The insulation needs support underneath, or it will begin to sag and may eventually pull loose from the joists; to prevent this, use chicken wire, wire mesh, or heavy single wire to support the insulation. (See Figure 7-4.)

Basements and utility areas are frequently the location of furnaces, water heaters, and other equipment that can cause insulation facings to burn if the insulation is placed too close to heat or flame without adequate protection. Check the building code or consult appropriate authorities for recommended fire precautions in such circumstances.

Installing insulation between wall studs requires special attention because of the many obstructions and openings: doors, windows, electrical boxes, and so on. The insulation must fit snugly against openings and around equipment that may be located in the walls without compressing the insulation. The fastest way to insulate around doors, windows, and other obstructions is to first cover the openings or obstructions with insulation, then cut around them with a sharp knife. The odd pieces of insulation created by this installation technique can be used in smaller areas. Avoid all gaps in insulation—any area that is not insulated causes a loss of energy and may provoke condensation. Insulation can be installed with power staplers or with nails.

If using a separate vapor barrier (a 4- to 6-mil polyethylene membrane is often used), install the unfaced insulation between the studs, then secure the vapor barrier over the studs (and over the obstructions too, then cut the vapor barrier from around openings and obstructions as described for insulation). Any cuts in the vapor barrier must be repaired. The interior wall material is installed over the vapor barrier.

Insulation Summary

1. Insulating a house or building requires adequate ventilation.

2. Choose insulation with an R rating that is appropriate for the particular building materials you are using and for the weather conditions in your locality. Do not grossly overinsulate; do not underinsulate.

3. Most areas of the country require the use of a vapor barrier, installed on the warm side of the wall, ceiling, or floor. This vapor barrier may be bonded onto the insulating material, or a separate vapor barrier of polyethylene or other suitable material may be installed separately.

4. Houses are typically insulated at the walls, ceilings, and floor (if there is a space beneath the floor). Floor-slab edges and foundation walls are usually insulated with rigid insulation. Water-supply and waste pipes are often insulated in cold areas and sometimes omitted in milder climates. Heating and cooling ductwork should be insulated.

FIGURE 7-4 Wire mesh or heavy single wire used to keep insulation in place between floor joists.

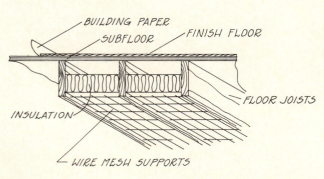

(a)

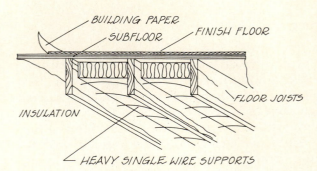

(b)

Doors, Windows, and Stairs

8

Doors and windows, almost without exception, are produced in factories. They may be bought prehung, which means they are already attached to a framework such that the builder has only to provide the manufacturer's required rough-framing opening; then, with ordinary carpentry, the doors and windows may be nailed in place. Prehung doors and windows vastly speed up the process of installing doors and windows and sharply reduce the number of on-job fabricating errors. Prehung doors and windows are almost always used in present-day residential construction.

Every manufacturer provides rough-framing dimension requirements with the door and windows available. Know the type of windows and doors you will use and provide the required framing opening when the wall framing is done; this will avoid rework in framing.

DOORS

The typical wood doors used in residential construction are flush, panel, folding, and sliding. (See Appendix D-2 for hinge information.) The overhead garage door has become standard. Flush doors are smooth-surfaced; they may be either solid or hollow. Flush, solid-core doors are usually used for exterior doors; as the name implies, they have a solid interior. This solid interior material often is made of blocks of wood glued together, although

other materials may be used; the interior is sandwiched between veneers. Flush, solid-core doors are generally 1¾ inches thick.

Flush, hollow-core doors are used for interior doors. They are usually made by spacing blocks of wood between the veneer faces, although other materials may be used. Hollow-core doors usually are 1⅜ inches thick. (See Figure 8-1.)

Panel doors are made by installing panels of plywood, hardboard, or solid wood within a larger framework. Glass is sometimes substituted for wood for the panels. The horizontal framing members are called *rails*, and the outside vertical mem-

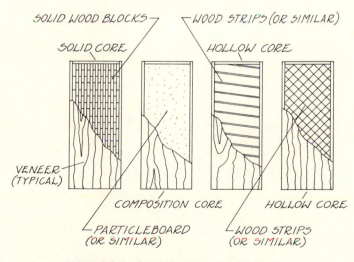

FIGURE 8-1 Four types of residential doors.

bers are called *stiles*. The interior vertical members are called *mullions*. Panel doors vary widely in design. (See Figure 8-2.)

Folding doors, as the name implies, fold. They are often made of two to four doors per opening. Folding doors may be either wood or metal. The most frequent use is for closet doors, dressing areas, and similar utility spaces. The most typical installation method is on one or more tracks, installed within a framed opening. The doors are equipped with pivot brackets and slide guides; when compressed, the doors open along the tracks. (See Figure 8-3.) Sliding doors work similarly to folding doors and are used for similar spaces.

Sliding glass doors are almost synonymous

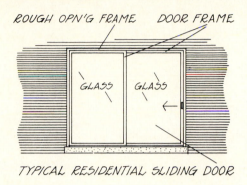

FIGURE 8-4   Typical residential sliding glass door.

with patio doors. These doors are ideal where inside casual areas such as dens or recreation rooms, or kitchens are related to outside areas such as patios, terraces, or swimming areas. Glass doors are an appropriate transition between such areas. Sliding glass doors are typically made of aluminum; they fit within head and sill tracks and slide on rollers. (See Figure 8-4.)

The overhead garage door, or "sectional upward-acting door," as some manufacturers call it, is standard in many areas of the United States. These doors may be built of wood, fiberglass, or metal; the wood doors are particularly heavy, and electrically powered drive components are recommended. Radio-controlled activators are easily installed and make it possible to open the garage door without getting out of the car. Power openers and radio controls are welcome conveniences and may be a necessity for the elderly or the handicapped.

Before the garage framing is begun, the door should be selected and the manufacturer's framing requirements studied. The framed door opening must be high enough and wide enough to accommodate the door and whatever hardware and casement are necessary to secure it. Also, appropriate overhead space must be allowed for the drive and other components. (See Figure 8-5.)

WINDOWS

The most commonly used windows in residential construction are the double-hung window and the single-hung window. The appearance of the two windows is the same; a walk down any American street will provide the viewer many examples of both. The double-hung window allows movement of both the top and bottom sections of the window. The single-hung window allows movement

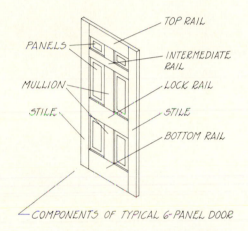

FIGURE 8-2   Typical six-panel door.

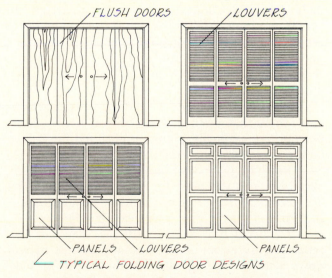

FIGURE 8-3   Typical folding door designs.

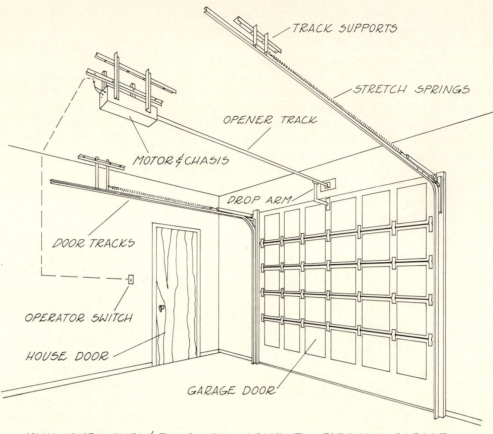

FIGURE 8-5 Electric garage door opener: main components and arrangement.

MAIN COMPONENTS & THEIR ARRANGEMENT—ELECTRIC GARAGE DOOR OPENER

of only one section, typically the lower one. (See Figure 8-6.)

Casement windows are another popular and handsome style common in residential construction. The casement window is equipped with cranks that, when turned, move the window out vertically. The casement window is available with a horizontal swing, also; such windows sometimes are called awning windows and often are used at the base of fixed glass windows (discussed later) for ventilation. (See Figure 8-7.)

Sliding windows typically have two movable sections that slide horizontally. Sliding windows are not especially handsome and are often used on back bedrooms, utility rooms, and so forth. (See Figure 8-7.)

Fixed windows are simply immovable sheets of glass installed in an opening. These windows are usually used to let in natural light or to allow views to the outside. (See Figure 8-8.) Where ventilation is needed, awning windows are often used at the base of fixed glass windows.

Windows, like doors, are factory-built units. Each manufacturer provides information concern-

ing rough-framing opening requirements and other installation details. Study these manufacturer requirements carefully before attempting to frame or install windows. Often, the rough opening is an inch or so wider and about ½ inch higher than the actual window; this extra space is used to level and plumb the window; wood shims are often used between the window unit and the rough framing to get the window level and plumb. (See Figures 8-9 and 8-10.)

Windows that will be used for ventilation and any others that open will, of course, need screens. Screens are typically aluminum-framed nowadays, and the screening material should be noncorrosive.

Insulating window glass is expensive, but its use eliminates the need for storm windows, which are also expensive, not to mention troublesome to put up and take down each year. Insulated glass is constructed at the factory with two pieces of glass separated by an air space. The appearance of windows with insulating glass is about the same as ordinary windows. Insulated glass, because of the extra cost, would have to be explained to potential buyers, but the energy savings possible with in-

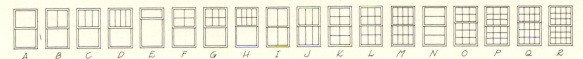

TYPICAL DOUBLE-HUNG WINDOW TYPES

STANDARD DOUBLE-HUNG WINDOW SIZES

Height of Opening	Width of Opening									
	1' 4"	1' 8"	2' 0"	2' 4"	2' 8"	3' 0"	3' 4"	3' 8"	4' 0"	4' 4"
2' 6"	ABFN	ABFN	ACGN	ACGJN						
2' 10"	ABFN	ABFN	ACGN	ACGN	ACGJN	ADHJN	ADHN			
3' 2"	ABFN	ABFN	ACGILN	ACGILN	ACGIN	ADHIJN	ADHIN			
3' 6"	ABFN	ABFN	ACGILN	ACGILN	ACGIN	ADHIJN	ADHIN	ADHN	ADHN	
3' 10"	ABFN	ABFN	ACGIKLN	ACGILN	ACGILN	ADHIJLMN	ADHIN	ADHN	ADHN	
4' 2"	ABFN	ABFKN	ACGILN	ACGILN	ACGIN	ADHIN	ADHIN	ADHN	ADHN	
4' 6"	ABFN	ABFKN	ACGIKLN	ACGIKLN	ACGILN	ADHILMN	ADHILMN	ADEHLMN	ADEHN	ADEHMN
4' 10"	ABFN	ABFKN	ACGIKLN	ACGILNO	ACGILN	ADHILNQ	ADEHIMN	ADEHIMN	ADEHN	ADEHN
5' 2"	ABFN	ABFN	ACHIKLN	ACGIKLN	ACGILN	ADHILMN	ADEHILMN	ADEHIMN	ADEHN	ADEHMN
5' 6"	ABFN	ABFN	ACGIKLN	ACGINOP	ACGILND	ADHILNOQR	ADHINOQ	ADEHIMNQ	ADEHN	ADEHNQ
5' 10"	ABFN	ABFN	ACGKN	ACGIKN	ACGIKLN	ADHILN	ADHILMN	ADEHIMN	ADEHN	ADHMN
6' 2"	—	ABFN	ACGN	ACGIN	ACGINO	ADHIN	ADHINQ	ADHN	ADHN	ADEHN
6' 6"	—	ABFN	ACGKN	ACGIKNOP	ACGIKLNOP	ADHILNOPQR	ADHILNOPQR	ADHMNQR	ADHN	ADHMNQR

Since the double-hung window is so popular with homeowners, it is available in many styles.

FIGURE 8-6 Typical double-hung window types and standard sizes. (Courtesy Creative Homeowner Press.)

sulated glass and the trouble-free aspect of the windows should be valid sales points.

STAIRS

Builders often buy stairs between the first and second floor as prefabricated units, because these stairs require time and skill to construct. However, some builders prefer to build the stairs on-site. And basement and other open stairs may be built easily on-site. Also, exterior stairs are often needed (for main entries, outside basement entry, patios, etc.), and these are usually site-built. Interior and exterior stairs vary somewhat in design, construction, and materials, but all stairs have many design and construction characteristics in common.

The following is a listing of typical design and construction requirements for stairs.

1. The main stairs between floors should have at least 6 feet 8 inches of headroom all along the stairs. Interior basement and service stairs, whether closed or open, should have 6 feet 4 inches of headroom all along the stairs. Exterior basement stairs with overhead cover may have as little as 6 feet 2 inches of headroom, if construction conditions are such that more headroom is impossible or impractical.

2. Main stairs between floors should be at least

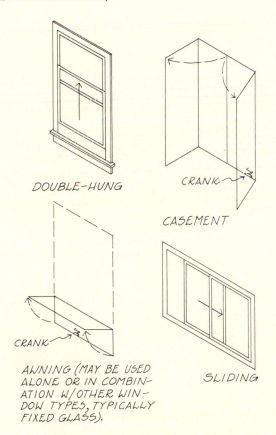

DOUBLE-HUNG

CRANK

CASEMENT

CRANK

AWNING (MAY BE USED ALONE OR IN COMBINATION W/OTHER WINDOW TYPES, TYPICALLY FIXED GLASS).

SLIDING

OPERATIONAL SCHEMATICS FOR COMMON WINDOW TYPES

FIGURE 8-7 Schematics for double-hung, casement, awning, and sliding windows.

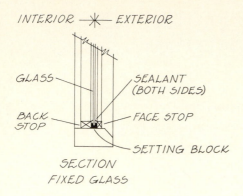

FIGURE 8-8   Fixed glass window installation.

2 feet 8 inches wide. Stairs usually need handrails, which take up space; the stairs have to be wide enough to have 2 feet 8 inches clear width after the installation of the handrail. Interior basement and service stairs should have at least 2 feet 6 inches clear width. Main exterior stairs should be the same width as the walk but should have not less than 3 feet of clear width.

3. The run (depth) of the treads for main stairs should be at least 10⅛ inches, including the tread nosing (front edge of the treads, as you face the stairs); this depth applies to both open and closed risers. Risers are the vertical components of the stairs; main stairs, for example, are generally closed; basement and service stairs are usually open. If basement or service stairs have closed risers, the run should be at least 10⅛ inches, including nosing (this is because the closed riser reduces the amount of usable foot space on the tread); if the risers are open, the run should be at least 9½ inches, including nosing. The run of main exterior stairs should be at least 11 inches. The run of open basement stairs also should be at least 11 inches; basement stairs with overhead cover should have a run of at least 9½ inches. The tread run should always be the same in any one flight of stairs, for safety.

4. The rise of main interior stairs and interior basement and service stairs should not exceed 8¼ inches; exterior stairs should not exceed 7½ inches, except that covered exterior basement stairs may measure 8¼ inches, if necessary. The height of risers for any one flight of stairs should always be the same, for safety.

5. Where a door opens out toward the stairs at the top of a flight of stairs, there should be a landing; the depth of the landing (measured from the approximate face of the door) should be at least 2 feet 6 inches. A landing is preferable to a winder, where the stairs must turn on their way up (a winder is the rotation of the treads and risers to make a turn—see Figure 8-11).

6. Handrails or railings should be installed for flights of three risers or more. (See Figure 8-11.)

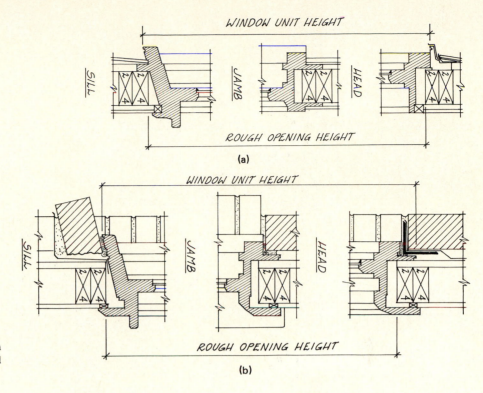

FIGURE 8-9 Sections through windows with (a) wood siding and (b) brick veneer.

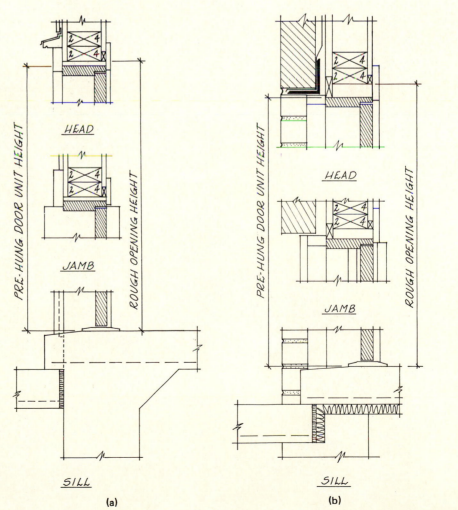

FIGURE 8-10 Sections through exterior doors with (a) wood siding and (b) brick veneer.

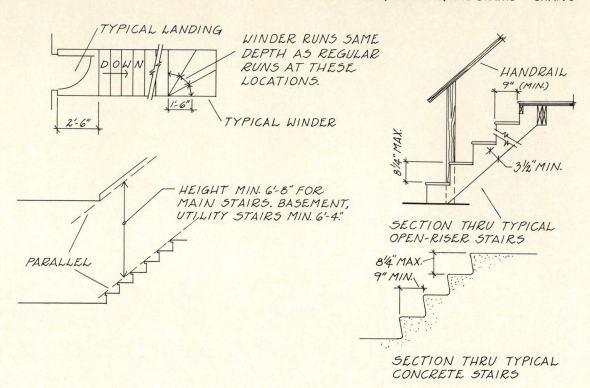

FIGURE 8-11 Typical stair designs.

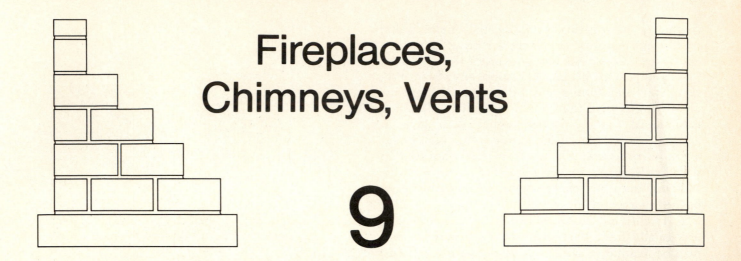

Fireplaces, Chimneys, Vents

9

FIREPLACES, CHIMNEYS, AND VENTS

Electric heating devices and cooking appliances can often be used without vents (with the exception of ranges, which should be vented regardless of the type of energy). The other fuel-burning devices—appliances, warm-air furnaces, hot-water heaters, wood- and coal-burning stoves and heating devices, and so on—require a flue-gas escape route, or vent to the outside, because the products of combustion are usually dangerous. Fireplaces, of course, are vented to the outside by chimneys. Safety is a prime consideration when dealing with fuel-burning devices; the proper selection of materials and proper installation are critical and must be done in strict compliance with applicable authority codes and with careful workmanship.

In most new houses, central heating systems are typical. These systems may be vented through masonry vents or prefabricated metal vents with outer metal casings (to protect wood framing and other flammable materials from the heat). Some prefabricated metal vents may be framed with

wood, if they are so designed and if codes permit. Sometimes prefabricated metal vents are used within masonry housings; such vent systems often approach the size of fireplace chimneys and are built similarly.

Smaller or supplementary heating units, such as wall furnaces, room heaters, and hot-water heaters, use smaller metal vent pipes. These smaller vents are usually routed and tied into the larger flue-gas vents, although they may be routed to the outside individually, if necessary and if proper precautions are taken. But there usually is no reason why the hot-water heater cannot be located in the same space with the heating system and thus share the same vent.

Prefabricated metal fireplaces and chimneys, where codes permits, offer some advantages to builders and homeowners alike. First, the installation of prefabricated units is much faster than the building of a traditional masonry fireplace and chimney. Prefabricated metal units are framed with typical wood framing members and sometimes may even touch the wood members. There is much

flexibility in locating prefabricated metal units; traditional masonry fireplaces are troublesome to construct unless placed at an exterior wall, whereas metal units may be placed almost anywhere. Prefabricated metal units may be bought with a heat-circulating system that makes the unit an effective, economical heating device rather than just an aesthetic "toy." The hearth, side trim, mantel, and other portions may still be constructed traditionally with brick, stone, or tile. Also, the chimney above the roof may be veneered with brick, rather than housed in a wood chase, which is typical.

Traditional masonry fireplaces and chimneys are usually built by specialists. They may or may not be equipped with a heat-circulating system, but this addition is certainly advisable with the present cost of energy, and energy costs will likely continue to rise over the long term. (See Figure 9-1.)

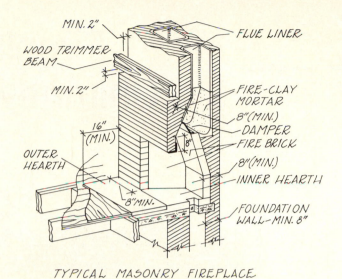

TYPICAL MASONRY FIREPLACE

FIGURE 9-1 Typical masonry fireplace.

Exterior Finish Materials and Installation

10

The most common residential exterior surfaces are brick and stone veneer, plywood siding, wood shingles and shakes, wood siding, stucco, and various synthetic materials.

BRICK AND STONE

Brick veneer is a protective and decorative skin around the house. It is attached to the exterior stud wall; the stud wall, and possibly some interior bearing walls, bears the weight of the house, not the brick veneer. Brick can be used as a structural material and often is in commercial construction, but it is seldom so used in residential construction.

Some definitions are necessary before discussing brick application. (See Figure 10-1.)

1. A single thickness of brick, such as the brick veneer skin of a house, is called a *wythe*. Where more than one thickness of brick is used, you have a *multiwythe* wall. Multiwythe walls are seldom used for houses, but they are often used outside for such structures as garden and patio enclosures.

2. Brick veneer is normally laid flat, with the longest dimension parallel to the face of the wall; bricks laid in this manner are called *stretchers*. Occasionally, bricks are laid flat but with the long dimension perpendicular to the wall; these bricks are called *headers*. Headers are sometimes used in brick veneer

walls as part of decorative patterns, and they are used to turn corners.

3. A brick laid on its edge is called a *rowlok*; when you lay the rowlok with the long dimension perpendicular to a stretcher course, it is called a *soldier*.

4. Often, brick or other masonry is laid by following a string stretched from corner to corner; the string gives the mason a guide to help keep the courses straight and level. This procedure is called laying *to the line*. (See Figure 10-2.)

5. *Patterns* in brick veneer may be achieved by varying the arrangement of the brick, the color of the bricks, the texture of bricks, and the plane of the wall (some bricks sticking out

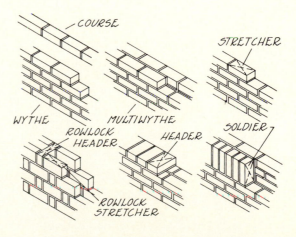

FIGURE 10-1 Basic masonry terminology.

137

somewhat and others recessed). There are many types of brick arrangement patterns. The most common pattern is *running bond*. (See Figure 10-3.)

6. Bricks are usually laid starting at the corners and working toward the middle. This is done because the corners require particular care, and adjustments are easier to make near the center than at a corner. The built-up corner areas are called *leads*. (See Figure 10-4.) When the bricks meet at the wall center, it is usually necessary to cut a brick to close with; this is a *closure* brick.

Mortar Joints

The most typical brick mortar joints are concave, vee, flush, raked, extruded, beaded, weathered, and struck. Each of these joints gives the brick veneer wall a certain appearance, and each joint has a best use or function. (See Figure 10-5.)

Concave and flush joints tend to emphasize the wall flatness. Vee joints emphasize shadows. Raked joints, because the mortar is recessed almost ½ inch, create dark shadows. Extruded joints allow

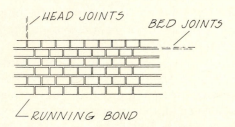

FIGURE 10-2 Laying to the line.

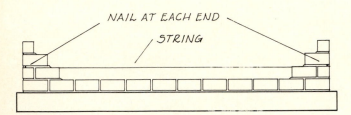

FIGURE 10-3 Running bond, one of many brick arrangement patterns.

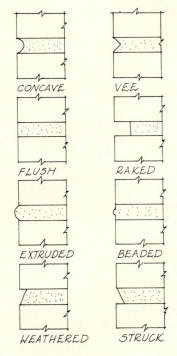

FIGURE 10-5 Brick mortar joints.

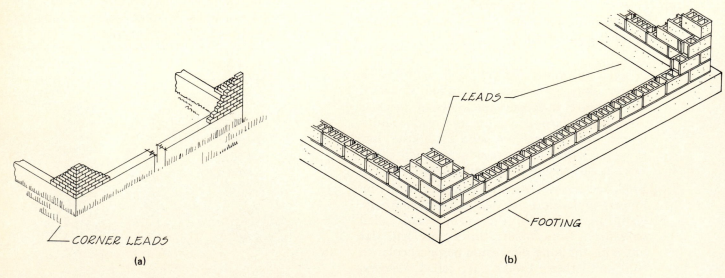

(a)   (b)

FIGURE 10-4 Corner leads for brick and concrete block walls.

the mortar itself to become part of the wall pattern. Beaded joints work similarly; they are just more formal in appearance. Weathered joints and stuck joints create shadow lines. All these joints contribute to the appearance of the wall, either by the shadows they make, by the pattern of the mortar itself (as with extruded and beaded joints), or both. Also, joint types may be mixed on the same wall, if desired.

As with bricks, mortar joints are identified by their location in the wall. Horizontal mortar joints are called *bed joints*, and vertical mortar joints are called *head joints*. (See Figure 10-3.)

Joints should not be selected for appearance alone, however. Concave and vee joints drain well, an important consideration in rainy areas and in cold areas where snow and ice may collect in the joints. Flush joints are easy to make, but they soak up water; freezing and thawing water can damage mortar joints. Flush joints are especially good where the brick will be covered with stucco because flush joints present a flat surface for the stucco to bond to. Raked joints collect water, snow, and ice— important in the North, irrelevant in the dry, sunny areas of the Southwest. Extruded, beaded, and struck joints collect water. Local codes and other authorities often have requirements for the types of joints used and the way they are tooled (finished).

Joint Tools and Joint Finishing

The type of tool used depends on the style of the joint and worker preference. The S-shaped jointer, as the name implies, is shaped like the letter *S*; it is used for concave joints. A ½-inch-square bar can be used to make vee joints; a bar edge is pulled along the joint, creating the V shape. Flush, struck, and weathered joints can be made with a trowel. A special tool is required for beaded joints.

Joint finishing makes a neat appearance, but that is not the only reason for finishing. When the brick is laid, mortar is pushed out of the joints; this is because of the weight of the brick and because a little more mortar is always put between the bricks than will be allowed to stay there to ensure complete contact between mortar and bricks. The mortar that is pushed out of the joint is not compressed. The jointing tool compresses the mortar, making it denser and bringing it in closer contact with the brick; this makes the joint more water- and weather-resistant.

There is, literally, a rule of thumb for telling when joints are ready to be tooled (mortar joints have to set somewhat before they are stiff enough to be tooled): After laying a section of wall, test the mortar at the joints by pressing it with the thumb; if the thumbprint remains but the mortar does not stick to your thumb, the joints are ready for tooling. Do not finish the joints too early or they will not hold their shape; do not let them set too much or the joints will be hard to finish and may be damaged in the process.

Most of the problems that occur in brick veneer walls are related to the mortar joints: Either the mortar itself is faulty (wrong type of mortar, improperly prepared mortar, etc.), or the installation is faulty (inadequate coverage of the bonding surface, poorly selected or finished joints, etc.). Mortar joints, when the mortar *is* properly prepared and installed, help produce strong walls that look good for decades.

Mixing Mortar

First, verify the mortar mix ingredients and proportions for your job in your locality; the building department or applicable authority can give you this information. (See also Appendix D-8.) Typical mortar, for general use including brick veneer walls, is about 2¼ to 3 parts masonry sand per 1 part masonry cement. Thorough mixing is important, either by manual methods or with a mechanical mixer. Mechanical mixers are usually required for a job as large as a house, but even on large jobs there is usually detail work (planters, drive edging, steps, etc.) to be done, and thus both mechanical mixers and manual methods may be used on the same house job.

Mechanical mixers are sometimes equipped with electric motors; sometimes gasoline-powered motors are used. In any case, be sure the mixer is suitable for mortar mixing; some mixers are for concrete only, some for mortar only; some will do both.

Check to be sure there is no debris in the mixer before using; then rinse out the mixer with water before starting the first batch.

If the mixer is gasoline-powered, be sure the tank is full before starting a mix.

The particular mixer used should have the manufacturer's instructions labeled somewhere on the machine, and these should be followed.

Generally, begin by emptying any water from the mixer; then add a portion of the dry ingredients (the amount you put in the mixer depends on the particular mixer). Add the desired proportion of water, preferably drinking water but in any case water free of debris, organic matter, and other pollutants. Let the machine run a few minutes, until the ingredients and water reach the desired consistency. The mortar may then be delivered to the masons.

The number of masons and helpers needed varies, of course, with the size of the job. It is usually more efficient to free masons to do the bricklaying and have enough helpers to do the mixing, mortar carrying, and brick carrying. Masons often work in pairs, one starting at each corner of a wall and working toward the middle.

Clean the mixer out with water after each batch is mixed. Empty the water out before starting the next batch.

Manual mixing requires a mortar box. These boxes are available at builder supply houses or may be built on the job. The size, as with mechanical mixers, depends on the size of the job. In any case, the box should have a waterproof bottom and sides; otherwise, the correct amount of water cannot be maintained in the mix.

Put the dry ingredients in the mortar box. Some masons layer the dry ingredients: one layer of sand, then a layer of masonry cement, and so forth; then they hoe the dry ingredients twice, once in each direction lengthwise.

Water is then added, and the ingredients are hoed together. Hoeing is continued until the ingredients are thoroughly mixed to the proper consistency and all dry pockets are eliminated. This can usually be done in about 5 to 8 minutes.

Protect the mortar on hot, sunny days by draping plastic or some suitable cover over the mechanical mixer or mortar box. Do not mix more mortar than can be used in about 2 hours. If the mortar is not used within 2½ hours, so much water will probably have evaporated that more will have to be added, and this should be avoided if possible. Water should not be added more than once after the initial preparation of the mortar.

LAYING BRICK AND OTHER MASONRY

The first step before laying brick, concrete block, or other masonry is to check the foundation wall or footing the masonry veneer will rest on; the surface must be level enough so that the first bed joint can maintain a maximum average thickness of ½ inch. No joint, including the first bed joint, should be more than ¾ inch or less than ¼ inch thick. If the surface is not level within this toler-

ance, level with load-bearing grout (within the thickness tolerance for the type of grout used). If the surface cannot be leveled with grout, the first course of brick may be custom-cut to fit; however, if the surface is this much out of level, the foundation contractor would probably be called back to repair or replace the foundation wall or footing.

Masonry veneer may be laid by starting at one corner and proceeding to the next, but the typical procedure is to build up what are called *corner leads*. (See Figure 10-4.) The corner leads serve as guides to use in placement of the intermediate courses. When all the intermediate brick is laid, two more corner leads are built, then the intermediate brick laid, and so forth, up the wall until it reaches the desired height. *Story poles* are used to measure the height of the courses and keep them uniform; factory-made story poles are available, or wood members (such as two-by-fours) may be used. (See Figure 10-6.) The courses may also be measured with a special device called a *mason's rule*. The corner leads must be built very accurately, since they serve as the guides for the intermediate brick. With the corner leads in place, a line is stretched from corner to corner. The line is typically pinned in a joint near the corner edge, close to the brick but not touching it (about $\frac{1}{16}$ inch out from the face of the corner lead faces). Each brick course is then laid to the line. Frequent checks for plumb and levelness should be made with a *spirit level*. (See Figure 10-7.)

If the brick veneer wall is long enough that the line will sag between the corners, a center lead, called a *rack-back lead*, is built in the center of the wall so that the line may be secured there. (See Figure 10-8.)

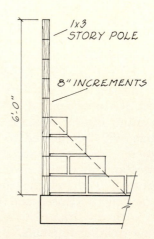

FIGURE 10-6 Vertical spacing of courses checked by using a story pole (concrete block shown here).

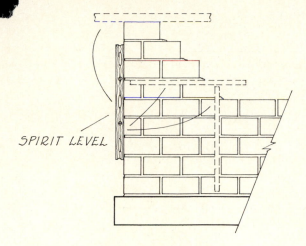

FIGURE 10-7   Use of a spirit level.

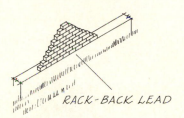

FIGURE 10-8   Rack-back or center lead.

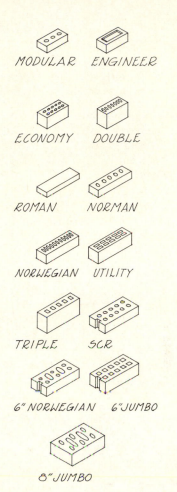

MODULAR   ENGINEER

ECONOMY   DOUBLE

ROMAN   NORMAN

NORWEGIAN   UTILITY

TRIPLE   SCR

6" NORWEGIAN   6" JUMBO

8" JUMBO

FIGURE 10-9   Some types and sizes of brick available for residential and light construction use.

Brick Ties

Brick veneer is usually anchored to the stud wall with corrosion-resistant, corrugated metal ties. The ties are L-shaped, and the longer, corrugated part of the tie fits in the brick mortar joint; the shorter end is secured to the stud wall, at the studs, not to the sheathing. (See Figure 5-6.)

The nominal dimensions of a standard modular brick are 4 by 2⅔ by 8 inches; thus, three modular courses are 8 inches high. Standard brick veneer is attached to studs spaced at 16 inches on center horizontally (at each stud) and at 24 inches on center vertically (nine courses between ties). Different sizes and types of bricks require somewhat different anchorage; check with appropriate authorities. Figure 10-9 shows some typical modular brick types; Figure 10-10 gives their nominal dimensions.

Brick Sizes

A nominal dimension of a masonry unit is equal to the actual dimension plus the thickness of the mortar joint to be used. Mortar joints are standard—usually ½ or ⅜ inch for brick veneer. Thus, the actual length of a nominal 4- by 2⅔- by 8-inch brick, using a ½-inch joint, is 7½ inches; actual length, using a ⅜-inch joint, is 7⅝, and so forth. The joint thickness is subtracted from each of the three nominal dimensions of a masonry unit to obtain the actual dimensions of the unit. Of course, the "actual" dimensions of the unit itself, as it comes from the factory, will vary within standard tolerances, just as the mortar joints will vary within standard tolerances.

Modular brick, however, is not always used. And if not, the typical brick for brick veneer walls usually has actual dimensions of 3¾ by 2¼ by 8 inches (within standard tolerances).

Flashing and Weep Holes

Condensation often collects on the back side of the brick veneer. The condensation runs down the back side of the brick to the top of the foundation wall. Flashing and weep holes are a method of collecting and directing the water out of the wall, before it can cause damage. The preferred flashing material is sheet copper, but 30-pound asphalt-saturated felt and other acceptable materials are

NOMINAL MODULAR SIZES OF BRICK

Unit Designation	Dimensions			
	Thickness (inches)	Height (inches)	Length (inches)	Modular Coursing
Standard Modular	4	$2\frac{2}{3}$	8	3C = 8[c]
Engineer	4	$3\frac{1}{5}$	8	5C = 16
Economy	4	4	8	1C = 4
Double	4	$5\frac{1}{3}$	8	3C = 16
Roman	4	2	12	2C = 4
Norman	4	$2\frac{2}{3}$	12	3C = 8
Norwegian	4	$3\frac{1}{5}$	12	5C = 16
Utility[a]	4	4	12	1C = 4
Triple	4	$5\frac{1}{3}$	12	3C = 16
SCR brick[b]	6	$2\frac{2}{3}$	12	3C = 8
6 in. Norwegian	6	$3\frac{1}{5}$	12	5C = 16
6 in. Jumbo	6	4	12	1C = 4
8 in. Jumbo	8	4	12	1C = 4

[a]Also called Norman Economy, General and King Norman.

[b]Reg. U.S. Pat. Off., SCPL

[c]Measurement given in inches.

Source: Brick Institute of America

FIGURE 10-10 Nominal modular sizes of brick. (Courtesy of Brick Institute of America.)

often used. The flashing is sized so that it will extend up the face of the stud wall at least 6 inches, from where it fits into the mortar joint at the top of the foundation wall. The flashing is installed continuously.

If sheathing paper is used over sheathing, the flashing should be run up under the sheathing paper at least 6 inches. If sheathing paper is not used, run the flashing up under the sheathing at least 6 inches.

The continuous flashing collects condensation at the foundation wall top. To get the water out, *weep holes* are made by omitting the mortar in the head joints of the brick veneer at 4 feet on center. (See Figure 5-6.)

Flashing is also required at the sills and heads of doors, windows, and similar openings. Sheet metal is the typical head and sill flashing material, but certain building papers and other materials may be locally acceptable.

For heads, install the flashing from the front edge of the lintel (lintels are discussed in the next section), over the lintel, and up at least 2 inches behind the sheathing paper. If there is no sheathing paper, extend the flashing up 2 inches behind the sheathing.

For sills, install the flashing from the edge of the brick veneer (under the brick, stone, concrete, or similar sills) up on the sheathing and then under the wood sill to inside of wall. See Figure 10-11 for head and sill construction details.

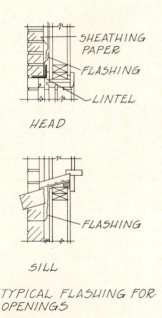

HEAD

SILL

TYPICAL FLASHING FOR OPENINGS

FIGURE 10-11 Typical flashing for openings.

Where wood head and sill members meet masonry, caulk the joints.

Lintels

Brick above doors, windows, and other openings must be supported in some manner. In residential construction, the most typical support is a steel angle that is fitted over the opening; angle lintels are barely visible when the brick is laid over them. Lintels should be designed to support the brick and other loads imposed on them; but, for typical single-entry doors, windows, and other openings not exceeding about 3 feet wide, the steel angle often used is 3½ by 3½ by ¼ inches and long enough to span the opening and provide adequate bearing surface on each side of the opening. (See Figure 10-11.)

Control Joints

Brick (or any material) expands when the temperature increases and contracts when the temperature decreases. The amount of expansion and contraction depends on the type material and on the temperature rise or fall. For the sake of example, a 100-foot-long brick wall will increase in length by about ⅜ inch if its temperature is increased 100° F.

In residential construction, especially with single-family detached houses, the walls are seldom long enough to require control joints. And when control joints are required, they should be located and designed by a professional. Figure 10-12 shows the typical locations where control joints are used and the typical types of joints.

Corbels

A corbel is a gradual projection of several courses of brick out from the normal face of the wall or brick surface. Corbeling may be used for decorative purposes—chimneys, for example are often corbeled at the top—or the corbeling may be for a functional reason, such as thickening a wall for some purpose. (See Figure 10-13.)

STONE VENEER

Stone has some of the technical requirements of brick veneer in its installation requirements: It must be flashed and tied to the frame wall, and weep holes and lintels must be provided where necessary. The special considerations of stone are discussed in the following.

The most typical stones used in the United States are limestone, marble, granite, slate, and sandstone; other stone varieties may be available locally.

Stone Patterns

Generally speaking, stone patterns range in sophistication between *rough rubble* and *ashlar walls*. Rough rubble walls are built of stone right

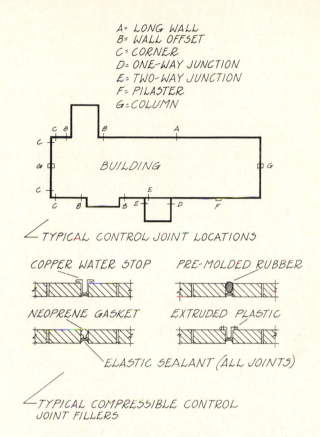

FIGURE 10-12 Typical control joint locations and types.

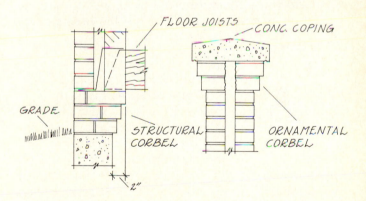

FIGURE 10-13 Corbeling.

out of the quarry or field; thus, rough rubble is very unspecific as to stone type, and there is no attempt given to formal patterns when installing the stones. The next step in formalization is the *coursed rubble* wall; the stones are laid with coursing in mind, and the joints are more uniform than with rough rubble. *Squared stone*, as the name implies, has been cut for a smoother installation; squared stone is laid in courses, and the joints approach uniformity. The *ashlar wall* is made of precisely cut stone, and the joints are as uniform as brick joints; the ashlar wall is the most precise of these patterns.

Mortar for Stone Walls

Ordinary portland cement stains most stone. Therefore, it is common practice to use nonstaining white portland cement when making mortar. Portland cement-lime mortar may be acceptable because lime usually does not stain.

One common mortar mix for stone consists of 1 part masonry cement (ASTM Specification C91, Type II) and 2¼ to 3 parts mortar sand in a loose, damp consistency. Another typical mix is 1 part portland cement, ½ to 1¼ parts hydrated lime, and 4½ to 6 parts mortar sand in a loose, damp consistency. Areas of high wind, earthquake zones, or areas where frost heave is a problem sometimes use 1 part masonry cement (ASTM Specification C91, Type II) plus 1 part portland cement with 4½ to 6 parts mortar sand in a loose, damp consistency. The building department or applicable authorities should be consulted for mortar mix requirements.

Stone absorbency varies with the type stone. The more absorbent types may have to be moistened before laying; otherwise, the stone can pull the moisture from the mortar and ruin the bond contact. Do not apply too much water, however, and nonabsorbent stone need not be moistened; the stone supplier should be consulted about moisture requirements and mortar requirements.

Joints

Ashlar stone walls, generally speaking, follow about the same joint procedure as brick walls. In rubble walls, the joints are often raked out ¼ to ½ inch with a jointing tool or trowel, to accent the stone shape and play down the mortar joint, especially where the mortar joints are quite wide; to further play down the mortar joints, a color similar to the stone is often mixed with the mortar. Others prefer to color the mortar in contrast with the stone, making the joints a dominate pattern.

PLYWOOD SIDING

One of the chief advantages of plywood to the builder is its ease and speed of installation. The manufacturers have steadily improved the durability of plywood over the years and are constantly adding new surface designs and textures; plywood siding is available in a variety of configurations, often duplicating the appearance of individually installed boards. (See also Appendix D-3.)

Every manufacturer varies somewhat in offerings, but the following designs are fairly typical:

1. *Plain*. Plain plywood does not attempt to disguise itself. When using plain plywood, the quality of the grain becomes more important. All-heart grain is a beautiful, high-quality plywood panel.

2. *Board on batten*. In this design, the plywood panels receive grooves approximately an inch wide and about ¼ inch deep at about 8 inches on center; the resulting appearance is like 8-inch boards mounted over narrow wood strips (battens). Sometimes the grooves are cut at varying centers, giving the panel the appearance of different-sized boards mounted on battens. Also, the grooves may be cut at a number of regular intervals—2, 4, 6, 8, 10, or 12 inches, for example, where the appearance of these board sizes is desired.

3. *Narrow grooves*. Approximately saw-blade width, narrow grooves are sometimes cut to give the impression of a board mounted edge to edge; these grooves typically are cut at 4 inches on center.

This is not an exhaustive listing, but the designs listed are typical of plywood seen in residential and light commercial use. Many textures are available in plywood panels: rough sawn, brushed, and striated, to name a few.

INSTALLATION OF PLYWOOD PANELS

The stronger panels may be nailed to studs installed at 24 inches on center; the strength of the panels depends largely on thickness and on whether or not the panels are grooved (grooves tend to weaken panels installed vertically). Also, more installation freedom is usually possible where plywood sheathing is used underneath. Generally speaking, ⅜-inch plain panels or ⅜-inch channel-grooved panels (the grooves are only 1/16 inch deep) may be nailed directly to studs (no sheathing required) if the studs are 16 inches on center. All ⅝-inch panels may be nailed directly to studs (no sheathing required) if the studs are 16 inches on center.

When the panels are installed vertically, the joints should fall at the studs or at comparable backup blocking. When the panels are installed horizontally, two-by-four blocking should be installed behind the panel joints.

Nails should be noncorrosive—stainless steel, aluminum alloy, or hot-dipped galvanized; otherwise, the nails will rust and stain the panels. The nails should be long enough to penetrate the panel and any sheathing and go into the studs or solid blocking at least 1 inch; 6d or 8d nails usually work. Typical nailing is 6 inches on center along the panel edges and 12 inches on center along intermediate members.

The panels expand and contract with temperature changes; $\frac{1}{16}$-inch to $\frac{1}{8}$-inch spacing between panels is typical, the former for grooved panels, the latter for ungrooved panels.

All corner joints (inside and outside) should be caulked. Other joints may be left uncaulked if they are shiplapped or covered with battens. Horizontal joints may be lapped, flashed with sheet metal, or waterproofed with a combination of metal flashing and horizontal battens (which form a drip). Figure 10-14 shows typical vertical batten joints, shiplap joints, horizontal joints, and corner details.

The foregoing discussion concerns typical installations. Local building conditions may require more or less stringent methods. Also, different manufacturers may vary slightly in their installation instructions.

It is important that plywood be properly stored at the job site. Where possible, indoor storage is best. Where indoor storage is not available, it is recommended that builders use factory-supplied covers for the plywood or equivalents of the factory covers. In all cases, keep the plywood

off the ground, keep the plywood neatly stacked to avoid warping, and provide adequate ventilation.

HORIZONTAL LAP SIDING

Horizontal lap siding, as the name implies, is installed such that each course of siding laps over the preceding one. Traditional lap siding is made of wood boards. Plywood also is being used nowadays, as well as a number of other materials, including metal and hardboard.

The installation of lap siding varies with each material and sometimes from manufacturer to manufacturer. However, there are several important construction detailing problems that have to be solved, regardless of the peculiarities of materials and manufacturers. These problems include the way the particular lap-siding courses relate to each other and to the house wall framing, corner treatment (of both inside and outside corners), and joint treatment.

Wood lap siding is *square-cut*, *beveled*, or *rabbeted bevel*. (See Figure 10-15.) Square-cut lap siding is the same thickness at both ends—a one-by-six board, for example. Beveled siding is cut so that the top edge is thinner than the bottom edge. Rabbeted-bevel lap siding is beveled and also has rabbets, or notches, on the bottom edges, which fit over the top edges of preceding courses.

Most wood lap siding is installed by first securing a cant strip at the base of the first course and proceeding up from there. The cant strip is a narrow strip, about $\frac{3}{8}$ inch by 1 or $1\frac{1}{2}$ inches, and its purpose is to put the first course at a slant. Succeeding courses get their slant by lapping the lower courses.

Square-cut lap siding emphasizes the lap effect

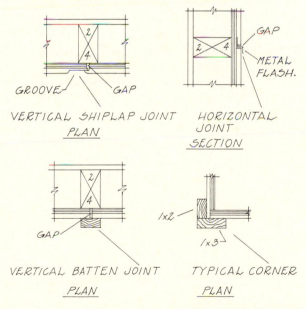

VERTICAL SHIPLAP JOINT PLAN

HORIZONTAL JOINT SECTION

GAP

METAL FLASH.

GAP

VERTICAL BATTEN JOINT PLAN

TYPICAL CORNER PLAN

1x2

1x3

TYPICAL SIDING JOINT & CORNER TREATMENT USED W/PLYWOOD SIDING (EXTERIOR).

FIGURE 10-14 Typical plywood-siding joint and corner treatments (exterior).

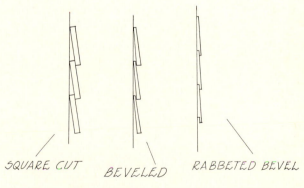

SQUARE CUT   BEVELED   RABBETED BEVEL

TYPICAL HORIZONTAL LAP SIDING TYPES

FIGURE 10-15 The most common types of horizontal lap siding.

most because the bottom edge is usually thicker than the bottom edges of beveled lap siding or rabbeted-bevel lap siding. This thicker edge, in addition to the noticeable greater thickness, creates the darkest horizontal shadow lines.

Beveled siding is typically about $\frac{3}{16}$ inch at the top and anywhere from $\frac{15}{32}$ to $\frac{3}{4}$ inch at the bottom, depending on the width of the boards. The wider the board, the thicker the bottom edge.

In square-cut and beveled siding, only the top portion of the siding courses touches the house wall framing, which leaves a space between the siding and the wall-framing surface.

Rabbeted-bevel lap siding is beveled, plus it has a rabbet, or notch, on the bottom edge, which fits over the top of the preceding course. The effect of the notch is to reduce the thickness at the bottom edges of courses, creating narrower shadow lines and a smoother overall surface. Also, the shape causes the courses to fit flush against the house wall framing.

The courses are usually lapped at least 1 inch. Generally speaking, the wider the board, the more overlap is required. The nails are usually 7d or 8d, depending on the thickness of the siding; in any case, the nails should be long enough to penetrate the siding and extend into the studs at least 1 inch. If the siding is under 12 inches wide (and it usually is), one nail is placed through the siding at each stud. The nails are placed near the bottom of the siding but should clear the top of the course below by $\frac{1}{8}$ inch; this space is needed to allow for expansion of the courses. If the siding is over 12 inches wide, an additional nail is used at the center of the course. All nailing should be done at the studs (consequently, all joints occur at the studs), and the joints should be staggered up the wall. Fill the joints with caulk.

If plywood sheathing is used, the stud spacing sometimes may be as much as 24 inches. If sheathing is not used, the stud spacing is normally 16 inches on center. Stud spacing is also influenced by the thickness (strength) of the exterior siding. Building paper is normally required where no sheathing is used but may sometimes be omitted where sheathing is used. Thus, no single component of the exterior walls should be selected without considering the other components. And, as always, local building departments and other authorities influence the selection and assembly of materials.

The outside corners of wood lap siding may be mitered (usually at a 45-degree angle), fitted together, and caulked. Or factory-made metal corner covers may be used; caulk the joint before installing the corner pieces. (See Figure 10-16.) Inside corners are built by first installing a corner post (usually about 2 by 2 inches), then butting the siding against the post; caulk the joints. (See Figure 10-17.)

The installation procedure for plywood lap siding is similar to the preceding.

Hardboard lap siding is well liked by builders and homeowners alike. There are good reasons for the acceptance of this material as a residential and light commercial siding material. Hardboard resists moisture and the effects of temperature extremes, it has good sound qualities, and it is easily cut with ordinary carpenter's tools. Another quality is particularly important for marketing: Hardboard imitates wood well (it does not look cheap or phony). The material also lasts well.

Hardboard is made from a mixture of wood fibers and natural and synthetic binding elements pressed together under heat. A variety of surface textures, colors, and thicknesses are available.

The exact installation procedure for hardboard lap siding varies with the manufacturer and with local building conditions. As with all siding, the stud spacing is dependent on the presence or absence of sheathing, the thickness of the siding, the type of interior wall finish, and similar factors

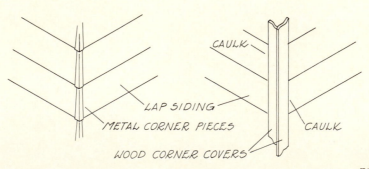

TYPICAL OUTSIDE CORNER TREATMENTS— LAP SIDING

FIGURE 10-16 Treatment of outside corners of lap siding.

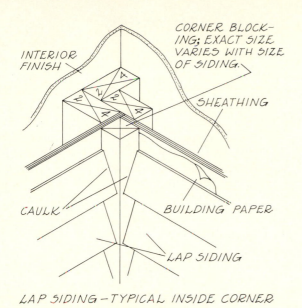

INTERIOR FINISH

CORNER BLOCK-ING; EXACT SIZE VARIES WITH SIZE OF SIDING.

SHEATHING

CAULK

BUILDING PAPER

LAP SIDING

LAP SIDING—TYPICAL INSIDE CORNER

FIGURE 10-17 Treatment of inside corners of lap siding.

that make up the total wall system. As a minimum, hardboard siding is usually ¼ inch thick.

The following is a *typical* installation sequence, not a definitive one:

The first step is usually the installation of a cant at or near the sill plate. This cant strip, for hardboard and metal siding, is usually metal. It is important that the cant strip be installed level; a chalk line usually is used to establish a line to build to. The cant strip causes the slant in the first course; each future course is lapped over the top of the preceding one. (See Figure 10-18.)

After the cant strip, install the inside corner posts. These posts may be wood (typically a two-by-two) or, typical with hardboard, metal. Metal inside-corner fittings have flanges that are nailed to

the wall framing at about 8 inches on center, vertically, with 8d nails. (See Figure 10-17.)

Install the first course of siding at the bottom. The starter strip, or cant, if it is metal, may be fitted with a special metal angle that joins with a similar angle attached to the first course of siding—this is a factory item and should be installed according to the manufacturer's instructions. Wood cant strips also may be used, in which case the first course of siding is nailed along the cant strip at a maximum of 16 inches on center.

The next course and all succeeding courses are lapped at least 1 inch and nailed through both courses. The nails should not exceed 16 inches on center, and end (butt) joints should fall on studs, if no sheathing is used. If sheathing is used, joints may be placed between studs. Joints should be staggered. Metal joint molding may be furnished with the siding; if so, install according to the manufacturer's instructions. Usually, the joint molding is slipped between the ends of the two siding pieces at the joint and the pieces nailed in place. They are easy to install and unobtrusive in appearance. (See Figure 10-19.) If there is no joint molding, the joint is usually nailed at the top and bottom of each siding piece, leaving about a 1/16-inch gap for caulk. The installer should never force the siding in place.

Metal outside corners are installed by fitting the pieces under the edges of courses and nailing in place. Wood outside-corner trim may be used, in which case the trim should be at least 1⅛ inches wide.

Where the siding butts door, window, and similar trimmed openings, leave about a ⅛-inch gap between the siding and the trim and caulk carefully. (See Figure 10-20.)

Building paper may sometimes be omitted when sheathing is used, but some localities require it nevertheless.

Matching color touch-up paint and matching caulk are available from the manufacturers.

Metal lap siding is installed similarly to hardboard siding.

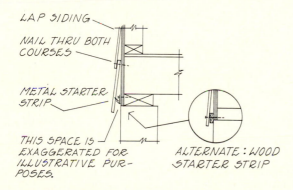

LAP SIDING

NAIL THRU BOTH COURSES

METAL STARTER STRIP

THIS SPACE IS EXAGGERATED FOR ILLUSTRATIVE PUR-POSES.

ALTERNATE: WOOD STARTER STRIP

LAP SIDING DETAILS

FIGURE 10-18 Installation of lap siding that requires a starter strip (cant strip).

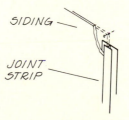

SIDING

JOINT STRIP

FIGURE 10-19 Metal joint moulding, often provided by siding manufacturers.

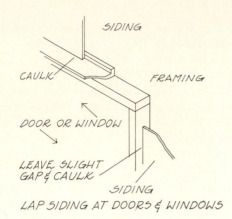

FIGURE 10-20   Lap siding at doors and windows.

STUCCO

Stucco is a versatile protective "skin" that can be installed over a variety of surface materials and shapes. In residential construction, stucco is typically installed over frame walls, and this is the application that will be discussed here. The greatest area of house use of stucco is the exterior wall surface, but it may be used on soffits and other surfaces as well.

Stucco is almost a "miniature concrete." The ingredients of stucco are portland cement, lime, an aggregate (sand), and water. Stucco is usually used with some type of metal reinforcement (wire lath or similar).

Portland cement may be one of three types, depending on use and local building conditions: portland cement (ASTM C-150 or F.S. SS-C-192), air-entraining portland cement (ASTM C-175 or F.S. SS-C-192)—both of which are used where freezing and thawing may be a problem, or plastic portland cement.

Lime also may be one of three types: hydrated lime (ASTM C-6) for normal finishing, hydrated lime (ASTM C-206) for special finishing, or quick-lime (ASTM C-5).

Aggregates may be natural sand, crushed stone, or crushed air-cooled blast-furnace slag. Aggregates are sized by standard sieve tests, and the builder must choose the most appropriate size of aggregate for the particular job. Figure 10-21 gives a typical range of sizes for stucco work.

The type of portland cement, lime, and aggregates used for stucco in a particular location tend to be standardized, so local usage should be studied. Proportions for typical stucco mix ingredients are given in Figure 10-21. Mix I may be used as a base course over porous surfaces such as concrete block

STUCCO MIX

Mix Type	Portland Cement[a] (parts)	Lime (parts)	Aggregate[b] (parts)
I	1	¼ max.	3 to 3¾
II	1	¼ to ½	3¾ to 4½
III	1	½ to 1	4½ to 6

[a]Where plastic portland cement is used, reduce lime by 25%.

[b]Proportioning of aggregate is based on a ratio of 3 parts aggregate to 1 part cementitious material (cement and line). Where aggregate is well graded with a good proportion of coarse particles, a ratio of 1 part cementitious material to 3½ or 4 parts aggregate may be suitable. The use of trial mixes for the particular aggregate available is recommended.

FIGURE 10-21   Stucco mix proportions. (Courtesy of Federal Housing Administration.)

or brick or where a dense, waterproof finish is desired. Mix II is a general-purpose mix for both base and finish coats. Mix III is a finish coat but should not be used where freezing and thawing is a problem.

Stucco lath is usually necessary to reinforce the stucco and help secure it to the supporting surface—wood-frame wall, masonry wall, or other. There are several common types of lath, including expanded metal lath, woven wire fabric, and welded wire fabric.

Expanded metal lath must be rust-resistant. It is usually made from galvanized sheets that are cut and then expanded to the diamond-shaped openings typical of the lath, or it may be made of copper-alloy steel treated with rust-resistant paint. The minimum weight for expanded reinforcing with large openings is 1.8 pounds per square yard; for corrugated flat lath (which has small openings), the minimum weight is 3.4 pounds per square yard.

Woven wire fabric looks like chicken wire; the holes are hexagonal, typically 1, 1½, or 2 inches. The 1-inch hexes should be 18-gauge wire; the 1½-inch, 17 gauge; and the 2-inch, 16 gauge. The wire should be galvanized.

Welded wire fabric should be galvanized *after* the welding so that the protective coating is not damaged and thus susceptible to rust. The openings are generally 1 by 1, 1½ by 1½, or 2 by 2 inches. Welded wire fabric can be bought with or without a backing material. If a backing material is woven into the wire, sheathing paper may sometimes be omitted; otherwise, a sheathing paper (typically 15-pound asphalt-saturated felt) is needed to protect wood wall framing. If the backing is included,

16-gauge, 2- by 2-inch wire is used; if there is no backing in the wire, use 17-gauge for 1½- by 1½-inch mesh and 18-gauge for 1- by 1-inch mesh.

Regardless of the lath used, it is secured to the wall frame with galvanized nails at 16 inches on center horizontally and 6 inches on center vertically, to the wall sheathing, or to the studs if there is no wall sheathing. Some lath is self-furring; that is, the lath is held away from the sheathing. If self-furring lath is not used, provisions should be made to hold the lath away from the sheathing at least ¼ inch. Lap woven wire and welded wire fabrics at least one mesh; the end laps should be at the studs or at other bearing and should be staggered.

Control Joints, Corner Reinforcement, and Screeds

The house wall frame, stucco reinforcement, and other building materials expand and contract differently; this can cause the stucco to crack. Differential settlement can also cause the stucco to crack. It is difficult, if not impossible, to prevent this type of cracking entirely. However, the cracking can be minimized by the use of control joints, which ensure that cracks occur where they will do no harm.

Control joints should be chosen and located by an expert—an architect, engineer, or other specialist. As a rule of thumb, control joints are recommended at least every 10 feet, horizontally and vertically. Control joints are typically made with an "accordion pleat" in 26-gauge galvanized steel.

Control joints may be spaced simply to perform their function of controlling cracks, in which case only the minimum number of joints would be used. However, additional joints may be added to create decorative architectural patterns on the wall surface.

In addition to control joints, specially formed metal corner reinforcement is usually called for on inside and outside corners. Other metal stucco trim pieces, reinforcement, and similar accessories also are available for use around windows, doors, and other openings. All such materials should be rust-proof. Where vertical and horizontal control joints intersect, the horizontal joint should be continuous; the vertical joint butts the horizontal joint.

Stucco is usually terminated at or near the house foundation wall. To create a neat edge and to protect the stucco from chipping, a *base screed* is used. Metal screeds are simple and fast to install,

or concrete and masonry walls may be indented to receive the stucco. (See Figure 10-22.)

Stucco Application

Stucco on a frame wall should be applied in three coats. The first coat is called the *scratch coat*; the second, the *brown coat*; and the third coat, the *finish coat*. Trowel on the scratch coat, thoroughly embedding the lath with stucco; this first coat should be a minimum of ⅜ inch and a maximum of ½ inch thick. Let the stucco set partially, then cross-scratch with a scratch tool. The purpose of scratching is to provide a good bonding surface for the brown coat.

If a wall sheathing is used, the scratch coat is sometimes strong enough and cured enough to accept the brown coat within 4 or 5 hours. If there is no sheathing, the scratch coat should be allowed to set for at least 48 hours, and the surface should be kept damp, but not wet, with a fine spray. Do not apply the brown coat over dried-out stucco; should the surface become dried (not desirable), dampen it with a fine spray before applying the brown coat.

The brown coat should be at least ⅜ inch thick. When applying the brown coat, plan the areas to be stuccoed carefully. Do not stop in the middle of a wall, and start work again after the stucco has set, because the stopping place may show through the finish coat. Plan the stops at corners, soffits, control joints, screeds, and similar areas. After the brown coat is applied and floated, damp-cure the stucco for approximately 7 days.

The finish coat is troweled on at least ⅛ inch thick and should be damp-cured for 48 hours. Because the finish coat is so thin, it should be allowed to set long enough (approximately 24 hours, depending on weather conditions) so that a fine mist will not damage the finish during curing.

The finish coat is sometimes called the *white*

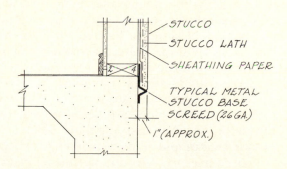

FIGURE 10-22 Typical metal stucco base screed.

coat, perhaps because white cement and white sand are typically used for traditional white stucco. However, a wide range of color additives is available. Also, the typical stucco finish is smooth, but rough, rustic surfaces or patterns may easily be made with hand tools. Aggregates may also be added, either by hand or with blower machines.

Stucco may also be applied with a machine. The instructions for the three coats still hold. The machine manufacturer's operating instructions should be followed, but generally speaking, the operation is very similar to spray painting, and similar operator skill is required. Keep the nozzle about 1 foot from the wall and perpendicular to it. Hold the nozzle closer for detail work (such as around window and door openings). Machines are faster than hand application if the operator is skilled.

Stucco Siding

Traditional stucco is a beautiful material, but it takes considerable time and skill to install. One faster, cheaper alternative worth considering is stucco siding. Stucco siding does for the house exterior walls what Sheetrock has done for interior wall partitions. Stucco siding is installed with essentially the same procedures as plywood and similar sidings. The appearance and durability of stucco siding rivals those of genuine stucco.

WOOD SHAKES AND SHINGLES ON EXTERIOR WALLS

Wood shakes and shingles differ principally in that shakes have one rough side, made by splitting the shake; this rough side gives the shake a more rustic appearance. Shingles are sawed and thus have a smoother surface. Rebutted and rejointed shingles have machine-trimmed edges, and the butts are straight and at accurate right angles to the edges, so joints accurately parallel joints, and the butts form very straight and parallel horizontal lines. Grooved sidewall shakes are rebutted and rejointed shingles with face grooves formed by a machine. The variances in these materials require different installation procedures, which are described by the various manufacturers and must be verified against local building and authority requirements. The following instructions are therefore only typical, not definitive. (See also Appendix D-4.)

Wood shakes and shingles may be installed by the single-course or the double-course method. Single coursing works similarly to the way shakes and shingles are installed on the roof, but more exposure is allowed because of the steepness of exterior walls versus roofs (roofs have three-ply construction; walls have two layers at any given point). Single-course shakes and shingles should be doubled at the foundation line.

Typical exposures are 7½ inches for 16-inch lengths, 8½ inches for 18-inch lengths, and 11½ inches for 24-inch lengths.

Nail the shakes or shingles with 1¼-inch 3d nails, two nails at approximately 1 inch above the butts, ¾ inches from the sides; drive two additional nails approximately 4 inches apart across the face of the shingle. Nail each course in this manner.

Outside corners may be made by overlapping alternating courses of shingles (called a *woven corner*) or by mitering the shingles at the juncture (usually at a 45-degree angle). (See Figure 10-23.) Inside corners may also be overlapped (woven), in which case a metal flashing is recommended, or they may be butted against a corner trim, such as a two-by-two. (See Figure 10-17.) If the shingles butt a corner piece, caulk carefully.

If double coursing is used, greater exposure is allowable. In double coursing, one complete layer of shakes or shingles is laid over another. The lumber grade also influences the amount of exposure. No. 1 grade is typically used for the finish layer and a lower grade for the under layer. Thus, typical shake and shingle exposures for double coursing are 12 inches for 16-inch shingles, 14 inches for 18-inch shingles, and 16 inches for 24-inch shingles.

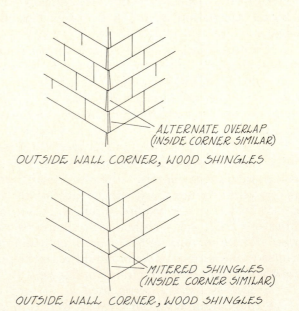

ALTERNATE OVERLAP (INSIDE CORNER SIMILAR)

OUTSIDE WALL CORNER, WOOD SHINGLES

MITERED SHINGLES (INSIDE CORNER SIMILAR)

OUTSIDE WALL CORNER, WOOD SHINGLES

FIGURE 10-23 Wood shingle outside wall corner treatment.

Double coursing requires larger nails (1¾-inch 5d). Place the nails approximately 2 inches above the butts, ¾ inches in from the sides; drive two additional nails approximately 4 inches apart across the face of the shingle. Nail each course in this manner.

All nails should be rust-resistant and should be driven flush with the shingle surface; never drive so hard as to embed the nail head in the wood or otherwise damage the shingle. When double-coursing, use triple shingles at the foundation line. If plywood sheathing is used, it is recommended that the plywood be covered with 15-pound felt before applying shingles; use threaded nails if plywood sheathing is used. If fiberboard or a gypsum sheathing is used, nailing strips should be used to provide backing for the shingles. Space the nailing strips (typically one-by-threes) according to the shingle exposure. Stagger shingle joints (usually 1½ inches or more).

GUTTERS

The roof collects water during a rain and disposes of it in concentrated quantities that may be harmful. Excess water at the foundation wall, for example, may result in wet basements or damp floor slabs and may even cause some structural damage. In some soils, the concentrated runoff can cause erosion. Plantings at the runoff points can be damaged. Where the roof overhang is less than 12 inches for a one-story house or less than 24 inches for a two-story house, unsightly exterior walls may result from runoff. These are some of the typical reasons for the use of gutters.

The purpose of gutters is to collect roof runoff water and route it to a disposal point.

The gutter system is made up of the gutter itself, downspouts, end caps, hangers, and whatever additional parts may be needed to install the system properly. There are a number of manufacturers of gutter systems, and exact assembly and installation procedures vary somewhat.

A typical installation procedure is as follows: The downspouts are first installed at their locations on the plans. The midpoints between the downspouts are found and the gutters sloped from that point to the downspouts; it is possible to install gutters dead level, but it is safer to slope them about $\frac{1}{32}$ inch per foot toward the downspouts. A chalk line is used to align the gutters at the desired slope. Hangers (installed before the gutters) hold the gutter in place.

End caps, as the name implies, cap the ends of the gutter and ensure that the water goes down the downspouts and not out the ends.

Gutter seal is used at all joints.

In the past, gutters were made of copper and of wood. Today, galvanized metal is much used, and vinyl gutters are seen more and more. Aluminum is also common.

In some cases, notably where the soil conditions are not subject to erosion or where the roof overhang is considerable, gutters may be omitted. However, if gutters are omitted, some type of water diverter (of galvanized metal, treated wood, vinyl, etc.) should be installed above entries and other areas where the runoff would be a nuisance or where runoff could be harmful to the house or grounds.

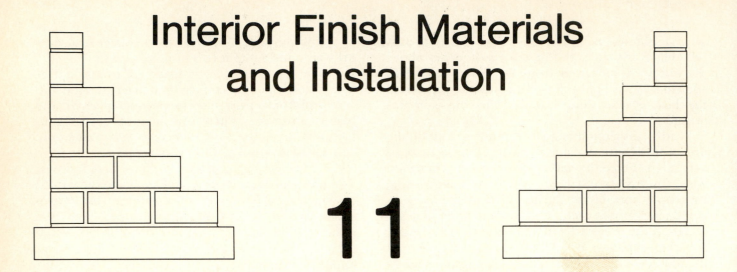

Interior Finish Materials and Installation

11

GYPSUM WALLBOARD

Gypsum wallboard is one of the most versatile and economical wall and ceiling materials and perhaps is the most used interior wall and ceiling finish material in residential construction. Gypsum wallboard panels are made by sandwiching gypsum between layers of paper. The back (the side applied to the wall studs) often has a vapor barrier bonded to it. The front (interior-room side) has a finish suitable for paint, wallpaper, and other decorative finishes.

The most used panel size is 4 by 8 feet, ½ inch thick; this size is quick and convenient for typical house construction with 8-foot finish ceilings and studs at 16 or 24 inches on center. These panels are typically installed in single layers, directly over the wall and ceiling framing. They may be nailed or screwed in place.

There are other panel sizes for special uses: 4 by 10, 4 by 12, and 4 by 14 feet. Thicknesses, besides the standard ½ inch, are ¼ inch, ⅜ inch, and ⅝ inch. The ¼-inch panels are often used to renovate existing surfaces of wallboard, plaster, wood paneling, and other materials. Where added fire resistance is needed, the ⅝-inch panels may suffice; if greater resistance is needed, two or more layers of wallboard may be installed (the ⅜-inch panel is often used as the finish surface in such fire-resistant, built-up walls). A side benefit of multilayer wallboard is quieter rooms (the extra layers damp sound).

Utility rooms, mechanical-equipment rooms, garages, and other spaces are sometimes required to have a specific fire rating by local or other authorities. Fire-rating requirements should be checked before the wallboard is selected. Type X fire-rated wallboard, ⅝ inch thick and installed on both sides of a two-by-four stud wall, has a 1-hour fire-resistance rating; a double layer of the same wallboard has a 2-hour rating. Some ceilings and floors may also require added fire resistance; check applicable authority requirements.

Wallboard must usually be kept dry, which means that standard wallboard cannot be used as a backing for ceramic tile in the bath, kitchen, or other locations where ceramic or other tile will be used. A special water-resistant type of wallboard is available as backing for tile.

Make sure that all required inspections are completed and approved before wallboard is installed (or before any wall-finish material is installed over framing, for that matter). Several local inspectors may be involved, several departments (plumbing, electrical, etc.) contacted, and numerous tests may be required.

You may install the wallboard parallel with the framing members or perpendicular to them. Wallboard has four commonly used shapes for the edges (long side of the panel): tapered, rounded, square, and beveled. The ends of the panels (short dimension) are square. Tapered-edge joints are the most frequently used in residential construction. The tapered edges provide a depression for the joint-treatment filler and tape. The panels are usually installed with the long side perpendicular to the framing members. Always provide solid backing at the joints. (See Figure 11-1.)

Check the wall and ceiling framing before installing wallboard. None of the members should be excessively warped or bowed. It is not the duty of the wallboard installer to replace poor framing, but minor adjustments may be made quickly with scrap two-by-fours or other wood blocking. In general, no single framing member should be out of the plane of the other members by more than ¼ inch.

If single-layer ½-inch wallboard is installed parallel to the framing members, the maximum stud spacing is 24 inches on center; the maximum for ceiling joists and for ⅝-inch wallboard is 16 inches on center. For single-layer ½-inch wallboard installed perpendicular to the framing members, the maximum stud spacing is 24 inches, and the ceiling joists can go up to 24 inches (the same is true for ⅝-inch wallboard).

Nailing typically is done one of three ways: by the single-nail method, the double-nail method, or the adhesive-nail method. In the single-nail

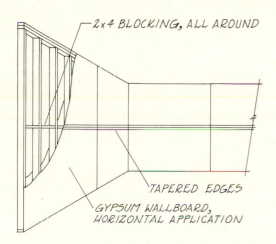

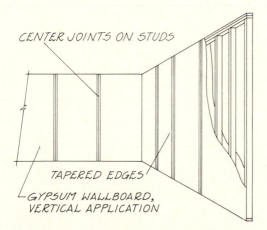

FIGURE 11-1 Tapered-edge wallboard may be installed horizontally but is usually installed vertically.

method, the edges and ends of the panels must occur over framing members or other solid backing. Nail with 12½-gauge smooth-shank bright drywall nails long enough to penetrate about ⅞ inch into the framing member or support (equivalent nails may be acceptable to applicable authorities); place the nails 7 inches on center around the perimeter of ceiling panels and 8 inches on center around the perimeter of wall panels. Do the same at framing members. Do not place the nails closer than ⅜ inch from the edges or ends and not more than ½ inch away. (See Figure 11-2.)

Double nailing is preferred by some drywall contractors, who think this method helps prevent nail pullout. The same type nails are used in this method as described in the preceding paragraph. The nails are placed around the perimeter in the same manner (7 inches on center for ceilings, 8 inches for walls). The difference is that double nails are spaced 12 inches on center at the intermediate framing members; this is done by first installing single nails at 12 inches on center at the intermediate members, then installing another set of nails 2 inches away from these. (See Figure 11-2.)

In the adhesive-nail method, a continuous bead of adhesive (available at any building supply house) is applied to all the framing behind a given panel. After the adhesive has set the length of time specified by the manufacturer, the panel is pressed against the framing members and nailed. The adhesive makes for a firmer contact, so wider nail spacing is allowable: single nails at 16 inches on center for the walls, 12 inches on center for the ceilings.

Screws also are used to secure wallboard.

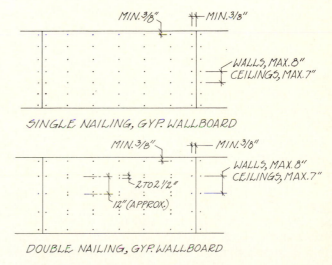

FIGURE 11-2 Wallboard may be single-nailed or double-nailed.

Typical screws are 1¼-inch, cadmium-coated, double-lead-thread drywall screws with contour head and diamond point (or equivalent screws acceptable to applicable authorities). If the framing members are 16 inches on center, place the screws at 12 inches on center on the ceilings and 16 inches on center on the walls. For framing at 24 inches on center, place the screws 12 inches on center for both the ceilings and the walls.

Wallboard installers normally place the ceiling panels first, then the wall panels. Do not force the panels into a space. Leave approximately a ⅛-inch space from panel to panel and where panels butt framing members or where the wall panels butt the ceiling panels. When hammering, drive the nail perpendicularly into the backing. Drive the nail flush with the panel, then give it another hit to form a slight indentation or "dimple" in the wallboard; do not drive the nail so far in that the paper backing is cracked, and do not drive at an angle. It is usually faster to install the panels completely in a room or given area before beginning the taping and finishing of joints. However, with large houses it may be efficient to keep installers working just ahead of finishers.

The typical finish ceiling height is 8 feet. The rough height is somewhat greater. Usually, the wall panels at the top of the wall are installed first, then the lower panels. There will be a gap of 1 inch or more between the bottom of the lower panel and the floor. This gap is blocked flush with the surface of the wallboard and will later be covered with the baseboard piece.

Help avoid loose nails by marking the location of framing or backing before nailing. Nail perpendicularly into the members so that the nails do not come out the sides of the members. If a nail misses a member or if it is obvious that a nail has barely pierced a member, pull it out.

Nail popping (loose nails) frequently occurs at inside corners due to otherwise insignificant settlement and structural movement in the house. To help prevent this condition, some of the nails or fasteners are strategically omitted near the intersection of walls and ceilings. Typically, if the ceiling joists are perpendicular to the line of intersection of the ceiling and wall, start the nails 7 inches from the line of intersection; if the ceiling joists are parallel to the line of intersection, start the nails at the intersection. Start the wall nails 8 inches away from the ceiling. Omitting the nails (or other fasteners in this manner allows the wallboard to slide slightly rather than shift nails loose. (See Figure 11-3.)

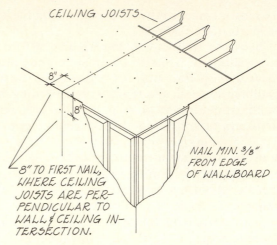

FLOATING INTERIOR ANGLE, GYP. WALLBOARD

FIGURE 11-3 Floating interior angles help to prevent cracking and nail popping in gypsum wallboard.

The vertical corners of interior wall partitions surfaced with wallboard must be protected. The most typical device used for this purpose is the metal *corner bead*. A corner bead may be an all-metal heavy-gauge hot-dipped galvanized steel angle or an angle of galvanized metal mesh. Sizes typically are 1 by 1 inch or 1¼ by 1¼ inches. The wallboard panels are installed first; then the solid metal or mesh angles are nailed on. A joint compound is then applied over the angle. Corner beads help to build straight, neat, vertical corners and protect the corners from ordinary household wear. (See Figure 11-4.)

Another vertical corner treatment is to omit the corner bead and instead use wood trim to cover the corner. This method usually presents a less desirable appearance, but it is possible to work

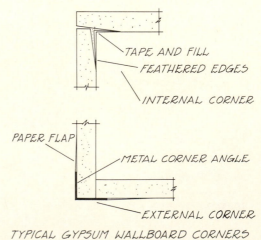

TYPICAL GYPSUM WALLBOARD CORNERS

FIGURE 11-4 Typical gypsum wallboard corners.

wood corner trim into the overall house-trim design and make it look good.

The corner bead for vertical corners is the most common metal wallboard accessory. However, a variety of metal wallboard edge trims and other metal accessories are available for special conditions in wallboard. Metal control joints also are available, but these are only needed for particularly large expanses of wallboard, which occur infrequently in residential construction.

In areas where two layers of wallboard are used, nail the first layer (base layer) with the same type of nails described earlier; space the nails 8 inches on center on walls and 7 inches on center on ceilings. Next, apply special adhesive (available at building supply houses) over the entire surface of the wallboard. Let the adhesive set according to the manufacturer's instructions. Then install the finish layer of wallboard at right angles to the first layer; first-layer joints and finish-layer joints should be offset a minimum of 10 inches. Nail the finish layer 16 inches on center on the wall and 12 inches on center on the ceilings. As usual with wallboard, keep the nails between ⅜ and ½ inch from the edges.

Taping and Finishing Wallboard

When the wallboard panels are secured and the metal corner beads and other accessories installed, it is time to tape and finish the joints.

First, "butter" the wallboard joints with the manufacturer's recommended joint compound. This coat of joint compound should have a uniform width, just a bit wider than the joint-reinforcing tape, and a uniform thickness. The buttering tool, held at about a 45-degree angle to the joint and used with moderate hand pressure, will give the desired result.

Next, press the special joint-reinforcing tape into the joint compound; center the tape over the joint. Holding the buttering tool as you did initially to butter the joint, pull the tool over the tape, squeezing out the excess joint compound. This should leave about a 1/32-inch thickness of joint compound under the tape if moderate hand pressure is used. Some installers add a very thin "skim" coat over the tape to ensure a firm contact between the joint compound and the tape, but this is still considered the first coat of compound.

This first coat of compound must now be allowed to dry. Meanwhile, place a small dab of compound over all the nailheads and smooth them. When the first coat of compound has dried

over the joints, apply a second coat. This second coat is considerably wider than the first coat. Allow the second coat to dry before applying the third and final coat.

While the second coat is drying on the joints, apply a second and slightly larger dab of compound over all the nailheads and smooth. As you can see, the nailheads are finished similarly to the joints, minus the reinforcing tape. Some installers use two coats of compound over the nailheads, and some use three; follow the manufacturer's and applicable local directives.

When the second coat of joint compound has dried, apply a third and still wider coat. This final coat will feather out to a very thin edge. (See Figure 11-5.)

When the joint compound is dry, sand the joints and the areas over the nailheads until smooth. Over a two-by-four, sandpaper or a vibrating sander may be used. In either case, sand the compound level with the surface of the wallboard, but be careful not to scuff the paper covering on the wallboard face.

Follow the manufacturer's instructions regarding precautionary steps to be taken when installing wallboard in adverse weather conditions. Generally speaking, weather extremes cause problems with the application of the joint compound. In extremely hot, dry weather, it may be necessary to raise the humidity (typically by sprinkling the subfloor with water); the effects of extreme cold can be neutralized by heating the house evenly to about 55° F or somewhat above (this is no particular inconvenience because the workers are more efficient when

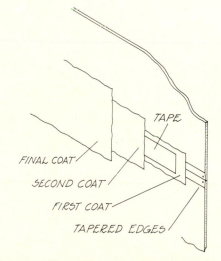

TYPICAL JOINT TREATMENT—GYPSUM WALLBOARD

FIGURE 11-5 Finishing gypsum wallboard joints.

the house is so heated anyway). Also, the house should be swept clean of dust and debris before joint compound is applied; if winds are a problem in this regard, the doors and windows should be kept closed during application of the compound.

Sanding wallboard may pose a health problem to workers unless precautions are taken. Some wallboard contains asbestos fibers, which become hazardous dust when the wallboard is sanded. Dust from any wallboard should not be inhaled. To avoid the dust problems, the joints may be wet-sanded or sponged rather than dry-sanded. In any area where construction dust is a problem, workers may gain protection through the use of acceptable protective eye goggles and respirators.

Predecorated wallboard joints are fast and simple to finish, compared to standard wallboard joints. The wallboard has excess covering material (flaps) at the edges that must be temporarily taped back to get to the joint. Spread the special joint compound over the joint and allow it to dry. Then coat the undersides of the flaps with the manufacturer's recommended paste or cement. Avoid getting paste or cement on the outside of the wallboard covering material. Then smooth the right-hand-side flap down over the joint; next, smooth the left-hand-side flap down over the right-hand flap. Hold a metal straight-edge centered over the joint and cut through both layers of wallboard covering material. Then remove the two excess strips, smooth the covering down with a roller, and sponge off any excess cement. (See Figure 11-6.)

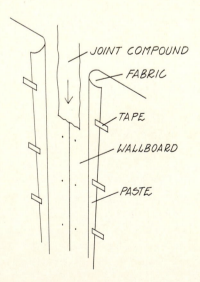

FIGURE 11-6 Preparation of gypsum wallboard when vinyl wallpaper is used.

Wallboard may also be purchased with radiant-heating components already installed in the panels. "Pigtails" extend from the panels to be connected to the house electrical system. These panels are installed substantially the same as standard panels, but care must be taken in storage, handling, and installation so as not to damage the heating elements.

PLASTER

Plaster is used much less in residential construction today than it was in the past. This is probably due to the fact that gypsum wallboard can duplicate most or all of the functions of plaster at less cost and with faster installation. Plaster also has another disadvantage: It is more difficult to alter or remove than is gypsum wallboard. However, plaster still sees considerable use in residential construction and is thought of by many as synonymous with high-quality work.

Plaster is similar in appearance, workability, installation, and function to stucco. Like stucco, plaster must be reinforced; plaster is applied in two or more coats, either by hand or by machine.

There are many different brand names for types of plaster that are mixed to fulfill specific building conditions. The need for strength, fire resistance, and sound resistance, for example, are qualities that affect the selection and installation of plaster. The type of lath (reinforcing) used is also a factor. The two typical plasters for most uses are gypsum plasters and portland cement plasters. As always, the function of the wall relative to the house must be studied and all the components—type of plaster, type of reinforcement, method of installation, and so on—chosen in light of that function.

Plaster is installed in two phases, the base coat and the finish coat. The base coat is usually installed in two or three applications (with a trowel or with a machine); the total thickness depends on the type of lath used, the finish coat, usage, and other factors but is usually about ⅝ inch.

Base-coat plaster comes as both gypsum plaster and portland cement plaster. Gypsum plaster is composed of gypsum rock (powdered) and hydrous calcium sulfate. Hydrous calcium sulfate is about 20% water; thus, the manufacturers remove about ¾ of the water, which is replaced in appropriate amounts on the job. Portland cement plaster offers resistance to moisture and thus may be used where high-moisture conditions exist.

The typical aggregates for base-coat plaster

are sand, perlite, and vermiculite. The type of aggregate used is influenced by the type of lath (metal, gypsum, or other) and by the type of plaster; check the manufacturer's instructions.

Finish-coat plaster is specially mixed to achieve certain characteristics of texture, color, hardness, or acoustics. In choosing a finish plaster, the base-coat plaster must be considered because some finish plasters require a strong base coat to achieve their desired functions.

The finish coat is usually about $\frac{1}{16}$ inch thick and is typically installed in two applications—either with trowel, which gives a smooth, dense surface, or first with a wooden float followed by a rubber float, which gives a somewhat rougher surface.

If this sounds complicated, it is. Fortunately, the manufacturers sell both base-coat and finish plaster that requires only the addition of water. Then, all that is required is a skilled plasterer to install the plaster.

In residential construction, plaster is typically installed over metal lath, wire mesh, or gypsum lath. The metal and wire laths form a mechanical bond with the plaster that offers resistance to cracks. Diamond mesh is good for contoured surfaces and all-around plaster work. Wire fabrics also are available.

Gypsum lath increases the fire rating, adds lateral stability, helps reduce sound transfer, and speeds up installation. Gypsum lath should not be used where the base coat will be portland cement or lime because portland cement does not bond well to this material. Gypsum lath is available plain or perforated.

Veneer Plaster

Special gypsum lath is available that makes it possible to apply a thin layer of plaster in two coats: After the first coat is applied, the plasterer may immediately begin the second coat; the total thickness is about $\frac{1}{16}$ to $\frac{3}{32}$ inch. Before application of the plaster, the joints are sealed with a special tape and corner beads are installed.

Radiant Heating in Plaster

Radiant-heating cables, which function the same as those described in the section on gypsum wallboard, may also be used with plaster. A special base is first installed on the ceiling. Then, a plaster base coat is applied. Next, the heating cables are installed. Last, the cables are covered with plaster. Be sure the plaster lath and plaster meet applicable code requirements.

Before plastering, the house should be kept at a constant temperature for one week. The temperature should not be below 55° F; keep the temperature the same throughout the house. Ventilation is also important for the plaster to set properly; fans may be used to increase circulation.

INTERIOR PANELING

There is an enormous range of choices in interior paneling. For the sake of illustration, the following text will consider three basic types of paneling: plywood veneer panels, hardboard, and wood planks.

Plywood veneer panels are available in virtually any wood desired and any texture desired, from rough-sawn to glassy-smooth, and may be installed in a wide range of configurations: board-on-batten, batten-on-board, butt joints—to mention but a few typical styles. Panels that give the appearance of 4-, 6-, 8-, 12-, and 16-inch boards are common. Plywood paneling is available unfinished or factory-finished.

Typical plywood-panel sizes are 4 by 8, 4 by 9, and 4 by 10 feet. Common thicknesses are $\frac{1}{4}$, $\frac{3}{8}$, and $\frac{3}{4}$ inches; other thicknesses may be available.

The minimum thickness recommended for direct nailing of plywood panels to wall studs is $\frac{1}{4}$ inch. All edges should be provided solid backing—typically two-by-fours. The maximum stud spacing for $\frac{1}{4}$-inch plywood paneling is 16 inches on center; if $\frac{3}{8}$-inch plywood is used, the studs may be 24 inches on center.

Installation is usually begun at a corner. To plumb the first panel, use a level or a plumb line, marking the position of the panel edge on the supports. Be very sure that the first panel is installed plumb: It is the guide for the remaining panels.

Nail the panels to the wall supports with 11- to 22-gauge casing or finishing nails, spaced 8 inches on center for the walls (if panels are used on the ceilings, space the nails 6 inches on center). Keep the panels at least $\frac{1}{4}$ inch off the floor. Place the nails not less than $\frac{3}{8}$ inch and not more than $\frac{1}{2}$ inch from the edges; the nails should be long enough to penetrate the support $\frac{3}{4}$ inch. The nails may be countersunk and filled with a filler that matches the panel finish, or color-coordinated nails may be used. Adhesives are also available. Follow the manufacturer's installation instructions, of course; usually the wall studs are beaded with adhesive;

the panel is pressed into place; the panel is then removed long enough (several minutes) for the adhesive to get tacky, and then the panel is pressed back in place and nailed at each corner while the adhesive sets. A combination of adhesive and nails may also be used, but simply nailing the panels as described is usually adequate.

Hardboard paneling is made from wood fibers, bonded together under heat. Hardboard paneling, like plywood panels, offers a wide variety of finishes and configurations (patterns). The outside dimensions of these panels are about the same as plywood, but the standard thickness of hardboard is ¼ inch, and ⅛-inch panels may be used over solid backing (such as gypsum wallboard).

The installation instructions for hardboard are about the same as for plywood paneling (except that ⅛-inch hardboard paneling may be installed over solid backing). Some manufacturers of hardboard provide tongue-and-groove joints with special nailing clips; if this is the case, follow the manufacturer's installation instructions, provided they meet applicable code requirements. Keep the panels ¼ inch off the floor.

It should be remembered that the manufacturers of plywood, hardboard, and synthetic wood paneling usually try to imitate the appearance of genuine wood planks while simplifying installation and lowering costs. Many of the manufacturers produce fine-looking products. But whether or not the panels look good in the house depends almost entirely on the quality of the installation. The panels must be installed plumb, and there must be no bulges in the panels. Bulges are especially noticeable in high-gloss finishes, where the bulges catch the light; this can spoil the whole desired effect of the panels and create a cheap-looking job. The installation instructions given in this section are minimum requirements. If the budget permits, it always is recommended that paneling be placed over solid backing. Gypsum wallboard makes an excellent backing, and the joints of the wallboard need not be finished: Simply nail the wallboard in place (see nailing instructions for gypsum wallboard)—horizontally if the paneling is to be installed vertically, vertically if the paneling is to be installed horizontally.

The gypsum wallboard will provide extra strength, insulation, and sound benefits to the wall and will ensure a flat plane of paneling, provided the panels are nailed and glued properly.

Wood-plank paneling is rarely seen anymore, except in very expensive houses and offices. It is, however, a classic, beautiful material.

Typical planks range from 4 to 12 inches wide by 8 feet long. Typical thicknesses are ½ inch to ¾ inch. Random widths and lengths are popular; in such an installation, all widths may be used, and approximately 20% of the boards are a full 8 feet long; the remaining boards are of varying shorter lengths.

Planks must have solid blocking; in frame walls, this is usually accomplished with two-by-fours turned with the wide side toward the room to provide a wider bearing and nailing surface. Typical plank thicknesses are ⅜, ½, and ¾ inch. For ½-inch planks, the nailing supports are typically spaced at 24 inches on center; the ¾-inch plank supports may be wider, but not over 48 inches.

The planks are usually nailed with two nails at each bearing if the board is 6 inches wide or less; if the board is wider, use three nails at each bearing. Use 11- or 12-gauge nails long enough to penetrate the supports ¾ inch or more. Apply 15-pound asphalt-saturated felt over the studs if the planks are to be installed directly to the framing members—that is, if no wallboard or similar solid backing material is to be used. If wallboard is installed, then furring strips, typically one-by-threes, must be installed as nail backing for the planks.

Trim for interior paneling may be necessary or desirable at interior and exterior corners and at the intersection of walls and ceilings. Plywood and hardboard paneling are typically butted at interior corners, and a trim piece is usually, though not always, added. Trim is the normal method of covering the exterior corners of plywood and hardboard paneling. Exterior-corner trim is sometimes omitted where plank paneling is used, because plank paneling can be mitered (however, exterior-corner trim protects the paneling); plank paneling is typically butted at interior corners (although it could be mitered), and trim may be used, if desired. Most paneling receives a trim piece at the intersection of walls and ceilings, but this piece may be omitted, if desired.

CERAMIC TILE

Ceramic tile is a much used material in bathrooms and kitchens. Some buyers do not even consider another material for the bath. Tile is used less in kitchens than in the bath, probably because the joints make the material harder to clean than jointless plastics when used as counter tops and counter and sink backsplashes. Actually, ceramic tile is not as functional in the bathroom as are some of the synthetic materials that are quick and easy to install and easy to keep clean and maintain after installation. Nevertheless, ceramic tile continues to be a very popular material and probably will continue to be used for many years.

Ceramic tile is available glazed and unglazed. Glazed tile has a high-gloss finish that is used primarily as a wall and counter surface. The smoothness of glazed tile makes it work well on bathroom walls and on counter tops; glazed tile is not a good surface to walk on because it is slippery, and foot traffic will break down the glaze.

Unglazed tile has a somewhat rougher nonglare surface. Unglazed tile may be used anywhere that glazed tile may be used, but unglazed tile is most frequently used for floors because it is safe to walk on and resists wear.

Both glazed and unglazed tile come in an almost limitless variety of colors and designs and in many sizes and shapes, the most typical being square, rectangular, and octagonal.

Ceramic Tile on Walls and Floors

The easiest and fastest way to install ceramic tile is with adhesive and, when performed properly, it gives good, long-lasting service.

Walls. The tub and plumbing should already be in place. Install special moisture-resistant gypsum board horizontally; nail or screw to the wall studs at 4 inches on center. Do not use a vapor barrier between wallboard backing and studs. Leave a ¼-inch space between wallboard backing and tub edge. Note the manufacturer's instructions for treatment of joints and fastener heads (screws, nails, or other); special joint tapes and cements are sometimes necessary.

Apply a tile sealer or tile adhesive that is compatible with the gypsum wallboard backing used and the adhesive to be used; in some cases a special sealer may have to precede the tile adhesive; in other cases, the tile adhesive may be applied in two coats. Check the manufacturer's instructions.

Apply the tile adhesive according to the manufacturer's instructions, usually over the entire wallboard backing surface. Use a notched spreader blade.

Next, install the tile by holding it against the adhesive and twisting it slightly so that a firm contact is made; do not "butter" the tiles and just stick them on. Let the tile set according to the manufacturer's instructions; a chemical reaction takes place, and usually some time must be allowed before grout is installed in the tile joints.

Fill the tile joints with grout. Grout may be installed with a trowel, a hand gun, or an air gun; the air gun is fastest. In any case, be sure the grout is forced fully into the joints; then clean excess grout from the tile with a heavy-textured cloth followed by sponging. Finally, caulk between the tile and tub with a waterproof caulk—this is usually done with a hand gun.

Floors. Ceramic tile is often used for bathroom floors and, somewhat less frequently, for kitchens and other areas where the floor may be subject to wetting or where water is needed to clean the floor. The underlayment for floors is especially important, because foot traffic and other weight are involved.

Ceramic tile floors need an underlayment that is very level, smooth, and rigid. Concrete is an excellent underlayment. Plywood may be used as an underlayment if it is sufficiently rigid. If plywood is used, it should be structural-interior type or exterior type and should be at least ¼ inch thick, regardless of how close the framing members are spaced underneath. The actual plywood thickness used depends on the rigidity of the framing members underneath (whose rigidity, in turn, depends on the size of the members, method of bracing, spacing, and other factors); ½-inch plywood is frequently used.

Hardboard and particleboard also are used as underlayment for ceramic tile floors. The installation is similar to that of plywood. In using plywood, hardboard, or particleboard as an underlayment, check the manufacturer's installation instructions and the applicable authority requirements.

The installation of ceramic tile on floors using an adhesive is similar to installing it on walls:

1. Clean the floor thoroughly and seal the entire floor with the tile manufacturer's suggested sealer or with a coat of adhesive, if acceptable.
2. Use a notched spreader blade to apply the adhesive.
3. Install the tile with a twisting motion.
4. Allow time for the adhesive to set, according to the manufacturer's instructions, before grouting.
5. Grout the joints.

Thin-Set Mortar

Setting tile in thin-set mortar is similar to setting it on walls and floors. However, because thin-set mortar is literally thin (about $\frac{3}{32}$ inch) and is applied in one coat, the installation surface must be very smooth and true; otherwise, a leveling coat of mortar will be necessary.

Portland Cement Mortar

Setting tile in portland cement mortar is slower, more difficult, and more expensive than setting the tiles with adhesive. It is, however, con-

sidered to be a very high quality method of installation.

Walls. The tub, shower, and plumbing connections should already be in place—see Figures 11-7 and 11-8 for tub and shower construction details.

The usual installation of tile over frame walls is as follows. First, install a vapor barrier (poly-

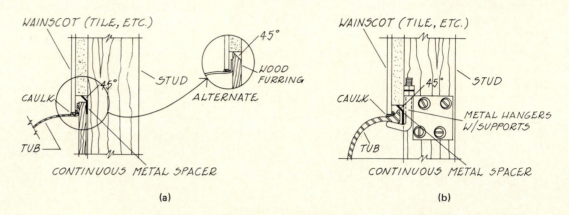

(a)   (b)

FIGURE 11-7 Tub installation. Tub may be supported by metal hangers or butt a wood furring strip. Tub is caulked all around, or special mouldings may be used.

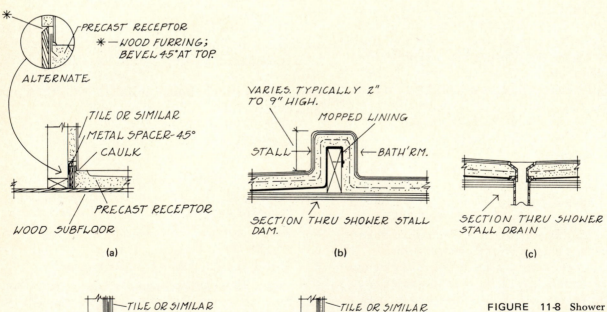

(a)   (b)   (c)

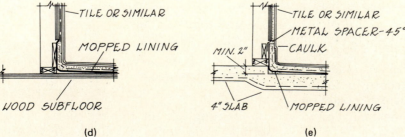

(d)   (e)

FIGURE 11-8 Shower installation. (a) Precast shower receptor installed over a wood subfloor. (b) Typical shower stall dam construction. (c) Typical shower stall drain through wood subfloor. (d) Typical job-constructed shower stall over wood subfloor. (e) Typical job-construction shower stall over concrete slab floor (precast shower stall installed similarly).

ethylene or building felt) on the studs to protect the studs and framing from moisture. A lath is installed next; if the lath is paper-backed wire fabric, the vapor barrier may be omitted because this type lath protects against moisture. Install the lath according to the manufacturer's instructions and applicable regulations. This completes the scratch coat of cement.

Next, install the setting bed of cement; this layer should be installed according to the manufacturer's instructions and will be about ¾ inch thick when complete. Allow proper drying time between coats.

The next coat is the skim coat, about ⅛ inch thick, into which the ceramic tile is set. Tiles have to contain a certain amount of moisture; check the manufacturer's instructions and local practice. The tile usually has to be dampened. Press the tiles in place, maintaining desired joint spacing. If sheet tile is used, make sure the joints between the sheets are the same width as the joints within the sheet; otherwise, the sheet outlines will be obvious and the job will not look good. Allow the tile to set properly, then install grout in the joints. Keep the grouted joints identical throughout.

Floors. First, install a water-resistant sheathing over the subfloor; if a paper-backed lath is used, this sheathing may be omitted. Next, install the lath.

Install the cement base according to the manufacturer's instructions, a minimum of 1¼ inches thick. Next, install a skim coat at least ⅛ inch thick.

Press the tile in place, allow proper setting time, and fill the joints.

When the installation is complete, damp-cure according to the manufacturer's instructions (this usually consists of keeping the tile damp and covered with waterproof paper for a period of time).

General Notes on Ceramic Walls and Floors

1. Use and install all products according to the manufacturer's instructions. If there is a conflict between the manufacturer's instructions and applicable authorities' directives, ask for clarification by applicable authorities (local building department, HUD, others).
2. Tile cartons should be dry when delivered and should be stored in a dry place.
3. Temperature should be at least 50° F where tile work is being done, and that temperature

should be maintained for at least a week after the work is complete. Provide adequate ventilation during tile work.

4. Temporary heaters should be vented to the outside of the house to avoid possible carton damage to tile work.
5. Use drinkable water for tile-setting materials.
6. Very little surface variation from level is acceptable: about ⅛ to ¼ inch in 8 to 10 feet for walls and floors, depending on the type settings used (conventional portland cement mortar, thin-set mortar, or adhesives). If the surface is not within the manufacturer's acceptable ranges of level, do not install ceramic tile—report to the appropriate authority.
7. Be sure the surface to be tiled is clean (free of dust, grease, labels, and other foreign matter).
8. Lay out the tile so that small pieces of tile (less than ½ the width of a given square) are minimized. Some working drawings show the desired joint placement very clearly, and some do not; make sketches, if necessary, so you will know what the best joint placement is. Locate key reference points and transfer them to the wall (corner joints, built-in soap dishes, towel racks, and other accessories).
9. Keep all joints the same width; keep the grout level; keep the joints straight. If tile sheets are used (each sheet typically contains 8 to 16 tiles), be sure the joints between the sheets are the same width as those within the sheet, and keep the grout looking the same between the sheets as within the sheets.
10. Smooth all exposed cut tile edges before installation; be sure the edges are clean before installation.
11. Size tile cuts carefully so that where the tile is covered by fittings (such as the chrome collars around faucet handles, towel racks, and other fixtures), the tile edges are completely covered.
12. After completion, clean the tile of grout film or haze, using a cleaner acceptable to the manufacturer. It is usually necessary to wash the tile with water after using a cleaner (acids are sometimes present in cleaners).
13. If construction still is going on after completion of the tile, cover the floor with heavy paper and secure it in place with masking tape; this protects the floor from dust, dirt, and worker traffic.
14. Follow the manufacturer's instructions per-

taining to traffic on newly tiled floors (usually 1 to 7 days should be allowed before tile is walked on, the exact length of time depending on the type of tile setting). If traffic on newly tiled floors is unavoidable before the end of the proper setting period, protect the tile by placing wide, very flat, rigid boards on the floor.

WOOD FLOORS

Preinstallation Requirements

Wood flooring should be brought to the house at least 48 hours before installation and stored in the house, which should maintain a temperature of 70° F during this period and during installation. After installation, the temperature should be maintained at a constant, normal house-interior temperature. The flooring should also be protected from dampness and should be well ventilated. Do not allow wood flooring to be delivered to the house site during rain or snow unless the wood is properly protected.

The two most common types of wood flooring are *strip flooring* and *block flooring*. Oak is the dominant wood because it is very durable, but other woods are used, typically beech, maple, birch, and even pecan. Strip flooring is typically $2\frac{1}{4}$ inches wide by $\frac{5}{16}$ inch thick if placed over a subfloor, $\frac{25}{32}$ inch thick if placed directly over the floor joists. It is placed perpendicular to the direction of the flooring. Strip flooring is usually tongue and groove on the edges; the ends may or may not be tongue and groove, but this makes for perfect installation because the matched ends help tie the floor together into a single plane. (See Figure 11-9.)

Wood blocks are typically either *unit blocks* or *laminated blocks*. Unit blocks are made by gluing together several strips of wood lengthwise; a common size is four $2\frac{1}{4}$-inch by 9-inch strips, glued together to form a 9- by 9-inch square. Laminated blocks are made by bonding together several layers of veneer to form 9- by 9-inch squares; when installed, the laminated veneers appear to be one solid square of wood. Both unit blocks and laminated blocks are tongued on two opposing sides and grooved on two opposing sides so that the floor is tied together to form a single plane. (See Figure 11-10.)

Discussed here will be the most common installations: strip and block flooring over wood subfloors, strip and block flooring over concrete subfloors, and strip flooring applied directly over floor joists. Literally scores of floor patterns are possible for creative designers and builders, but any arrangement chosen will be installed similarly.

Wood Strip Flooring over Wood Subflooring

The thickness and width of strip flooring, the joist spacing used, and the type of subfloor are all interrelated. Hardwood strip flooring, at least $\frac{25}{32}$ inch thick and at most $2\frac{1}{4}$ inches wide, may be installed over $\frac{3}{8}$-inch plywood over joists spaced at 16 inches on center. The flooring should be installed at right angles to the floor joists. This is a fairly typical installation.

Lesser thicknesses of flooring (typically $\frac{5}{16}$-inch hardwood) may be used with thicker plywood subflooring. If diagonal plank subflooring is used and the strip flooring is $\frac{25}{32}$ inch thick or greater, the joist spacing may be extended to 24 inches on center; the flooring should be installed at right angles to the floor joists.

The direction of the flooring is a design consideration; thus, where aesthetics determine the direction of the flooring and where the direction happens to be parallel to the floor joists, the strength of the subfloor must be increased—typically

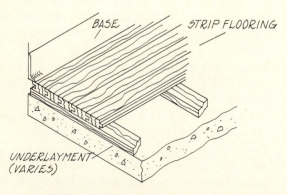

FIGURE 11-9 Typical hardwood strip flooring.

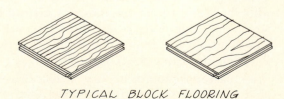

FIGURE 11-10 Wood block flooring: (a) unit blocks; (b) laminated blocks.

by increasing the thickness of the plywood. If a plank subfloor is installed at right angles to the joists, the strip flooring is usually installed at right angles to the planks (if, however, the planks are 2 inches thick or greater and a suitable underlayment is used, the flooring may be installed parallel to the planks).

If plank subflooring is laid diagonally, the strip flooring may be laid either parallel to or at right angles to the joists; this is a very typical solution where the house floor is wood framing because it provides a choice of flooring direction as well as a very rigid subfloor.

Tongue-and-groove flooring should be blind-nailed at an angle of about 40 to 50 degrees (in blind nailing, the nailhead cannot be seen; in face nailing, the nailheads are countersunk and the holes filled). Thin flooring ($\frac{5}{16}$ inch thick or less) usually does not offer enough thickness for blind nailing and thus must be face-nailed. Where plywood is used for the subfloor, threaded nails are typically used because plywood is not as dense as solid wood and thus does not hold nails as well. Provide a $\frac{1}{2}$-inch space between flooring and wall.

The following are typical nailing requirements:

1. *For $\frac{25}{32}$-inch tongue-and-groove flooring:* 6d nails (minimum), screw-thread or spiral-thread type, spaced at 10 to 12 inches on center, or 1d cut nails at same spacing

2. *For $\frac{15}{32}$-inch tongue-and-groove flooring:* 5d screw-thread or spiral-thread nails at 8 to 10 inches on center, or 6d, 12-gauge cut or steel-wire nails at 8 to 10 inches on center

3. *For $\frac{11}{32}$-inch tongue-and-groove flooring:* 4d screw-thread or spiral-thread nails at 6 to 8 inches on center, or 5d, 14-gauge cut or steel-wire nails at same spacing

4. *For square-edge flooring (no tongue and groove):* 1¼-inch fully barbed brads, or 4d screw-thread or spiral-thread nails. In either case, use two nails in each strip at 8 inches on center.

Wood Strip Flooring Installed Directly over Joists

Obviously, wood strip flooring placed directly over joists must be placed at right angles to the joists. The flooring must be at least $\frac{25}{32}$ inch thick when installed over joists spaced at 16 inches on center. If there is no basement, some type of vapor barrier must be installed at the bottom of the strip flooring; no. 15 asphalt-saturated felt may be used, or an insulation with a built-on vapor barrier. Allow ½ inch space between flooring and wall.

End joints for flooring that is not end-matched must fall on a joist. If the joints are end-matched, the end joints may fall between joists, but no two adjoining flooring strips may have end joints falling within the same joist space. Every flooring strip must be secured to at least two joists. Secure the flooring according to the manufacturer's instructions, typically with 8d threaded or screw-type nails at each joist, blind-nailed at a 50-degree angle.

Wood Strip Flooring over a Concrete Slab

A common installation technique for wood strip flooring over a concrete floor slab is as follows:

1. Clean the floor slab thoroughly and seal it with a primer that meets the manufacturer's requirements.

2. This technique uses wood sleepers—two-by-fours laid broad side down. The sleepers are secured to the slab with an adhesive suitable for bonding wood to concrete. First, mark the locations for the sleepers on the slab; the sleepers should be a range of lengths between 2 and 3 feet, approximately. The flooring strips will be nailed at right angles to the sleepers using a nailing technique similar to the one discussed for nailing flooring direct to joists (the sleepers act as joists). Apply the adhesive where the sleepers will be laid according to the manufacturer's instructions, usually in beads using a hand gun. If the slab is to be heated, the adhesive must be heat-resistant.

3. Press the sleepers onto the adhesive. Lap the ends of the sleepers 4 inches or more (lap them horizontally, of course, not on top of each other). Install the rows of sleepers at 12 inches on center, unless a subfloor is used over the sleepers (not typical), in which case the sleepers may be installed at 16 inches on center. The varying length of the sleepers allows you to stagger the joints from one row to another. Leave 1 inch clearance between the sleeper ends and the walls.

4. When the sleepers and the adhesive have had the proper bonding time, install the wood strip flooring. Every flooring strip should rest on at least two sleepers. Where end joints occur between sleepers, the ends should be

matched; do not allow two end joints between sleepers for adjoining wood strips.

The foregoing has been, and still is, a typical installation of wood strip flooring over concrete slabs. The sleepers (sometimes called *screeds*) may also be embedded in the concrete, if desired, or secured to the concrete with metal clips. Some manufacturers offer metal runners and clips or other devices for the installation of their flooring.

The preceding discussion focused on typical installations for wood strip flooring. Wood strip flooring is usually 2¼ feet wide. Some manufacturers produce wide plank flooring that can be installed directly to concrete slabs with adhesive—no sleepers are required; decorative nails (they have no function in the installation) are available, if desired, or simulated countersunk nails may be used.

Obviously, installing the wood planks directly to the concrete with adhesive, where the wide planks are appropriate to the overall design of the house, is a cheaper, faster installation method. However, the walking surface is "harder" than on wood sleepers, which offer some "give." There is also a difference in the sound of footsteps on flooring laid directly on concrete, since there is no space underneath the flooring. It is also likely that wood flooring laid directly on the concrete slab would be somewhat colder to the touch than floors laid on screeds. Where the budget permits, most people probably would prefer flooring installed on sleepers.

Wood Block Flooring over Wood Subflooring and Concrete Slabs

Wood block flooring may be installed over wood subfloors or concrete slabs. Over wood subfloors, the block may be either glued or nailed. Wood block flooring is glued to concrete slabs, and this is a popular installation because it is fast and cheap to install and looks good.

The typical wood subfloor for wood block flooring is plywood or tongue-and-groove boards. Install with at least 4 nails per block, driving the nails through the tongued sides at an angle of 40 to 50 degrees. The nails should be 7d threaded for ²⁵⁄₃₂-inch-thick blocks. For ¹⁵⁄₃₂-inch blocks, use 6d threaded nails or 7d, 12½-gauge cut or wire nails. Threaded nails or the equivalent must always be used where the subfloor is plywood. Allow ½ inch space between wood flooring and wall.

If gluing to a wood subfloor, follow the manufacturer's instructions for installing the adhesive. The adhesive must be the type for bonding wood to wood, of course. The blocks are pressed onto the adhesive, and bonding time is allowed before use.

To install block flooring over a concrete slab, first clean the slab and apply a primer-sealer. Install adhesive according to the manufacturer's instructions. The adhesive should be the type for bonding wood to concrete; if the slab is to be heated, the adhesive must be heat-resistant. A surface damp-proofing may be necessary, depending on the manufacturer's recommendations for the products used.

Finishing Wood Floors

Unless they are prefinished, wood floors usually have to be sanded smooth. Sanding should be done only by experienced workers, or the floor can be ruined or seriously damaged. The particular wood producer's sanding instructions should be followed; four sandings is a fairly common method, working from a no. 2 paper down to a no. 00.

There are a number of ways to finish floors. The following methods are common:

1. Apply one coat of sealer, then two coats of wax.
2. Apply stain, then two coats of wax.
3. Apply one coat of shellac, then one coat of wax (varnish or lacquer may be substituted for shellac).
4. Apply two coats of shellac or two coats of varnish.
5. Apply two coats of paint (as recommended by manufacturer), then one coat of wax.

The second method is often used. There is a wide range of stain colors. Dark stains are very popular as of this writing. After staining, the wax is applied and machine-buffed between applications. Stained and waxed floors give a low sheen that is easy on the eyes, and the floor retains a rich, natural-grain appearance. This is also a simple, fast, cheap finishing method.

CARPET

Carpet is a major residential floor-finishing material. It is fast and simple to install, fits easily into most construction budgets, and is relatively easy to maintain after installation. These are some of the main reasons for carpet's popularity with both builders and homeowners. The word *carpet*, however, can be ambiguous relative to use, quality, appearance, and other aspects. Like the word *car*, *carpet* is very unspecific: Volkswagens, Cadillacs, Hondas, Chevrolets, and Mercedeses are all "cars," but each serves different needs and sometimes different markets.

Carpet can be any of a wide range of materials, both natural and synthetic. It may be the natural color of the material (such as wool), or it may be any of thousands of other shades. The selection of design patterns available is equally broad. To confuse this whole area of house-floor finishes further, manufacturers are almost daily evolving new materials, new methods of installation, and new ways to construct carpet.

The carpet surface material is typically either *knitted, woven*, or *tufted*. Knitted and woven carpeting are similar because they are constructed by machines that tie the carpet material into the backing material in one operation. Tufting, however, has become the dominant method of manufacturing carpet. In the tufting method, thousands of needles are used in a sophisticated manufacturing process that forces the carpet material through the backing to produce shag, sculptured, plush, and pattern carpet that are so common in residential construction. For many years, the carpet and the pad or cushion underneath were clearly distinct components: The carpet was the finish material, and the cushion was a separate layer of natural or synthetic materials that helped protect the carpet from wear, increased the depth and thus the softness of the walking surface, and provided a degree of floor insulation. Many carpet professionals still consider this the best way to handle a high-quality carpet installation.

But over the years, carpet manufacturers have produced carpet that is bonded to various materials, so that carpet installation is now a one-step operation. This development speeds the installation of carpet and helps cut costs.

Carpet is usually installed by means of *tackless strips* or by gluing the carpet directly to the subfloor; the subfloor may be either wood or a concrete slab. Tackless strips are fitted with pins such that, when the strips are secured to the subfloor pins up, the carpet may be hooked over the pins, thus avoiding the old tack-down method. (See Figure 11-11a.)

Different types of strips are available to match the job need: They can be nailed or cemented to wood or concrete subfloors. The pin length varies with the carpet type and depends on the type and thickness of the carpet backing. Typical pin lengths are $\frac{1}{4}$, $\frac{3}{16}$, and $\frac{7}{32}$ inch; follow the manufacturer's recommendations.

Installing Carpet on Tackless Strips

Before installing carpet on tackless strips, thoroughly clean the subfloor. Remove all grease, oil, debris, and other materials that interfere with installation or may damage the carpet. Concrete

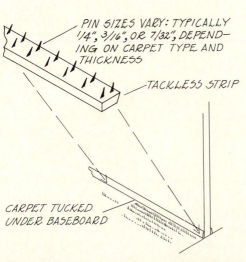

PIN SIZES VARY: TYPICALLY 1/4", 3/16", OR 7/32", DEPENDING ON CARPET TYPE AND THICKNESS

TACKLESS STRIP

CARPET TUCKED UNDER BASEBOARD

TACKLESS STRIP, INSTALLED AT WALL EDGE

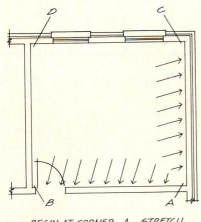

BEGIN AT CORNER A. STRETCH ALONG WALL AC, THEN WALL AB. STRETCH BD AND DC SIMILARLY.

INSTALLING CARPET ON TACKLESS STRIPS

FIGURE 11-11 (a) Tackless strips installed around the edge of the room. (b) Installing carpet on tackless strips.

(a)

(b)

subfloors should be tested for alkaline moisture; this may be done by wetting the concrete, then testing with pH test paper; the paper reveals by its color whether the concrete is acidic, neutral, or alkaline. If alkaline, wash the concrete with muriatic acid (readily available at building supply stores); a 5% acid solution is typical for this use. Rinse with drinkable water. Allow the concrete to dry thoroughly.

The tackless strips are secured at the walls. If a cushion is used, lay it next, with the minimum amount of sections. Cushions are usually stapled to wood floors and glued to concrete floors to prevent their shifting about. If a bonded-cushion carpet is used, the cushion may be omitted (although cushions still may be used, if desired and if the budget permits).

The carpet is best installed using a *power stretcher* and *knee kickers*. The power stretcher is used initially to stretch the carpet over the tackless strips at the wall; the installer stands while using this tool. The knee kicker is somewhat larger than an ordinary hammer, with a head that has rows of protruding teeth on one end and a pad on the other end; the installer works on hands and knees, placing the head against the carpet and "kicking" the pad of the tool with the knee. This provides a finer adjustment of the carpet than the initial setting.

The carpet is cut about a foot longer in each direction than the room to be carpeted. This gives an extra 6 inches of length in every direction. Run the extra length up the side of the wall while stretching the carpet over the tackless strips.

To stretch the carpet in place, begin in any convenient corner, hooking the carpet over the tackless strips (use the knee kicker for this), about 18 inches in each direction from the corner along adjacent walls.

Next, use the power stretcher to hook the carpet over the tackless fasteners along one of the walls from the starter corner. The stretcher should be pushed at about a 15-degree angle to the wall. Now stretch the carpet of the adjacent wall, moving from the same starter corner, and, again, pushing at about a 15-degree angle to the wall. Follow the carpet manufacturer's recommendations for the amount of stretching required; typically, this will be somewhere between 1½ to 3 inches per 12 feet of carpet. Maintain a constant pressure on the power stretcher so that the carpet will be stretched evenly over the tackless fasteners. To judge the amount the carpet has been stretched, measure the distance the carpet ran up the wall before stretching, then again afterwards. The difference is the amount of stretch.

Now, use the knee kicker to stretch further the same two walls. Kick the tool at the same 15-degree angle to the wall that you used with the power stretcher. Stretch evenly. Measure the stretch at the wall.

Hook the opposite walls over the tackless fasteners, first using the power stretcher, then the knee kicker. Maintain constant pressure on the tools. Push at about a 15-degree angle to the wall. Some adjusting may be necessary to get the carpet smoothly stretched over the subfloor.

When the carpet is smoothly in place, trim the edges with a carpet knife and tuck the carpet over the tackless fastener, against the wall. (See Figure 11-11b.)

Installing Carpet with Glue

Most carpet manufacturers produce a line of carpet specifically designed to be glued down. Glue-down carpets are popular for at least two reasons: Gluing is usually the cheapest way to install carpet, partly because the cushion is usually omitted, and glue holds the carpet firmly and permanently in place, eliminating the need for restretching, which must often be done with carpet laid over a cushion.

There are, however, certain drawbacks to glue-down carpet: If caster-wheel chairs (such as desk chairs) are to be used, chair pads are required. And glue-down carpets are troublesome to remove for replacement.

The proper adhesive must be used when installing carpet by the glue-down method. Latex adhesive is a typical glue, but because of the variety of backings and combinations of backings available, it is important to use only the manufacturer's recommended adhesive.

Generally speaking, glue-down carpet is easy to install. If the manufacturer does not specify the amount of glue and method of application, it is wise to experiment with a carpet scrap of the type to be used to determine the correct amount of carpet glue needed to bond the carpet firmly in place without soaking through the carpet. Carpet or linoleum rollers are typically used to push out air bubbles and smooth the carpet over the glue; it may be necessary to add weights over trouble areas until the glue bonds the carpet in place.

As with all carpet installation, clean the subfloor thoroughly before installing the carpet. Check concrete subfloors for alkaline moisture. Pits, cracks, and other openings in concrete floors greater than about ¼ inch should be patched; the concrete may have to be sealed before carpet installation, depending on applicable regulations and the carpet manufacturer's recommendations.

Edges and Seams

Keep carpet seams to the absolute minimum. In house construction, most rooms can be done with a single piece of carpet; seams should rarely be necessary. Where seams are necessary, they may be secured with a *heat seam* or by stitching, depending on the type of carpet and the manufacturer's recommendations. A heat seam is usually made by inserting a special heat-activated tape under the seam; a carpet iron is then used to bond the tape to the underside of the carpet.

If a seam is to be stitched, the edges may have to be reinforced by installing a seam tape with adhesive before the stitching is done. The type of thread used and the method of stitching should follow the manufacturer's instructions; this often consists of a waxed linen thread used to make at least three stitches per inch near the edge of the carpet seam.

The carpet edges are protected at the walls by trim. Where carpet butts a fireplace or similar vertical surface, the carpet is generally stapled down firmly. Unprotected edges, such as at room openings or where the material changes (from carpeted living room to tiled entry hall, for example), the carpet edge is usually protected with a metal *binder bar* or *gripper bar*: the carpet fits into the metal housing of the bar, which is secured to the floor. (See Figure 11-12.)

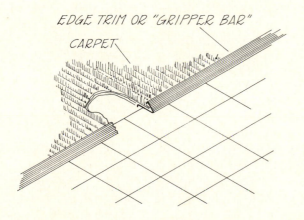

FIGURE 11-12 Gripper bar.

RESILIENT FLOORING

Some of the most commonly used resilient flooring materials in residential building are sheet vinyl, vinyl tiles, vinyl-asbestos tiles, and asphalt tiles. These materials are often used in entry halls, kitchens, bathrooms, utility areas, and recreation rooms.

Sheet vinyl is made of vinyl resins with an asbestos-fiber back. Sheet vinyl may be obtained with a foamed-vinyl or similarly cushioned backing, for a softer walk and to help reduce the noise level. Special "no-wax" surfaces are sometimes available. Thicknesses and widths vary somewhat, but ⅛-inch thicknesses and 6- to 12-foot widths are common. Vinyl tiles are of the same composition, but tile-shaped.

Vinyl-asbestos tiles are made of vinyl resins and asbestos fibers. Asphalt tiles contain asphaltic resins and asbestos fibers. All resilient tiles are available in 9- by 9-inch and 12- by 12-inch sizes. Coloring pigments and some inert filler material are usually ingredients in resilient flooring materials.

All these materials may be installed over concrete slab or wood subfloors. However, the underlayment for resilient flooring must be appropriate for the type of flooring used, and the installation of the underlayment must be performed carefully. Similarly, the adhesive used must be appropriate to bond the flooring material to the underlayment.

There is variation among manufacturers, but it is a typical requirement that the house temperature be maintained at about 70° F for at least 48 hours before, during, and after installation of tile.

Concrete subfloors must have a vapor barrier underneath; otherwise, moisture will cause a breakdown of the bonding adhesive used to secure the flooring material. The concrete-surface requirements vary somewhat with flooring material and manufacturer, but usually the concrete should be steel-troweled to a very smooth surface; there should be no score marks or depressions. Concrete should be dry and clean before the installation of these finish flooring materials; as a technical matter, the concrete is probably not completely dry inside when the installation begins because this process may take months to complete, but the more drying time, the better.

Concrete subfloors, if installed over a gravel bed and vapor barrier as discussed elsewhere in this book, should not present a moisture problem for resilient flooring materials.

Always use the manufacturer's recommended sealers, fillers, adhesives, and other materials—otherwise, the warranties may be invalid. In the loose-laid method for sheet vinyl, adhesive is used only at the seams. Seams in sheet vinyl should always be kept to a minimum and should be planned to fall where they will be least conspicuous or where the traffic is lightest. To loose-lay sheet vinyl, cut the vinyl so that it is somewhat larger than the room or area to be surfaced; lap the excess vinyl up the walls. Then press the vinyl down at the walls and cut away the excess vinyl, gradually, at each wall, until the vinyl fits neatly. To secure the seams, fold the vinyl sheet back from each side of the seam and install the manufacturer's recommended adhesive on each side of the seam with an adhesive spreader (the size varies, but approximately a 4-inch band on each side of the seam). Fold the sheet vinyl back in place over the seam—the seam should be barely visible, and the pattern, if any, should be matched. Roll the seam, typically with a 100-pound roller.

Another way to fit the sheet vinyl to the wall is to use dividers. First, move the sheet vinyl as close to one wall as possible. Set the dividers so that one leg moves along the wall and the other leg scribs the shape of the wall on the vinyl; then cut the vinyl to fit. Follow the same procedure around the room. This may be a good procedure for beginning workers, but experienced installers can work accurately and swiftly without the dividers.

To cut the seam accurately, lap the sheet vinyl at each seam, then cut through both layers; this produces a very tight seam. It is wise to use a chalk line to locate the seam and to use a metal straightedge when cutting the seam.

The amount of adhesive used and its location vary somewhat with different manufacturers and with the particular material used; you may have to use adhesive at the seams only, at the seams and perimeter, or uniformly under the entire material. Check the manufacturer's instructions.

Resilient flooring can usually be installed over existing resilient flooring. Because resilient flooring contains asbestos fibers and the existing flooring may have to be sanded, the installers always should wear approved respirators and goggles and take other precautions as required to avoid health hazards associated with asbestos and other minute particles that may be breathed or work their way into the skin while working. Occupational Safety and Health Administration (OSHA) publishes guidelines concerning proper equipment to wear while performing construction tasks.

OTHER FINISH-FLOOR MATERIALS

The flooring materials listed thus far are those most commonly used in residential construction. There are many other flooring materials that may be used, including terrazzo (usually precast), marble, slate, flagstone, and brick.

KITCHEN AND BATHROOM CABINETS

Kitchen and bath cabinetry may be handcrafted on the job, but this process is usually too expensive for most home buyers and too time-consuming for most builders. Thus, most cabinetry is factory-made, then installed in standard modules in the house.

Cabinets are generally base cabinets or wall cabinets. Base cabinets are secured to the wall and floor with screws or nails; counter tops are secured to the base with wood or metal braces. Wall cabinets are secured to the wall at some point above the floor. (See Figures 11-13 and 11-14.)

INTERIOR TRIM

Interior trim may be used as sparingly or as abundantly as desired, but the typical trim in most houses falls under one of the following categories: base trim, located at the joint of wall and floor; ceiling trim, where the walls meet the ceiling; outside corners for wall partitions; and door and window trim.

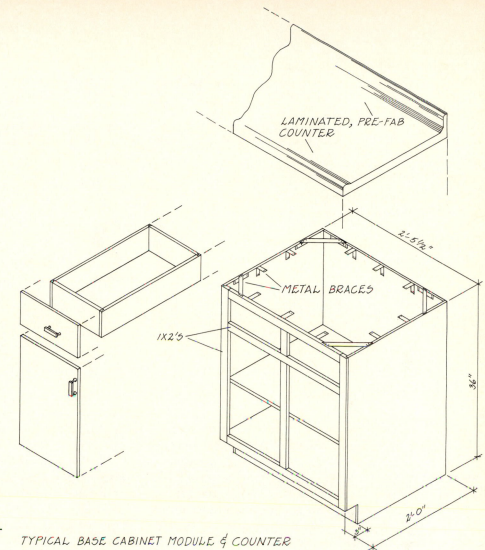

LAMINATED, PRE-FAB COUNTER

METAL BRACES

1X2'S

2'-5½"

36"

2'-0"

9"

FIGURE 11-13 Typical prefabricated base cabinets.

TYPICAL BASE CABINET MODULE & COUNTER

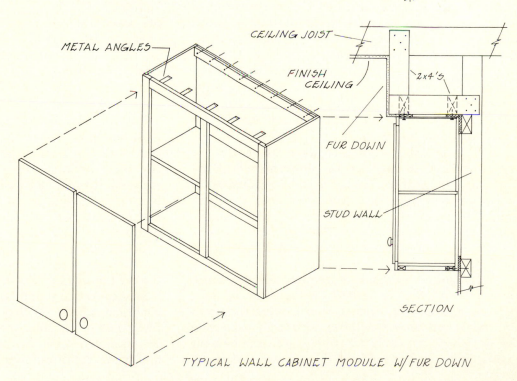

METAL ANGLES

CEILING JOIST

FINISH CEILING

2x4'S

FUR DOWN

STUD WALL

SECTION

FIGURE 11-14 Installation of prefabricated wall cabinets.

TYPICAL WALL CABINET MODULE W/ FUR DOWN

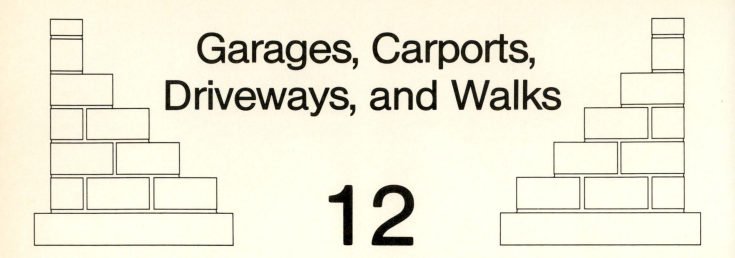

Garages, Carports, Driveways, and Walks

12

GARAGES AND CARPORTS

Garages and carports may be attached to the house, or they may be detached and located some distance away. Attached two-car garages have become almost standard over the years, for several reasons: With the automobile so much in use today, the garage entry is often used by the residents more than the main entry. In the winter, the car may be started and left to warm until the resident is ready to leave. All enclosed garages, both attached and detached type, offer more security to automobiles, tools, appliances, and storage items than do open carports. There are materials savings with attached garages, because the house forms one wall of the garage; this can be a fairly significant saving, especially if the house is brick-veneered. Also, the attached garage offers some energy savings by insulating the house against the weather. The savings from building a one-car garage are usually not great enough to sacrifice the convenience of the two-car garage; thus, like the attached garage itself, the two-car garage has become almost standard. There is a marketing aspect to the attached garage, also; many home buyers like the fact that the attached garage makes the house look larger than it really is.

If limited weather protection for the automobile is all that is desired and security is not a particular problem, open carports offer savings in materials and labor. Carports may be either attached or detached. An enclosed storage area can be built at the back or at one side of the carport, if desired.

In the 1940s and through much of the 1950s, the detached garage (often a one-car garage) was typical. Detached garages, especially one-car garages, are used much less today.

The construction techniques used for garages and carports are, for the most part, the same as those used for the house. Framing design and construction are about the same. The roof structure and covering are the same. The foundations of garages and carports are sometimes simpler and lighter-duty than the house foundation; for example, carports in some areas may be built with a concrete slab on grade.

Double-car attached garages usually have the same foundation system as the house, except in some areas the foundation at the doorway may be constructed by turning the slab down at the edge, thickening it up, and taking that edge below the frost line. (See Figure 12-1.)

Attached garages and carports may be required (by local building codes) to have a fire-resistant material at the house wall. The typical requirement is a 1- to 1½-hour fire-resistance rating. This rating can often be met with ⅝-inch fire-resistant gypsum-board panels or a similar material. Check the local codes to be sure for your area.

Garage interiors vary widely, largely depending on the market. The interiors may, for example, have exposed framing with minimum lighting and electrical supply. Or they may be finished out with Sheetrock or paneling and have quality lighting and be equipped with laundry rooms, workbenches, and game facilities; much of the interior depends

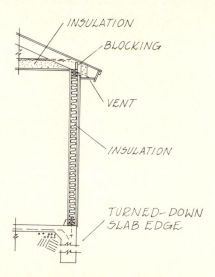

INSULATION

BLOCKING

VENT

INSULATION

TURNED-DOWN SLAB EDGE

FIGURE 12-1 Turned-down slab edge, section view.

on what the prospective buyers want—and what they are able and willing to pay.

DRIVEWAYS AND WALKS

Typical residential driveways and walks at the street are concrete; the double driveway has become almost standard. The following discussion will first focus on building the concrete driveway and the concrete front walk. Then, drives and walks will be divided into two types, hard and soft, and several variations of each will be discussed.

Driveway Specifications

Driveways serve as access to the garage or carport and sometimes as parking space for residents and guests. The following specifications are typical:

1. Minimum length for parking purposes is 22 feet.
2. Minimum width is 10 feet for a single driveway, 18 feet for a double.
3. Maximum gradient is 5% (⅝-inch drop per horizontal foot).
4. The drive may be crowned (high in the middle) or cross-sloped (tilted). Crowns and cross-slopes range between a maximum of 5% (⅝-inch drop per horizontal foot) and a minimum of 1% (⅛-inch drop per horizontal foot); slopes and crowns facilitate drainage.
5. Driveways typically extend from the garage or carport slab (or parking slab if there is no garage or carport) to the street. As a mini-

mum, there must be an isolation (expansion) joint where the driveway slab meets the garage, carport, or parking slab and another isolation joint where the slab meets the public sidewalk. In general, isolation joints are needed wherever the driveway butts other slabs or construction.

6. Contraction joints are required every 10 feet, as a minimum. Thus, if the driveway is 20 feet wide, a contraction joint down the middle will be required in addition to contraction joints across the driveway at 10 feet on center. Contraction joints are simple and inexpensive to make and should be used generously; these joints may be increased for decorative purposes, and they may be slanted to help facilitate drainage in areas subject to heavy snows.
7. The driveway entrance (apron) must meet local and appropriate authority standards; these standards are usually available from the local building department.
8. Where the driveway changes slopes, the vertical transition should not exceed 14% (1¾ inches per foot). The purpose of this standard is to avoid scraping the car undercarriage on high spots in the driveway created by the changes in slope that result in "hills" on the driveway or scraping the bumpers where slope changes result in "valleys"; most residential drives are not long enough to require changes in slope.
9. Turnarounds may be required in instances where it is not safe to back automobiles out of the driveway. The turnarounds are usually additional slab widths located at or near the garage, carport, or parking space.
10. The type of concrete, mix, and installation should comply with local engineering practices. Ready-mix suppliers as well as the building department are familiar with these requirements.
11. The concrete driveway should be at least 4 inches thick for ordinary use. It is recommended that the edges be thickened to 8 inches, 1 foot wide, all along the driveway.
12. Steel reinforcement is recommended, typically 6- by 6-foot no. 10 wire mesh; 4 inches of gravel is recommended under the driveway.
13. A broom finish is recommended, after the concrete has been troweled smooth.
14. The concrete should be kept moist for approximately 3 days to ensure adequate curing.

Double driveways are recommended. For typical residential situations, the drive is usually not long enough to gain significant savings by building a single driveway, and the convenience of a double drive is demanded by most buyers.

Ribbon driveways (concrete runways) are not recommended either. However, on suburban or rural lots of 5, 10, or more acres, ribbon driveways may be used for economy. Still, full-size driveways are recommended for at least 22 feet from the house, garage, carport, or parking space. Ribbon driveways are typically two slabs of concrete, each

a minimum of 2 feet wide, 5 feet on center. If the ribbons double as walks, they should be at least 3 feet wide. The concrete apron at the street should be the same as described earlier, and the driveway should be full-width at least 12 feet prior to reaching the apron. There should be contraction joints at least every 6 feet on ribbon driveways; on full-width driveways and full-width portions of ribbon driveways, at least every 20 feet. (See Figures 12-2 and 12-3.)

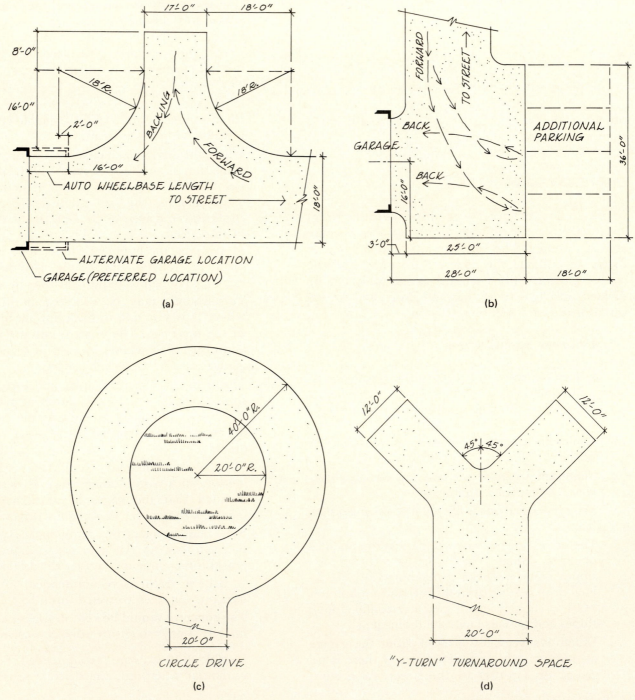

(a)

(b)

CIRCLE DRIVE

(c)

"Y-TURN" TURNAROUND SPACE

(d)

FIGURE 12-2 Typical driveway layouts.

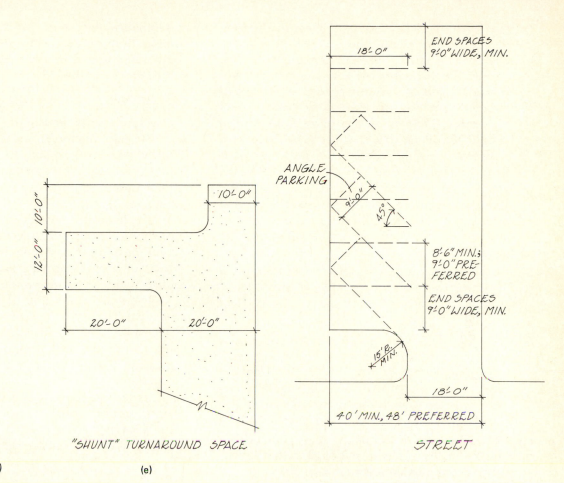

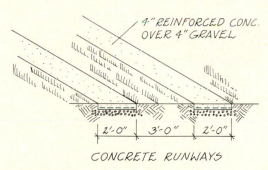

"SHUNT" TURNAROUND SPACE

STREET

FIGURE 12-2 *(Continued)*
(e)

FIGURE 12-3 Concrete runway design.

Construction of Concrete Drives and Walks

The construction sequence for both concrete driveways and walks is as follows:

1. Excavate the area for the driveway. The top edges of the driveway should be at grade; the subgrade will be the depth of the concrete (usually 4 inches) plus the depth of the gravel underlayment (usually 4 inches); thus, the subgrade under the driveway slab will be 8 inches below grade except at the edges, if the edges are thickened. If the edges are thickened, the excavation must be deeper by the extra amount of depth desired; if the edges are 8 inches thick by 1 foot wide, dig down another 4 inches, making the depth at the edges 1 foot. Use string and stakes to mark off the excavation area, allowing an extra foot or so of excavation on each side of the driveway to work in. The house foundation walls, curbs, and gutters will likely already be in place; thus, there will be ample reference points both at the house and at the street from which to locate the excavation. The stakes themselves can be used as references to ensure that the excavation slope is correct.

2. Install the form boards. First, remove the excavation stakes and string. Locate and stake the exact location of the four corners of the driveway or walk. Stretch string between the stakes (if the driveway or walk is long and the string sags too much, it may be necessary to use intermediate stakes). The string should be installed to represent the exact outside edges of the length of the driveway or walk. Check the string layout to be sure it is square. Come out from the string 4 inches and drive two-by-four stakes with the wide side facing the string; you can do this by holding two

two-by-four stakes together, then hammering the farther one into the ground (this leaves 2 inches of space between the stake and the string; the form board, a two-by-six or two-by-eight, will go in the space. It is recommended that the form stakes be 4 feet or less on center. Next, mark the drive slope on the inside face of the stakes (this can be done with a string and line level) and nail the form boards in place. The top of the form board should be the exact height of the top of the slab. Check the forms with a spirit level and plumb bob to ensure that the slope is correct. Double-check the forms to be sure they are square.

3. Install isolation joints between the slab and any construction such as walks at the street or foundation walls or garage slabs.

4. Place the 4-inch gravel underlayment.

5. Place the wire-mesh reinforcement. Keep the wire mesh up off the gravel by using small reinforcement rods driven in the ground and wired to the mesh to hold it up; small stones may also be used to hold the mesh up. The mesh should be located in the lower third of the slab.

6. Oil the inside of the form boards or use special lubricants from supplier.

7. Saw the form stakes level with the top of the form boards.

8. Place the concrete, being sure the base is moist; screed it level, then float it smooth. Next, trowel the concrete and add the required contraction joints. Round the edges with an edger.

9. Final-finish the concrete—a broom finish is recommended for driveways and walks.

10. Remove the forms when the concrete has hardened sufficiently (usually a few hours).

11. Keep the concrete moist for approximately 3 days while it cures.

HARD WALKS AND DRIVES

Hard walks and drives are concrete, asphalt, or similar consolidated or semiconsolidated materials. High usage or heavy usage usually demands some type of hard surface. Driveways, public walks, parking areas, patios, and service areas are typical examples. Sometimes a combination of hard and soft materials will work for a given situation: For example, concrete pavers may be laid a foot or so apart with the spaces filled with gravel (a "soft"— that is, unconsolidated—material); such a combination might work as a garden walk. Function and maintenance determine whether hard, soft, or a combination of hard and soft materials is used. Hard walks and drives are usually easier to maintain than soft walks.

Asphalt

Asphalt is second perhaps only to concrete as a utilitarian, high-use, low-maintenance material. Typical uses for asphalt include streets, parking areas, driveways, bicycle lanes, service areas, and similar uses. Used creatively, it makes good garden paths, pavers, even patios and similar uses.

The black color of asphalt is often overwhelming when it is used in large areas. This problem can be minimized by pressing in pea gravel, crushed brick, or other aggregates. Specially prepared asphalt paints in a range of colors are also available.

There are two types of asphalt, hot and cold. Hot asphalt is made by coating crushed rock with hot asphalt cement. The bonding action between aggregate and asphalt cement is similar to the bonding action that takes place in concrete. Hot asphalt is usually used for the heavier-duty, high-use areas such as streets and parking lots (asphalt also has roofing applications, discussed elsewhere in the text). Hot asphalt is purchased from asphalt plants and typically carried to the job in dump trucks. The trucks dump into spreaders, which in turn are used to place the asphalt. Hot asphalt is about 350° F when placed.

Cold asphalt may be bought in premixed form and, as the name implies, may be placed cold. Cold asphalt may be used for paths, pedestrian pavers, and similar light-duty uses.

Asphalt is placed similarly to concrete. The ground surface must be excavated to the proper subgrade, forms should be used, and special attention should be paid to the edges of the asphalt because the edges often crumble if not supported properly. When laying cold-asphalt walks, the forms used may be left in place to support the edges. The permanent frame or form may be 2-inch wood members of the appropriate depth, crossties, brick, stone, or any other material substantial enough to hold the asphalt edges firmly in place. If wood is used, it should be treated, preferably pressure-treated.

A sand base is recommended for asphalt, typically a minimum of 2 inches thick for residential walks.

The cold mix is dumped within the forms in small mounds and then raked level. The forms serve as a guide in keeping the asphalt level. Asphalt may be sloped to drain one way or the other or may be crowned to drain to both sides.

If forms are not used at the edges, the asphalt

should be thickened at the edges, similarly to concrete drives: about 8 inches thick and 1 foot wide at the edge. This practice is often omitted, which is why crumbled edges are often seen.

Power and manual rollers and tampers are available for compacting the asphalt. Rollers in combination with tampers usually give the most level surface; small areas and pavers may be leveled with manual tampers. Rollers should be brushed with water immediately before using to minimize asphalt buildup on the rollers. Several rollings are usually needed, and asphalt may be added to low areas as required.

When the asphalt is leveled and sloped as desired, finish aggregates may be added. Simply sprinkle the aggregates over the asphalt in the desired pattern and final-roll or tamp the asphalt until the aggregates are level with the surface.

Asphalt, both hot and cold, has some distinct advantages and disadvantages, compared to concrete, brick, and similar hard surfaces. First, asphalt is very malleable; this makes it easy to work with, and in the case of cold asphalt, less skill is required in its placement. Also, mistakes tend to be easier to correct with asphalt; once laid, if low spots are discovered, they may be leveled by adding more asphalt and retamping. Asphalt can be shaped to fit land contours more easily than concrete. As a supplementary material to concrete and similar hard surfaces, asphalt is an adequate, cheap solution.

Asphalt, especially cold asphalt, has disadvantages over the harder surfaces, too. It typically supports less weight; it sometimes develops ripples if the base is not well prepared; mud and moisture will seep up through asphalt if a minimum 2-inch sand base is not used; and asphalt does not support concentrated weight—such as garden furniture legs —as well as concrete or brick. Also, oil and gasoline dissolve asphalt, which means that provisions must be made to protect the asphalt where automobiles will be parked on it. The typical solution here is to leave a grassy strip where the auto will be parked. Other solutions to protect against auto gas and oil include paving strips of brick or concrete or specially manufactured paints that protect the asphalt.

Asphalt will cure enough for foot traffic and similar light use within a week or so. However, asphalt contains an agent that must evaporate before curing is complete; this may take months. Once fully cured, asphalt is almost as hard as concrete—for its intended uses.

Soil Cement

Soil cement uses the soil itself as an ingredient. In appearance, soil cement is like gritty, hard-packed soil. In function, it is somewhere between concrete and asphalt. It looks best where some support more substantial than earth is required but where concrete or asphalt or some other hard surface would interfere with the appearance of the overall design (garden walks, curbs, drainage swales, stepping stones, and so forth). Mixed properly, it may even be used for driveways where a more natural appearance than concrete or asphalt is desired.

To install soil cement, the soil is first rototilled or otherwise dug up 4 inches deep. Sandy or granular soil is required. If such soil does not exist on the job, it must be imported. Forms should be used, similar to concrete forms, and as with asphalt, special attention to the edges is required. The edges should be thickened, and permanent forms of wood, brick, concrete, or similar materials should be used. (Where permanent forms create a negative appearance, thickened edges approximately 8 inches deep by 1 foot wide may be used.)

The dry cement is mixed into the soil according to the manufacturer's instructions. Hoes and rakes are the typical tools. Rocks and organic matter should be removed from the soil. The soil and mix are worked until the cement is evenly blended in the soil. The soil with cement should be leveled and compacted with tampers or rollers. As with asphalt, more soil (mixed with cement) may be added to low spots and tamped level. High spots may be removed with a screed. The soil-and-cement mixture may be worked as required to drain—sloped one way or another or crowned, in the case of walks or drives. When the mixture is formed as desired, a fine spray of water is used to initiate the curing process. The spraying is continued until the mixture is thoroughly wet but not soupy—it should be similar to poured concrete.

When the soil cement has dried enough so that it will not stick unduly to the roller or tampers, it should be compacted. After compaction, keep the surface moistened for several days to a week to aid the curing process.

Flagstone

Flagstone is a natural stone, usually available in earth colors: yellow, brown-yellow, orange, red, gray, or some mixture of these colors.

Flagstone, for paving purposes, should be a minimum of 2 inches thick, unless it is laid on a firm base, such as concrete; the typical base for outside use is sand. There is some variance in the thickness of flagstone, and this adds to the problem of laying the stone levelly. Furthermore, the edges are always more or less irregular, creating difficulty with the joints. Sand is a typical joint filler for outside use, and sand does not work well in wide joints. Similarly, mortar joints break down if the joints are wider than those commonly used for brick. Thus, flagstone is best used for informal, outdoor surfaces: garden walks, pavers, and similar uses.

Because of its beauty, flagstone has been used for more formal areas such as patios and terraces, entry walks, and even interior entry halls. It is not recommended for these uses unless the budget can bear it and strict attention is paid to the base preparation and the laying of the material.

Where a very smooth surface is of no particular value—garden pavers, for example—flagstone may be laid directly on the earth, simply by scooping away enough earth to allow the stone surface to be at grade.

For smoother surfaces, more formal surfaces, and surfaces where people congregate, a base of sand and gravel is required. Sand and gravel is always required in areas of extreme cold, where frost heave is a problem. To prepare such bases, excavate the subgrade such that 4 inches of gravel and 2 inches of sand over that may be laid, leaving enough depth to install the flagstone (typically 2 inches thick) so that its surface will be at the desired finish-grade elevation.

For still smoother surfaces, flagstone may be laid on concrete. Where concrete is the base, the flagstone may be less than 2 inches thick. The concrete is formed and placed as discussed elsewhere in the text, and the flagstone is laid over the concrete, preferably with mortared joints.

Tile

Ceramic and quarry tiles are made of fired clay. Common paver sizes range from 4 by 4 to 12 by 12 inches, and tiles are available in octagonal and other geometric shapes. Unglazed tiles assume the color that their fired natural ingredients produce. Glazed tiles have ceramic coatings in a wide range of colors. The surface texture of both glazed and unglazed tiles varies considerably, depending on usage and desired appearance. A discussion of the outdoor usage and installation of tile follows.

Tiles vary in thickness, typically from $\frac{1}{4}$ inch to 2 inches. The thickness chosen depends on usage and installation method. Generally speaking, the firmer the base, the thinner the tile can be.

Tile surfaces should be planned so that a minimum of tile cutting must be done. Tile is brittle and requires skill to cut by hand. Tile saws cut down on cutting time, decrease waste, and ensure the desired cut. Saws are usually necessary when many tiles must be cut; they may be rented or purchased from building supply houses. Areas where cutting may be difficult to avoid include patio edges (especially when the patio is of a curvilinear design), areas where walks or other surfaces change direction, and areas where horizontal surfaces join vertical ones.

Joint sizes must be carefully planned before the tile is laid (typically somewhere between $\frac{1}{2}$ to $\frac{3}{4}$ inch). Also, contraction and expansion joints must be provided, where needed; these joints usually require engineering consultation or strict adherence to the manufacturer's recommendations when large areas are involved.

Bases for outdoor tile include sand, mortar, wood, concrete, or some combination of these materials. Sand bases are typically for informal outdoor areas—patios, cooking areas, and the like. A mortar base, laid directly over the soil, is possible where weather conditions are mild. Tile may be laid on wood decks, walks, and similar surfaces, but extreme caution must be taken to ensure that the wood members do not sag, breaking the tile joints. Tile on a concrete base is expensive but provides the best assurance of a level, long-lasting surface where more formal use and appearance are important.

Tile on sand. Tile to be laid on sand should ordinarily be about 2 inches thick. Thinner tile tends to break or tilt in the sand base. Prepare the subgrade, keeping it level and free of rocks and organic matter. Compact the soil with a manual or power tamper. Where the soil is unstable, dig down another 4 inches and add 4 inches of gravel fill.

Use treated form boards, as in placing concrete slabs, and plan to leave the boards in place to protect the edges; if another border is desired, such as concrete or brick, place the border first and use it as the form. Spread a 2-inch bed of sand over the base material and screed it level; then compact the sand with a tamper. Moisten the sand with a fine spray of water and retamp.

When the sand is smooth and compacted, place the tiles with the desired joint spaces. Spread sand lightly over the tile surface and sweep it into the joints; repeat this process several times, compacting the sand joints with a thin board or similar tool. Moisten the joints with a fine spray and compact again. The joints may settle somewhat in several days and have to be swept with sand, moistened, and compacted again.

With a firm border and with a well-compacted base and joints, such tile surfaces will remain level and stable. Some maintenance will be required to keep weeds out of the joints; this problem may be reduced somewhat, if desired, by planting a small, hardy ground cover in the joints.

Tile on mortar. Tile on mortar requires a mild climate and a smooth, stable base. If used in areas of extreme cold, which are subject to frost heave, the shifting earth will break up the joints and throw the surface out of level.

To lay tile on mortar, first stake out the area of work as if placing concrete. Tile on mortar should have a firm edge material: permanent wood forms, left as the edge, or brick, concrete, or similar materials. The earth must be smooth and level and free of rocks and organic matter.

Place the mortar in small heaps and work it into place with a wood float or other convenient tool; then screed it level. The mortar bed, for light usage such as pedestrian walks on private property, should be approximately 1 inch deep; if it is less thick, it may crack with use. Mortar tends to crumble if thicker than 1 inch.

When the mortar is smooth and level at 1 inch thick, place the tiles, maintaining ½- to ¾-inch joints. When laying the mortar bed or filling the joints, do not mix more mortar than can be used in approximately 1 hour; otherwise, it will begin to harden and lose its functional qualities. The tiles should be soaked in clean water—typically for at least 15 minutes—before laying; otherwise, the tiles will pull moisture from the mortar, weakening the bonding of the tiles with the mortar. Store the tiles on edge long enough to drain off excess water before laying, or the bonding may be weakened.

Place the tiles, keeping them level; a large spirit level is a typical tool for this job. Use a length of two-by-four and a hammer to tap high tiles level; do not tap tiles into place by hitting them directly with a hammer or similar tool. The hammer handle may be used to tap down single problem tiles.

When the tiles are level, fill the joints with mortar. Generally speaking, the smaller joints are "poured" with a rather fluid mortar mix. Larger joints may be troweled with mortar, if this is convenient. When the joints are filled, finish them with the appropriate tool (joints for walks, patios, and similar surfaces are usually best flush with the tile surface, but other joints may be used). The joints, like the mortar bed, must be scheduled so that the mortar does not become too hard before it is finished. Clean up mortar spatters immediately, using a damp cloth.

The mortar should be allowed to cure for approximately one week before it is used. Tile on mortar beds is suitable for foot traffic and light usage. Tile on mortar is relatively cheap and quick and easy to install, it gives a more level surface than tile on sand, and it is easier to maintain than tile on sand.

Tile on concrete. Tile on concrete lasts longer, requires less maintenance, is stronger, and provides the most finished, formal appearance of the various ways of laying outdoor tile. It is also one of the most expensive outdoor surfaces because it involves first placing a concrete slab and then veneering the slab with tile—two distinct jobs.

First stake out the slab (walk, patio, terrace, or other area) as for a concrete driveway or walk. Dig the subgrade, taking into consideration the thickness of all the materials plus any base-preparation materials such as gravel. The concrete base (minimum 2 inches thick, recommended 4 inches) should be allowed to cure before the tile is placed.

The edges of tile on concrete generally require no borders, because concrete supports well. Patios, terraces, and walks, however, may have a more finished appearance with a border (such as brick), where this extra expense seems worthwhile. Thickened slab edges are not a requirement either, but they are recommended.

If frost heave will be a problem, the slab may have to be reinforced, more gravel used under the slab, and possibly other local engineering safeguards taken.

The concrete slab should be rough-finished so as better to receive the mortar bed. If applying tile over an existing slab and the surface is smooth, treat it with a dilute muriatic-acid wash, then clean, wash, and rinse the concrete. Brush on a light coat of cement paste.

Mortar should not be placed on dry concrete; thus, the slab should be wet down, then excess water removed before applying the mortar. This keeps the concrete from pulling the moisture out of the mortar (the tile itself must be moistened before placing, as described in the instructions for placing tile over mortar).

Lay the mortar about ½ inch thick over the concrete slab. Screed level.

Apply the tile according to the instructions for tile on mortar.

Tile on wood. Laying tile on wood is possible, but care must be taken to ensure that the wood does not sag, especially if the joints are mortared; engineering consultation is recommended. Typical uses include wood decks and balconies, walkways, steps, and ramps.

In mild climates or where a waterproof surface is not a necessity, the tile may simply be laid over the wood with closed joints; if this solution is used, a substantial border material should be installed flush or slightly above the surface of the tile to hold the tile in place.

In cold climates, water between the tile and the wood base and within the tile joints can cause the tile to move. In such cases it is desirable to provide waterproof joints and a waterproof contact between the tile and the wood base.

For a waterproof surface, install building paper or similar waterproofing over the wood, then install a stucco mesh over the waterproofing.

Apply the mortar over the mesh and then place the tile.

Brick on concrete

Brick over concrete is a very expensive surface. It will, however, last for generations, and it is one of the most beautiful surfaces.

The brick is placed on a ½-inch bed of mortar (follow the instructions for tile on concrete and tile on mortar).

The joints may be mortared or made of sand, or they may be omitted entirely (closed joints). Fully mortared joints, flush with the brick surface, are recommended for areas of extreme cold.

SOFT WALKS AND DRIVES

Soft walks and drives are loose aggregate, bark and wood chips, and similar unconsolidated materials.

Soft walks of loose materials are best used for informal areas of fairly light and infrequent traffic: patios, garden walks, and so forth.

Loose Aggregate

Loose aggregate may be whole or crushed stones, typically from ¼ inch to approximately 1 inch; for border material, much larger stones are available. Crushed brick, granite, and similar materials may also be used.

Loose aggregates drain well, absorb little moisture, and require less base preparation than concrete and other consolidated materials. All loose surface materials, however, must be contained by some substantial border, such as wood or masonry; otherwise, the loose materials are a continual maintenance problem due to scattering. The sizes of aggregate should be varied to aid compaction.

First, build the form (border) for the desired material. Keep the borders level. Then place the aggregate, level with a screed, and compact it with manual or power tampers. The borders should be approximately 2 inches above the finished elevation of the aggregate. Although loose aggregate is one of the cheapest ways to build walks, parking areas, patios, and so forth, it will have a finished,

professional appearance *if* the borders are constructed with care. The borders, in addition to keeping the aggregate in place, form a visual edge to the aggregate surface; otherwise, the aggregate will appear sloppy and temporary and grow more so in time.

In addition to these more functional uses of aggregate, aggregate may be used effectively as a ground cover where poor soil, drainage, or other conditions make planting difficult. Some home-owners, who want only minimal lawn maintenance, opt for aggregate "landscaping" over plantings.

Weeds can grow through loose materials but can be easily removed by hand. Traffic also cuts down on weeds. Maintenance can be reduced further by using a plastic membrane under the loose materials; soil sterilizers may also be used, but care must be taken in their application.

Bark and Wood Chips

Bark and wood chips are probably the least formal of the various loose-surface materials. Typical uses include play areas for children (especially where there is climbing equipment that may result in potentially injurious falls) and walks for kitchen and flower gardens (the chips decompose in time, forming a useful mulch). Bark and wood chips may also be used as landscape material—as a backdrop for trees, shrubs, and other plantings or in combination with rock gardens and the like.

Bark, wood chips, other organic materials, and generally all loose materials must be contained with some type of form or substantial border for functional, maintenance, and aesthetic reasons.

Wood

Wood surfaces fill needs somewhere between the hard and soft materials. Wood is, of course, a classic building material. But it often has been used outdoors primarily to save on initial building expenses; the savings over, say, brick, were lost over the years due to maintenance and replacement costs. However, improved chemical treatment, especially pressure treating, has resulted in increased use of wood where brick or concrete or similar materials formerly would have been preferred—and the savings incurred are not lost to eventual maintenance and replacement costs *if* the appropriate treatment is matched with the wood use.

Treated wood may, for example, be used for walks, patios, and other ground-contact applications. It is even being used as foundation material in some areas, as discussed in the text on foundations.

The proper usage of wood is, however, essential. Treatment quality is standardized; when purchasing lumber, the builder should be certain of exactly what quality is being purchased. This information is typically stamped on the lumber. Lumber marked "LP-22 ABOVE GROUND USE ONLY .25," for example, would be suitable for above-ground decks and fence panels but not for deck or fence posts; lumber marked "LP-22 GROUND CONTACT .40" is suitable for limited ground contact—walks, patios, fence posts; "FDN FOUNDATION .60" stamps indicate the wood is suitable for underground use. The numbers after the usage (.25, .40, .60) refer to the amount of preservative treatment the lumber has absorbed; without going into specific details, suffice it to say that the more preservative the lumber absorbs, the more weatherproof it will be. And pressure treatment results in the maximum absorption (as opposed to soaking in preservatives).

Treated wood surfaces offer advantages over more traditionally used materials. There is great freedom of scheduling and method of installation in building with wood: projects can be done on-site, over as long a time as needed, or the projects may be prefabricated off-site and installed very quickly. A walk, for example, can be built on-site or assembled in modular section off-site and brought to the site, where it can be installed in a few hours. Decks, patios, and other building components can be handled similarly.

Wood may be used with more aesthetic freedom than many other materials. Wood looks good in combination with almost all other materials. It is also cheaper and easier to install than many other materials suitable for a given job.

Wood sizes offer wide design possibilities. Treated wood is available in the typical residential lumber sizes, plus the larger landscaping sizes. Designers can add to the versatility of the sizes by cutting and by building up sizes; a walk, for example, could be made by laying flush four-by-four members across the width of the walk, or the four-by-fours could be cut into 4- by 4- by 4-inch blocks and laid like brick. To get a larger-size member for strength or aesthetic purposes, two or more of the members may be bolted together.

Even if using the most durable wood-treatment grade, wood should be protected where it touches the ground. The typical protection is to lay the wood on a bed of sand or gravel; this helps prevent the accumulation of moisture and thus increases the life and serviceability of the wood; it also helps to maintain its original appearance. (See Figure 12-4.)

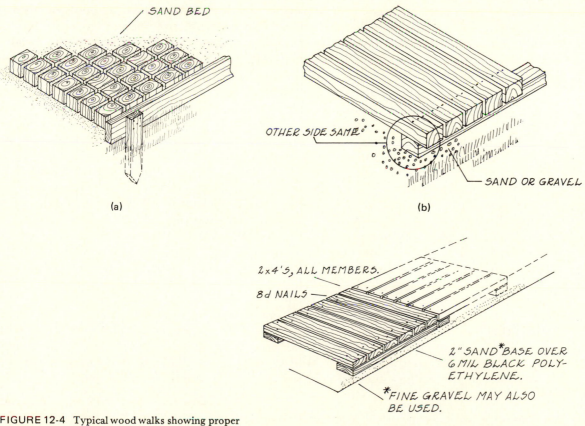

FIGURE 12-4 Typical wood walks showing proper construction.

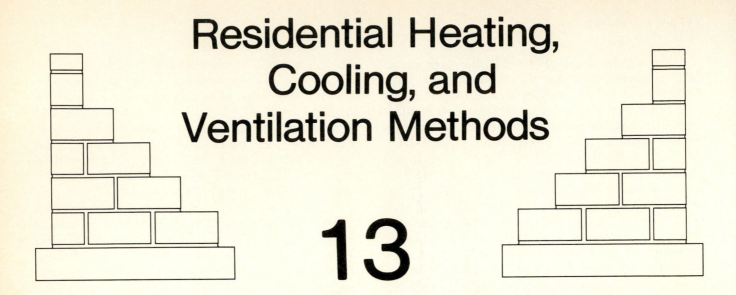

Residential Heating, Cooling, and Ventilation Methods

13

At least three methods of heating and cooling residences have been used over the years: hot water, radiant electric, and central forced air. Central forced air is probably the most used system in new American housing today, so it will be discussed in more detail than the other systems.

HOT-WATER SYSTEMS

Hot water systems were used a great deal in the past and are still used and maintained in older homes. These systems consist of a network of pipes, a heating device (often a gas furnace), and radiators or similar devices in the rooms to transmit the heat. The pipes contain water, which passes through pipes in the furnace, which heats the water; the water, in turn, moves through the room radiators, where the heat is given off to the rooms. The pipes are usually routed in the house crawl space or basement. Hot-water heating systems are very good systems, but they are usually too expensive to be attractive to home buyers or builders. Contractors are often called on to install central cooling systems to older homes that use hot-water heating systems. (See Appendix D-10.)

RADIANT ELECTRIC HEAT

Radiant electric heat is transmitted to the rooms by means of cables installed in the ceilings and/or floors or may be installed as baseboard units. The heat is caused by resistance in the cable, which heats the cable and gives off heat to the rooms. Baseboard electric-heating units are easier to install than cables embedded in plaster ceilings or within concrete slabs.

CENTRAL FORCED-AIR SYSTEMS

Central forced-air systems heat or cool the air in a centrally located unit or units, then push the air with fans through a duct system to the various rooms. The furnace may be fueled with gas, oil, or electricity. Electric furnaces offer very clean heat, are simple to install (the venting system is usually not required because there are no toxic combustion gases that must be expelled to the outside), and usually require little maintenance after installation. However, electricity is so expensive in some regions of the country (especially the northern areas of extreme cold) that electric-furnace systems are not used. In the warmer southern areas, electric furnaces may be practical.

The ductwork is a network of round or rectangular ducts that route warm or cool air from the furnace or cooling unit to the supply outlets; *return registers* are employed to take more air back to the unit to be heated or cooled. Ducts are typically made of galvanized iron or of aluminum. Round aluminum ducts are very popular with residential builders. The ducts are thin and lightweight, their exact thickness depending on the size and shape of the duct. Metal ductwork often must be insulated.

Glass-fiber ductwork may also be used; fiber ductwork eliminates the need for insulation around ductwork that might otherwise be necessary. As with metal ductwork, fiber ducts are available in a variety of sizes, both round and rectangular. Fiber ducts may also be combined with metal ductwork, and fiber connections are sometimes used within metal ductwork runs and at the furnace.

Asbestos-cement ducts are often used where the ducts are installed in concrete slabs. Concrete corrodes some materials, but not asbestos cement; also, asbestos cement is resistant to the heat generated by concrete. Asbestos cement simplifies the installation in concrete slabs because it is heavy enough so that it does not "float," as the ligher ducts do, and is less subject to damage during installation. Asbestos cement is more expensive than metal or fiber ducts, however, so they are seldom used except in or under concrete slabs. Lengths of asbestos-cement ductwork are tied together by means of a rubber sleeve, which covers the joint, and two stainless steel straps, which secure the sleeve in place.

In residential construction, much or all of the metal ductwork will have to be insulated. Temperature differences between the duct and the surrounding air can cause condensation and loss of heating or cooling capacity. Ductwork is typically placed in house attics, which have significant air-temperature differences from the air in the ducts; thus ductwork in attics must usually be insulated. In some cases, the surrounding air temperature and the air temperature within the ductwork may be about the same—in some basements not exposed to outside air, for example. Sometimes, in such cases, insulation may be omitted; when in doubt, insulate.

The typical insulating material for residences is fiberglass with a built-on aluminum-foil vapor barrier.

Ducts often are run within walls and floor joist spaces. Ducts are the largest items that have to be framed around (plumbing is the second largest; electrical wiring presents the fewest problems in framing). Thus the heating and cooling system should be planned before house construction begins, so that framing conflicts with the ductwork and with plumbing can be avoided. Ductwork is usually installed before the furnace and cooling unit.

The actual selection and sizing of ducts should be done by an engineer or by the heating and cooling subcontractor. It is usually best to size the ducts so that the same ducts may be used for both heating and cooling, even though cooling equipment may not be installed until later. In residential construction, the subcontractor usually knows the technical and code requirements and may select, supply, and size the ducts. However, the actual selection of major equipment, such as the furnace and cooling unit, should be done by the owner, working with an engineer or architect (the owner in tract housing is understood to be the general contractor, unless the house is being custom-built).

Again, the ductwork is usually completed before the furnace is installed. There are two categories of ductwork: the ductwork that is within the wall or other framing members, of which the framing contractor must be aware to avoid framing conflicts, and the ductwork that is in the attic or other open spaces, where little installation preparation is necessary.

The routing of ductwork varies considerably with different houses. Generally speaking, small houses (basically rectangular in shape) may use a simple rectangular duct down its length with smaller branch outlets to the particular rooms or spaces. These branch outlets may be rectangular or round. The main rectangular duct will likely be the same size along its entire length.

For larger houses, the main supply duct may be too long to be of uniform size. Such long ducts are typically narrowed in depth or width at designed intervals; the purpose of narrowing the duct is to maintain a satisfactorily uniform supply of air all along the duct. The branch duct lines angle or curve off the main supply duct to the particular spaces.

The ductwork must often be suspended from joists or other framing with metal straps. The ducts are equipped with flanges at the ends and secured together with sheet-metal screws. Joints must usually be wrapped with a special heat-resistant tape to guard against air loss. Insulation is used more often than not, not only to prevent heat loss and gain from the ducts and to help prevent condensation but also to reduce noise, which can be an annoying problem with ducts.

The return-air ductwork is similar (almost a duplicate) to the supply ductwork. At the furnace, there is a return plenum. From the plenum, there is a main return line, located to receive branch lines from central hallways and rooms; these branch lines are equipped with grills at their ends in room walls and floors.

One of the most typical heating and cooling methods for residential construction is to select a furnace with a cooling coil in the supply plenum. The cooling condenser is then located outside the house, usually at ground level on a concrete pad.

This arrangement allows the owner to purchase the heating system first, then install the cooling coils and condenser at some later date, thereby lowering the initial cost of the house somewhat.

Before the furnace is installed, the electrical-service entrance and distribution wiring should be in place, be inspected by the appropriate authorities, and be either "hot" (hooked up) or ready for hookup to the utility line. Also, the furnace location should be prepared: a finished concrete basement floor or other space prepared according to authority requirements. It is not wise to bring the furnace to the job site and let it set around for days before installation. With the ducts, wiring, and furnace location completed, the installation of the furnace and controls is a quick job.

The furnace includes four main components: the fuel-injection or power-connection unit, the firebox or heat-exchanger unit, the blower compartment, and the control devices. When these components are arranged vertically within a metal housing, it is called a *highboy furnace*. Highboy furnaces are usually located in specially built spaces in basementless houses; the highboy conserves the amount of floor space devoted to the furnace. The highboy furnace receives air near the bottom of the unit and delivers air through the top. Consequently, the ducts often are placed in the attic.

The *lowboy furnace* places the blower compartment alongside the heat exchanger. The lowboy also delivers air through the top. Lowboys are typically used in basements.

The components of the furnace may also be arranged horizontally (the *horizontal furnace*) so that the unit may be used in narrow crawl spaces or similar locations. The air enters the unit at one end and is delivered from the other.

In the *counterflow furnace*, the air enters the unit at the top and is delivered from the bottom. These units are typically located in first-floor spaces in basementless houses or used when it is not desirable to locate the furnace in a crawl space.

These are typical furnace arrangements and locations. There are other methods. And there are additional combinations of these. However, most new houses today use either a lowboy or a highboy in the basement or a highboy on the first floor if there is no basement. (See Figure 13-1.)

Before deciding on a furnace, its component parts, features, and accessories should be analyzed in light of owner or market needs. Heat exchangers are usually made of steel, surfaced to resist the high temperatures; the heat exchanger is the heart of the system, and owners should be advised to seek high quality on this item.

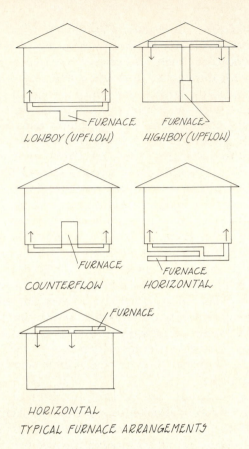

TYPICAL FURNACE ARRANGEMENTS

FIGURE 13-1 Typical furnace arrangements.

With gas or oil furnaces, the burner assembly is the most essential component part. These assemblies must all do the same job: provide a uniform flame surface and ignite quietly. Also, they should not develop hot spots. Some manufacturers offer electric-spark ignition devices that cut down on the energy lost to pilot lights. Safety features should include an automatic shut-off device, should the fuel flow be interrupted; and limit switches, which shut off the burners should the unit overheat. Some manufacturers offer two-stage burners; in mild weather, the burners operate at a lower level than they do in times of extreme cold.

The fan motors (blowers) are all very similar, but some are more electrically efficient than others. The shaded pole motor, for example, is traditional but is less efficient than the permanent split-capacitor type.

Two accessories, while not essential to the operation of furnaces, offer considerable additional comfort and convenience to future owners: humidifiers and air cleaners.

Humidifiers add moisture to the air, as needed; they are equipped with humidistats, which turn the humidifiers on when the moisture in the air falls below the desired level. Humidifiers relieve and help prevent dry throats, noses, and skin and help prevent sore throats. Humidifiers help promote indoor plant growth and reduce static electricity; they also help provide a better atmosphere for furniture and some interior materials. Humidifiers vary somewhat in design. They require a cold-water supply line.

Air cleaners reduce the amount of particles in the air. All furnaces are equipped with air filters, but these filters are not as effective as the air cleaners available. Air cleaners may be either electric or nonelectric. Electric air cleaners remove particles by using two oppositely charged plates, which trap air particles. Nonelectric air cleaners utilize some type of filter. Manufacturers claim that ordinary furnace air filters stop only a small portion of the total amount of particles in the air; air cleaners, they claim, reduce housework and cut down on allergic reactions. Air cleaners are typically installed in a return-air plenum or may be installed at the return-air flange connection at the furnace.

The thermostat measures temperature and turns the heat or cooling equipment on or off, as required to maintain the desired temperature, which is set by hand on the thermostat. The thermostat should be located in a hall, family room, or similar open, centrally located space. There should be no direct air currents, heating or cooling devices, or other equipment that would interfere with the thermostat's measuring the general temperature of the house. Install the thermostat on a wall within comfortable reach. The thermostat location should be determined before the wiring is done, so that the thermostat may be wired at that time; thermostats are wired to the furnace with low-voltage wire.

Air is moved from the blower compartment (with fans) through ducts to the various rooms and spaces to be heated or cooled. The air enters the rooms through *registers*. Registers are typically located near the exterior walls and preferably at windows because the exterior walls, and especially the windows, are the coldest part of the house in winter and the hottest part in summer, so air from the registers counters the heat loss or gain at these critical areas and helps keep the temperature uniform within a given room. Also, air coming through the registers circulates the air at the windows, and this helps prevent condensation.

The volume of air coming through a given register is adjusted by means of a damper at the register (typically a louvered section whose blades may be completely opened or completely closed). The air is directed similarly by means of louvers that can be turned to direct the air where it is de-

sired. If the registers are under the windows, the blades typically are adjusted to direct the air over the window and upward.

The registers are secured with sheet-metal screws at the duct ends. Registers are usually not installed until the finish floor is in place. When the finish floor is in place and with the furnace installed, the system should be ready for balancing and adjusting, which will be discussed later.

The actual installation of the furnace is a fairly simple but cumbersome job. The furnace is heavy and is generally moved to its resting place with some type of dolly and lots of muscle power. If the furnace is to be installed on a concrete slab, little preparation is needed; the furnace is set in place and leveled with shims. On wood floors, the furnace must have a noncombustible surface underneath, such as asbestos millboard or metal. Some furnaces have built-in fire protection on the bottom so that they may be installed directly over wood floors. When the furnace is secured and leveled, the supply and return plenums may be installed.

The plenums are a transition between the furnace and the duct systems; they are usually built by the heating subcontractor or by a sheet-metal fabricator for the heating subcontractor. The plenums must be sized such that one end fits the flange of the particular furnace and the other end fits the type duct used.

The heating subcontractor will make the final hookup to electrical and gas lines, adjust the blower and the limit switch, make the final thermostat connection, and probably make an initial test for balance—that is, check to see that the furnace is receiving proper return air, the rooms are receiving proper supply air, the various controls function properly. The heating subcontractor also installs the water-heater vent, although the installation of the water heater itself will be performed by the plumber. All this may take several visits by the heating subcontractor. The heating subcontractor's job often requires another visit after several weeks to recheck the equipment.

The cooling equipment often used for new homes nowadays consists of a set of cooling coils, which are placed next to the furnace bonnet, and a condenser, which is typically located outside the house (usually at ground level on a concrete pad). This equipment may be purchased and installed as an integral part of the total heating and cooling system, or it may be bought as a package and installed later, provided the ducts are designed for both heating and cooling (this usually means making the ducts somewhat larger).

If cooling is necessary, central cooling is

recommended over individual room air conditioners because the central system is more efficient and provides more uniform cooling. However, room air conditioners can be practical where, for some reason, it is not desired to cool the entire house. Also, some areas of the country have relatively short periods of truly hot weather, and the initial cost of a central cooling system may not be worth the expense to owners.

In many areas of the country, the installation of an attic fan is a practical consideration. Attic fans may be arranged to draw air by way of house windows up through a vent (usually through a hallway ceiling); this cools the house when the outside temperature is lower than the inside (typically at night) and prevents the air in the attic from becoming excessively hot. If it is not desirable to draw air through the windows (if, for example, dust or other airborne particles are a problem to residents), the ceiling vent may be omitted and the soffit vents increased so that the hot attic air will not be allowed to build up. It should also be noted that in areas where cooling equipment is not needed, the furnace blowers may be operated without the heat on, thereby providing ventilation and fresh air.

Other ventilation and exhaust devices include bathroom exhaust fans and range exhaust fans. Both these exhaust devices are frequently required by various authorities. The bathroom exhaust fan is usually installed in the ceiling and draws the warm, moisture-laden air up through a metal vent that exits at the roof. The range fan works similarly; the fan is usually within the hood that is placed over the range to gather smoke.

Plumbing

14

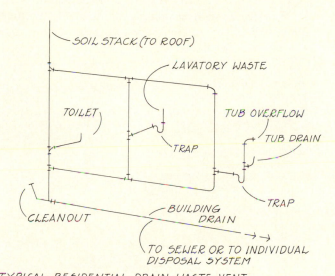

SOIL STACK (TO ROOF)
LAVATORY WASTE
TOILET
TUB OVERFLOW
TRAP
TUB DRAIN
TRAP
CLEANOUT
BUILDING DRAIN
TO SEWER OR TO INDIVIDUAL DISPOSAL SYSTEM

TYPICAL RESIDENTIAL DRAIN-WASTE-VENT SCHEMATIC FOR 3 FIXTURES

FIGURE 14-1 Drain-waste-vent system.

BASIC COMPONENTS

The most basic components of a house plumbing system include the drain-waste-vent system and the water supply system. The drain-waste-vent system is a network of lines that collect human and other wastes and transfer them to the building drain, which in turn transports the wastes to some point outside the building (this distance varies, but 5 feet or so is typical). From that point, another line takes the waste to the public sewer or to an individual disposal system (septic tank and disposal lines; individual disposal systems will be discussed separately). (See Figures 14-1 and 14-2.)

The water supply is usually provided by the municipality, but individual wells are sometimes used.

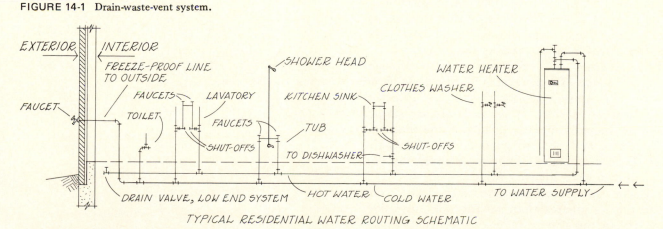

EXTERIOR   INTERIOR
FREEZE-PROOF LINE TO OUTSIDE
FAUCETS   LAVATORY   SHOWER HEAD   WATER HEATER
FAUCET   TOILET   KITCHEN SINK   CLOTHES WASHER
FAUCETS
SHUT-OFFS   TUB   SHUT-OFFS
TO DISHWASHER
DRAIN VALVE, LOW END SYSTEM   HOT WATER   COLD WATER   TO WATER SUPPLY

TYPICAL RESIDENTIAL WATER ROUTING SCHEMATIC

FIGURE 14-2 Typical residential water-routing system.

In typical house-building or subdivision situations, the landscape shape is manipulated such that storm runoff water runs away from the house to the street, where it is collected by gutters and transported to catch basins and thus into the public storm-drain lines. Roof runoff is usually collected by gutters, transported down by downspouts to splash blocks (usually concrete pads that prevent erosion), then onto the landscape and thence to the gutters, and so on. In the past, storm water (runoff) was routed into the sewer lines. But that practice resulted in overworking public water-purification plants, so now separate storm lines and sewer lines are the rule, although local requirements may vary.

Where the landscape cannot be manipulated to handle storm runoff water adequately, underground drainage lines may have to be used. Builders avoid these lines whenever possible because of the extra expense; further, such lines often require maintenance by homeowners at a later date (cleaning out tree roots and other stoppages, etc.). These lines typically use some type of catch basin, located at a low point, where the runoff enters and is then transported to the public storm line.

The purpose of the interior house drain lines is to dispose of wastes. Lines that dispose of human waste are called *soil pipes* or *soil lines*. These are, of course, polite terms used in place of less pleasant terms. Lines that dispose of sink water, bath water, and the like are called simply *waste lines*. All these lines are more or less horizontal and are routed to vertical lines called *soil stacks*, which transport the wastes down to the main building drain, then out to the sewer. In small houses, this system is quite simple, similar to that shown in Figure 14-1. In larger houses, the system is more complex, with various branch soil and waste lines and perhaps several soil stacks. The size of the lines depends on the number of fixtures—toilets, bathtubs, showers, sinks, and so forth—and should be designed by a competent professional. (See Appendix D-9.)

Sewer gases are unpleasant at best and dangerous at worst. To prevent soil-line gases from entering the house, a very simple but effective method, called a *trap*, is used. Traps are simply horseshoe-shaped connections in the line that stay full of water and prevent gases from entering the house. Look under any sink in your house and you will see a trap. (See Figure 14-3.)

If the atmospheric pressure on the fixture side of the trap is greater than on the discharge side, the water in the trap can be pushed through the pipe and out the system; if this happens, sewer gases may enter the house. To prevent such a discrepancy in pressure, *vents* are used. Vents are lines

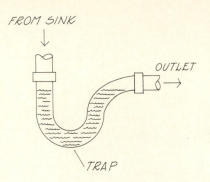

FIGURE 14-3   A trap.

installed above the water traps to equalize atmospheric pressure. (See Figure 14-4.)

Water in the trap can also be lost by evaporation if the sink or other fixture is not used for a long period of time. Also, hair and other debris in a trap sometimes act as a wick and draw the water from the trap; the lines should be kept clean.

The water supply system includes all the lines, valves, and fixtures. There are two sets of lines, one

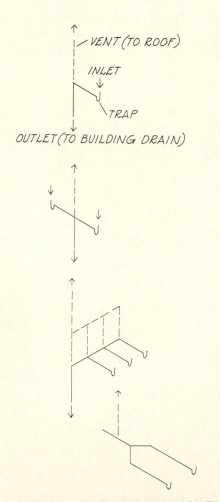

TYPICAL VENT ARRANGEMENTS

FIGURE 14-4   Waste-line venting methods.

for cold water and one for hot water. The cold-water supply line typically branches off to the hot-water tank, as shown in Figure 14-2; in larger houses, however, pumps may have to be added to push adequate amounts of hot water to the various fixtures.

All supply and waste lines require supports, typically metal straps at regular intervals. Plumbing must be coordinated with the framing contractor—and with the concrete contractor if a concrete slab is used (rough-in plumbing must be done before the slab is poured, provisions made to bring the plumbing through foundation walls, and so on). Plumbing is second only to heating and cooling ductwork in the amount of space required; most plumbing can be brought through the framing without undue cutting of members, but when excessive cutting is required, the framing must be reinforced properly. When plumbing lines cannot be incorporated in the normal thickness of the house walls, *chases* are used. Chases are simply thicker wall cavities, usually made by using two-by-six studs instead of two-by-fours or by staggering the two-by-four studs on each side of the wall's longitudinal center line so that the overall depth of the wall cavity becomes adequate to incorporate the pipes. The house floor plan shows the location of all fixtures, and from this plan the plumber can do any sketches or drawings required to ensure that the lines are installed at the proper location within the framing.

Plumbing fixtures vary greatly in configuration, material, and quality. The quality of the pipes, valves, and other functional considerations should take precedence over "flashy" fixtures, but the reverse happens too often, with accompanying higher maintenance costs to the eventual owner. A visit to a plumbing supply house or studying a plumbing supply catalog will provide a better knowledge of plumbing fixtures than any text could provide.

PIPE MATERIALS

Plastic

Plastic pipe is being used more and more. Plastic pipe is light, easy to cut, and simpler to install than metal pipe. Plus, it generally takes less skill to install plastic pipe than it does to install metal pipe.

The basic types of plastic pipe at this writing are polyvinyl chloride (PVC), chlorinated polyvinyl chloride (CPVC), acrylonitrile-butadiene-styrene (ABS), and polybutylene (PB), which is the latest type. Besides the advantages mentioned in the first paragraph, these plastic pipes are smoother than metal pipes and thus offer less resistance, so smaller pipes may sometimes be used for a job that would require larger metal pipes. Further, plastic pipes have better insulating qualities than metal pipes.

All the pipes mentioned here can use solvent adhesives for connections, with the exception of PB; PB cannot be solvent-welded and is instead joined with compression fittings.

Cast Iron

Cast-iron pipes are heavy and more difficult to handle and install than plastic pipe, but cast-iron pipe is very strong. It is often used for building sewer lines, sewer lines to the street, public sewers, waste-stack lines, and the like.

Cast-iron pipe is typically bought in five foot lengths, but longer lengths are available. Standard cast-iron pipe has a "bell" shape at one end and a "spigot" at the other. The smaller spigot end is placed inside the larger bell end, then oakum, a treated ropelike material, is packed in the joint, almost to the bell end. Molten lead is then poured over the oakum, and when the lead has cooled, the lead is tamped into the joint with a special tool.

Cast-iron pipes may also be jointed with neoprene gaskets. The gasket is placed in the bell end, then mopped with a special lubricant. Next, the small end of the connecting pipe is forced through the gasket. This second method is perhaps not as high-quality an installation as molten lead, but it has been proved to provide a satisfactory joint. There are other cast-iron pipe shapes and methods of connecting joints.

Vitrified Clay Pipe

Vitrified clay pipe is often used for sewers (house and public) and drain lines. It is a fired-clay product. The shape is similar to that described for cast iron. Joints may be sealed by first packing in oakum, then filling the remaining space with a mixture composed for portland cement, sand, and water. Vitrified clay pipe should be laid on a firm foundation—typically concrete, stone, or a similar base; soil is suitable only if it is stable, because sagging in the pipes can cause the joints to crack.

Steel Pipe

Steel pipe is strong, can be threaded for very tight joints, and is, of course, very expensive. Steel pipe may be used for water supply lines and gas lines. The only joint preparation necessary is to coat the contact areas with a pipe-joint compound before screwing the pipes together.

Copper Tubing

Copper tubing is expensive, but it is a high-quality material that is still much used for a variety of house plumbing lines, especially drain-waste-vent systems and hot- and cold-water supply lines.

Copper tubing is available in three wall thicknesses: type K, L, or M. K is a relatively thick wall, L is medium, and M is thin. Typical lengths are 10 and 20 feet. K, L, and M are available in hard-tempered straight lengths; hard-tempered tubing may be joined by soldering, by brazing, or by using capillary fittings. Only K and L tubing are available in soft or annealed-temper, straight-length, or coiled tubing. Soft-tempered tubing may be joined by soldering, or brazing or by using flare-type compression fittings.

INDIVIDUAL SEWAGE DISPOSAL SYSTEM

An individual sewage disposal system works as follows: A sewer is extended from the house to a septic tank. From the septic tank, an effluent sewer line is extended to a distribution box, from which the effluent sewage travels to a subsoil disposal field. The foregoing is the most typical individual sewage disposal system, although not the only one. It should be noted that local authorities usually require the owner to have considerable land on which to build a disposal system—typically 5 acres or more. Check the applicable authority requirements for land area required, soil type, disposal system design, and possible other requirements before attempting to build a disposal system. (See Figure 14-5.)

The House Sewer

The house sewer carries raw sewage from the house to the septic tank. Cast-iron pipe with leaded joints is the typical material, sometimes required near a potable water supply or where tree roots or similar growth might penetrate other pipe materials. Other sewer-pipe materials sometimes used are glazed clay tile and cement pipe. Regardless of the type of pipe material used, cast-iron pipe is typically used for the first few feet out from the house foundation—usually 5 or 6 feet or so. The size of the pipe and the slope at which it is laid depend on the amount of sewage to be transported.

Grease traps may be installed with a separate line to separate grease from the other household waste, thus relieving the house sewer of the burden of providing an easy means of cleanout. Grease traps are seldom used for residences, however. (See Figures 14-5 and 14-6.)

The *septic tank* is a simple boxlike structure, usually built of concrete or steel but sometimes of other materials. It may be built on the job or be a manufactured unit. Whatever the material, the tank must be sized to match the house sewage volume.

The action that takes place in a septic tank requires no chemicals or additives. The tank is a kind of natural processor. Raw sewage flows into the tank with the aid of gravity, and the heaviest matter settles to the bottom. Anaerobic bacterial action causes three layers of sewage to form in the tank. The bottom layer is a heavy sludge; the middle layer is a liquid; and the top layer is a scrum that seals off air and aids the bacterial action.

As the house sewer line brings more raw sewage to be processed, the effluent sewage is pushed up and then spills over into a siphon tank or directly

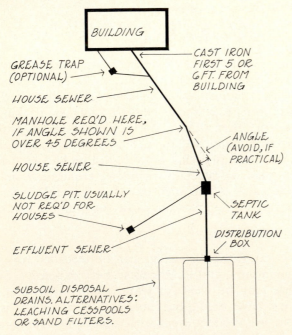

FIGURE 14-5   An individual disposal system.

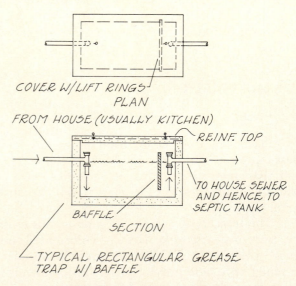

FIGURE 14-6   Typical grease trap.

into the effluent sewer line, which extends to the disposal system. If the septic tank holds 1,000 gallons or more, there usually will be a siphon tank (a siphon tank may be a local requirement). A siphon tank is always used where the disposal field uses sand filters. The siphon tank is just a separate compartment in the septic tank that collects over-flow from the septic tank and later discharges it to the effluent sewer line through the use of an automatic siphon. The function of the siphon tank is to regulate the flow of effluent to the disposal system, thus preventing overloads which may saturate the disposal system with accompanying horrific smells. (See Figure 14-7.)

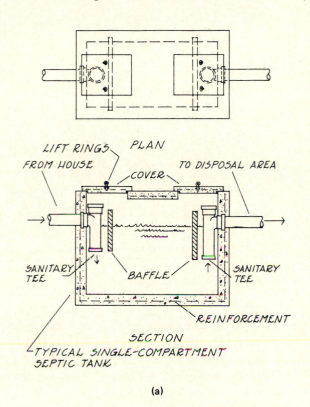

(a)

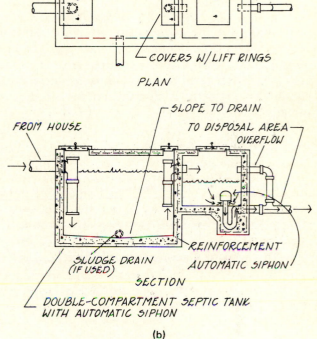

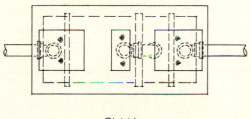

(b)

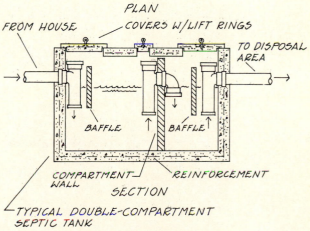

(c)

FIGURE 14-7 Typical septic tanks.

The *sludge pit*, if used, is connected to the septic tank with a separate line. The purpose of the sludge pit is to drain off the sludge at the bottom of the septic tank, thus reducing the number of cleanouts normally required of the septic tank. The sludge pit itself, however, requires periodic cleanout and is usually regarded as optional for residential use. (See Figure 14-8.)

The Effluent Sewer

The effluent sewer line is similar to the house sewer line, but the effluent line does not carry raw sewage. The effluent line is a closed pipe (without open joints) that transports effluent from the septic tank or siphon tank to a distribution box, through which the effluent is distributed to the disposal system.

The Distribution Box

The distribution box collects effluent from the effluent sewer line. The box is equipped with a number of outlets, through which the effluent runs to its final destination, the disposal system. The number of outlets from the box depends on the size of the disposal system. (See Figure 14-9.)

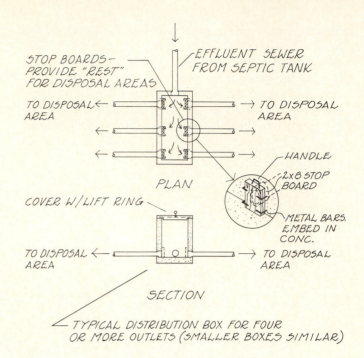

FIGURE 14-9 Typical distribution box.

The Disposal System

The disposal system is called a system because there are several typical methods of disposal: leeching cesspools, subsoil disposal beds, and sand filters. Subsoil disposal beds are the most typical method used in residential construction.

Subsoil disposal beds. The subsoil disposal bed is composed of a network of drainpipes; spread your hand and your fingers take on the shape of a typical disposal-bed pattern, though there are other patterns. The drainpipe is laid with open joints (the pipes do not quite touch at the ends), and the effluent is thus slowly absorbed into the soil. The exact layout used depends on topography, available spacing and other factors. One aspect of this system should be obvious at this point: The soil under the drainpipes must absorb well. Also, the groundwater level must not be too high; subsoil beds require that the water level be more than 2 feet below grade. Subsoil disposal beds may be used on level slopes by sloping the drain lines somewhat; on fairly steep slopes, the lines would be laid at a slope less than the ground slope. (See Figure 14-10.)

Leeching cesspools. Leeching cesspools are typically round masonry structures; they look something like a manhole and shaft. The joints are

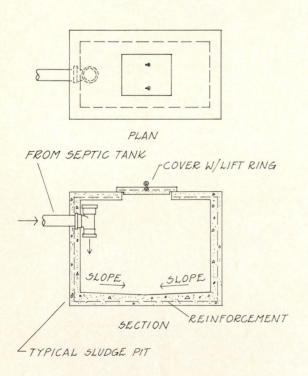

FIGURE 14-8 Typical sludge pit.

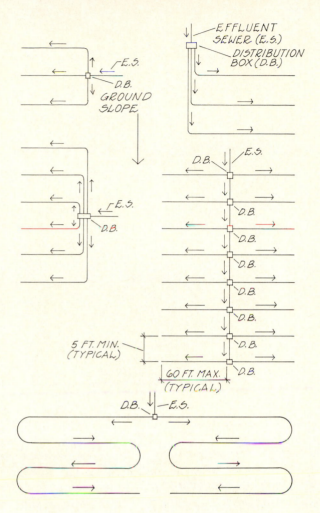

TYPICAL SUBSOIL DISPOSAL LAYOUTS

FIGURE 14-10 Subsoil disposal layouts.

systematically laid open so that the effluent can pass through into the surrounding soil; the bottom is some kind of absorptive surface. Typically, a 4-inch layer of graded stone is installed around the cesspool to aid absorption of the effluent. The cesspool, remember, is a disposal system; it receives effluent from the distribution box.

Leeching cesspools take up less land area than the other systems, and slope is of little consequence. But the soil must have excellent absorptive qualities, the ground-water level must be 8 feet or more below grade, and the cesspool must be located well away from the house and water supplies. The number of cesspools used depends on the amount of waste to be processed. The arrangement is chosen to provide continuous operation of the system during maintenance and cleanout. (See Figure 14-11.)

Sand filters. Sand filters are expensive, troublesome to build, and seldom used unless the other two systems are impracticable, as in the event that the soil is impervious.

To install sand filters, the impervious soil where the disposal field will be must be excavated and replaced with sand (other suitable filter material could be used, but sand is the most usual). (See Figure 14-12.)

FIGURE 14-11 Leaching cesspool and typical arrangements.

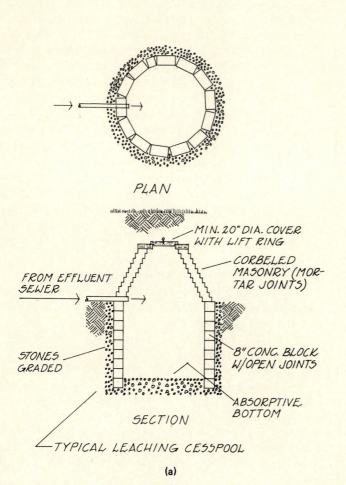

PLAN

MIN. 20" DIA. COVER WITH LIFT RING

CORBELED MASONRY (MORTAR JOINTS)

FROM EFFLUENT SEWER

STONES GRADED

8" CONC. BLOCK W/OPEN JOINTS

ABSORPTIVE BOTTOM

SECTION

TYPICAL LEACHING CESSPOOL

(a)

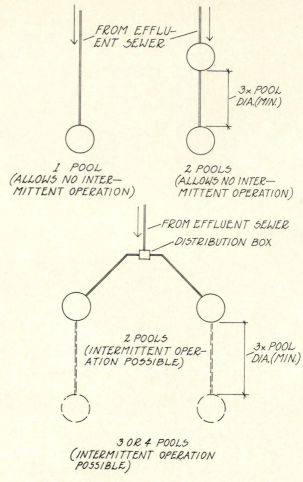

FROM EFFLUENT SEWER

3× POOL DIA.(MIN.)

1 POOL (ALLOWS NO INTERMITTENT OPERATION)

2 POOLS (ALLOWS NO INTERMITTENT OPERATION)

FROM EFFLUENT SEWER

DISTRIBUTION BOX

2 POOLS (INTERMITTENT OPERATION POSSIBLE)

3× POOL DIA.(MIN.)

3 OR 4 POOLS (INTERMITTENT OPERATION POSSIBLE)

TYPICAL LEACHING CESSPOOL ARRANGEMENTS

(b)

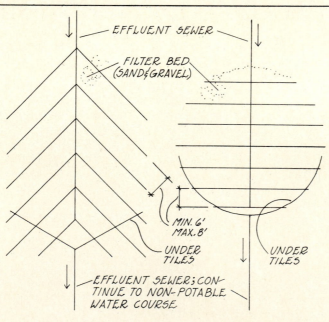

EFFLUENT SEWER

FILTER BED (SAND & GRAVEL)

MIN. 6' MAX. 8'

UNDER TILES

UNDER TILES

EFFLUENT SEWER; CONTINUE TO NON-POTABLE WATER COURSE

TYPICAL SAND FILTER LAYOUTS

FIGURE 14-12 Typical sand filter layouts.

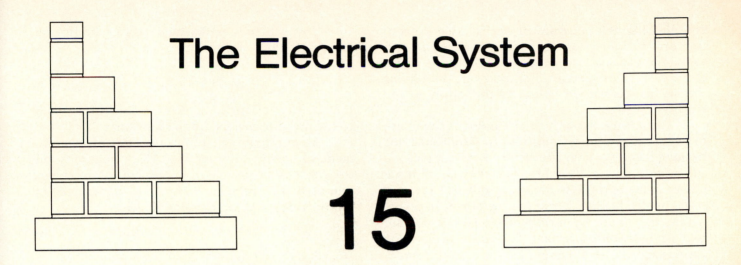

The Electrical System

15

All houses must be provided with electrical power for lighting, appliances, and other equipment. The power is typically alternating current (AC), provided by the municipality by way of a distribution transformer somewhere at or near the street; a line may be run overhead or underground (many localities are moving to total underground installation of electrical service lines) to the house service entrance. From there, the line goes to the house service panel, which further distributes the power throughout the house, as required for the various equipment, by way of the various circuits. (See Figure 15-1.)

BASIC ELECTRICAL TERMS

The electrical system is typically designed and installed by specialists. The general contractor or owner is most concerned with seeing that the drawings show the required switches, receptacles, lighting fixtures, and so forth and, later, that the installation be done according to the plans and specifications and that it meet the owner's particular requirements and the requirements of applicable authorities. To meet even these basic objectives, it is necessary for most contractors—and certainly the general contractor, who usually is responsible for

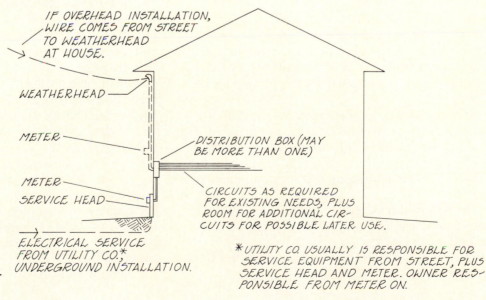

IF OVERHEAD INSTALLATION, WIRE COMES FROM STREET TO WEATHERHEAD AT HOUSE.

WEATHERHEAD

METER

DISTRIBUTION BOX (MAY BE MORE THAN ONE)

METER
SERVICE HEAD

CIRCUITS AS REQUIRED FOR EXISTING NEEDS, PLUS ROOM FOR ADDITIONAL CIRCUITS FOR POSSIBLE LATER USE.

ELECTRICAL SERVICE FROM UTILITY CO.* UNDERGROUND INSTALLATION.

* UTILITY CO. USUALLY IS RESPONSIBLE FOR SERVICE EQUIPMENT FROM STREET, PLUS SERVICE HEAD AND METER. OWNER RESPONSIBLE FROM METER ON.

FIGURE 15-1 Providing electrical power.

coordinating the activities of the various subcontractors—to know basic electrical terminology and to know, at least in general terms, how the house electrical system operates. The following definitions should be understood before further study of the electrical system:

The *service entrance* is the point at the house to which the utility company is responsible for running its power line. The service entrance receives the utility power line. From there, one or more power lines are run into the house and into a *service panel*, which further distributes the power to points within the house by means of *circuits*, which are two or more wires arranged so as to service the various appliances or equipment for a given area.

Receptacles are the outlets into which appliances, lamps, and other equipment are plugged; receptacles also are commonly called *convenience outlets*.

The term *fixtures* usually refers to the non-movable house lighting devices, which may be located on the walls or ceilings or both.

Switches are controls that break or connect circuit power lines, thus turning on or turning off the various equipment; light switches are an example.

Circuit breakers work similarly to switches, in that they break the circuit power lines. But circuit breakers are safety devices that automatically stop power through a circuit, when the circuit is overloaded and becomes too hot. Fires could start more easily if circuit breakers were not used.

Any wire or substance that carries electricity is called a *conductor*. Conductors are run in *conduits*, tubes or channels, often made of metal, that protect the house wiring.

A special wire, called a *ground*, is connected to the earth to protect users from electrical shock. Most appliances, for example, are now equipped with *ground wires* that protect users from shock. The house itself is grounded for protection against lightning.

Electric power is measured in a number of ways. The *ampere* (also called *amp*, abbreviated *A*) is a unit, or measure, of the flow of electrons through a circuit. *Amperage* refers to the number of amps a circuit carries or certain equipment requires.

The *volt* (abbreviated *V*) is the unit of pressure required to drive the electrons. *Voltage* refers to the number of volts.

The *ohm* is the unit of resistance to the flow of electrons.

You will now understand, if you did not before, a very basic formula in electricity:

$$\text{Amps} = \frac{\text{Volts}}{\text{Ohms}}$$

that is, the current in an electrical circuit is equal to the pressure divided by the resistance.

A *watt* (abbreviated *W*) is a unit of measurement of energy consumption. Your electricity bill is based on the number of kilowatt hours of energy you used. The number of watts used is obtained by multiplying Amps times volts; that is:

$$\text{Watts} = \text{Amps} \times \text{volts}$$

A *kilowatt* (abbreviated *kW*) is 1,000 watts. This is simply a convenient way of expressing a large number of watts; 28,000 watts, for example, equals 28 kW.

THE SERVICE ENTRANCE

From the distribution transformer, a number of wires are extended to the house service entrance. The utility company must be consulted concerning the location of the house service entrance. Often, the utility company pays for the line to the service entrance, so it usually wants the service entrance at the shortest distance from their service lines. If the utility company does not bring the line to the closest reasonable point for a service entrance, it will probably want to bring it to the point of greatest power usage in the house; this point is typically the kitchen. The greatest usage point is where the larger, more expensive wires will be used, and the utility company wants to use as little of this wire as feasible. However, there is some latitude in exactly where the service entrance is located, and it should be located where it will be least offensive in appearance. If the service entrance is an overhead installation, the wiring usually comes in through a weatherproof metal conduit (or *weatherhead*) installed at the eaves. From there, the wiring is run through a rigid metal conduit to the meter, typically located at a convenient height on the wall for reading. Underground installations, however, are being used more than overhead installations nowadays, even though they are more expensive, because the appearance is better. As with overhead installations, the underground wiring is run through a conduit to the meter, then to the house service panel.

The wiring—or *service conductors*—that is extended to the service entrance varies with the needs of the particular house. Two wires usually indicate a single-phase, a 120-volt system; one wire is neutral (the ground), and the other is hot. Three wires to the service entrance may indicate a straight three-phase system or a single-phase, three-wire system. Four wires typically indicate a three-phase system with one neutral wire. Although the house may be wired for the exact electrical needs shown on the plans, it is not uncommon to carry more power to the house than is initially needed, in anticipation of added future needs.

THE SERVICE PANEL

From the meter, the wiring is run to the house service panel. This service panel, or *distribution box*, is literally a metal box into which the wiring is run from the outside source. From this box, the wiring is fed through individual circuits that serve the house needs—kitchen ranges, lighting fixtures, and so forth. Also in the box are circuit breakers and/or fuses. The fuses protect the lines by burning out and opening the circuit before any serious damage may occur to the lines by overloading. Circuit breakers do the same thing by "tripping"—that is, by automatically interrupting (opening) the current when they sense an overload. After a breaker trips, it may be flipped back on without replacing it; burnt fuses must be replaced. If the breaker immediately trips again, the overload is still present. Overloads are typically caused by hooking up too much equipment on a given circuit, more equipment than the circuit was designed to handle. Also in the service panel is a *main disconnect switch*, which is used to turn off all the house circuits.

The size of the service panel depends on the number of circuits and amount of power the house needs. Panels are sized, or *rated*, in amperage and *poles*. Each 120-volt circuit requires one pole; 120-volt circuits are typical for lighting, some appliances, and convenience outlets. Ranges, air conditioners, electric hot-water heaters, clothes dryers, and similar heavy-duty appliances usually require 240 volts; thus, two poles are needed. The panel should be large enough to accommodate several more poles, even if the wiring is not brought to the poles at the time the initial wiring is done.

The panel box should be located where it can be conveniently serviced and where it will be safe from weather damage and away from water that might get to the box or on the floor around the box. Utility rooms, basements, garages, and similar protected areas are good places for the service panel.

The panel box may be secured on the wall surface or may be recessed. Assuming there are no code requirements in particular localities, it does not matter which way is chosen. In locations exposed to view, the recessed installation usually presents a more finished appearance.

CIRCUITS

From the service panel, circuits are run out to serve the various electrical devices. A general-purpose circuit, for example, may provide 20 amps of energy; 20 amps $\times$ 120 volts = 2,400 watts of energy. Thus, this line could serve appropriate appliances up to 2,400 watts; generally, however, not more than six outlets are connected to a given circuit. Some circuits may serve only one electrical device; for example, a water heater, range, air conditioner, or clothes dryer may have a single circuit.

RECEPTACLES

Receptacles (convenience outlets) are arranged in the various rooms and spaces to accept the plugs of lamps and other small electrical appliances. There is some variation in the location of receptacles, but, generally speaking, there should never be less than one receptacle per wall, usually located near the floor; also, the wall outlets should not be greater than about 10 feet apart, although applicable authority regulations may vary somewhat. In the kitchen, outlets are placed both along the wall at the floor and above the counter top of base cabinets. Beyond the very basic considerations of receptacles just given, the house and room designs often make additional receptacles desirable, if not absolutely necessary.

Receptacles are typically the *duplex* type; that is, each outlet has a place for two plugs. There are, however, variations in the design of receptacles for particular usages. Electrical devices such as clothes dryers and ranges, which require 240 volts, use special receptacles built such that 120-volt appliances cannot be plugged into them—this is a safety precaution. Outside receptacles are built to be waterproof.

LIGHTING FIXTURES

Lighting fixtures, like receptacles, must achieve certain functional requirements. Beyond that, there is much latitude for individual needs. A typical minimum residential lighting-fixture requirement, for example, is that each room have at least one surface-mounted fixture, usually installed in the center of the ceiling. Basics such as this ensure that each room will have enough light for safety and at least the minimum light to do ordinary housecleaning and similar maintenance. Incandescent lighting (light bulbs) are typically used.

Custom houses may feature indirect lighting by way of recessed fixtures. Table lamps may be used to supplement the general lighting or may be used alone, where a softer light is desired. There has been more use of fluorescent lighting tubes in the last 10 years or so, perhaps because the tubes have been improved and do not give off the dreary light they did at one time. Also, fluorescent tubes are more efficient and thus more economical to use; kitchens are favorite places to use fluorescent tubes. Lighting is an art, worthy of separate, concentrated study, and some schools are beginning to offer courses in lighting theory and design.

SWITCHES

The most typical household switch is the wall switch, from which the lighting fixtures may be turned on and off. Mercury switches are usually considered the most desirable, because they are noiseless; mercury, in a sealed tube, makes or breaks the circuit when the switch is activated. Toggle switches give the familiar click when used; these switches, though noisy, are usually cheaper than mercury switches.

Besides switching lighting fixtures on and off, switches may be used to control receptacles; such switches typically control lamps plugged into the receptacles, although appliances could be controlled the same way.

Where a room has two entrances, three-way switches are used at each door or opening. The three-way switch allows the lights to be turned on or off from either switch. Other types of switches and combinations allow one or more lighting fixtures or receptacles to be controlled from several locations, for larger than usual houses or unique circumstances. Dimmer switches have become popular over the years; typically, the light is turned on and off by pushing a round button, and the light intensity is controlled by twisting the button. Still another switch type is the delayed-action switch; you turn the switch off, but the light remains on for a minute or so before it goes off. Other switch types, including remote-control switches, are available.

CONDUCTORS

A number of conductors are available, each with a particular use or range of uses that is most appropriate. Further, codes may require the use of a particular conductor or conduit (conduits will be discussed later). Conductors may be either single wires, called *single conductors*, or several wires attached together, called *cables*. The following is a discussion of wire and cable types often used in residential wiring.

Single-Conductor Electrical Wires

Types RH, RU, RW, and RHW are rubber-insulated conductors. They are much less used nowadays than the improved plastic-insulated wires.

Types TW, THW, THWN, THHN have thermoplastic insulation around them. The *W* in the wire type designation means the wire may be used in some damp or wet areas. The *H* means the wire can stand high-heat locations. *N* refers to a special type of insulation (nylon coated), which is about ⅓ less thick than the other insulations; this wire is good for rewiring existing work, where space may be more critical than in new construction.

Type USE-RHW and *type XHHW* are resistant to oil and water. The first type may be used for underground service entrances or similar use but should not be buried in concrete or aggregate (the chemical action of concrete curing can damage the wire); USE-RHW also may be used for overhead service entrances.

Cables

Type NM is a nonmetallic sheathed cable. This cable has several conductors, each insulated, then covered with a paper wrap and finally an outer plastic sheathing. *Type UF* cable is similar, but the outer plastic sheathing is made of thermoplastic, making underground use possible. Nonmetallic sheathed cable is easier and faster to install than wiring through metal conduits, and most building codes now accept nonmetallic sheathing.

Type NMC and *UF* cable are similar to NM cable, but NMC cable is especially resistant to corrosion, and UF is suitable for underground use, having a special outer sheathing of thermoplastic. Neither of these cables should be buried in concrete or aggregate.

Type SE-U cable is used for overhead service entrances and for individual circuits to heavy-duty appliances such as clothes dryers, ranges, and water heaters. If codes permit, the cable may be used to connect the meter or entrance box to distribution panels.

Where electrical cable is strung through holes in the studs, metal plates should be used as a safety precaution, to guard against nail or screw puncture when interior-finish materials are installed later. (See Figure 15-2.)

CONDUITS

Conduits are special metal tubes or pipes through which electrical wiring is run. The metal tubes make for a very high quality, safe electrical wiring job. But they are expensive and more difficult to install than nonmetallic sheathed cable. In general, nonmetallic sheathed cable is replacing wiring brought through metal conduits. Some codes, however, still require metal conduit, and even when nonmetallic sheathed cable is used, there still may be particular instances where the additional safety of metal conduit is desirable.

There are several types of conduit, including rigid conduit, thin-wall conduit, and flexible conduit (also called *Greenfield conduit*). Rigid conduit, as the name implies, is stiff; special tools are available for bending conduit to follow the particular route of the circuits, turn into outlet boxes, and so on. Thin-wall conduit is less stiff and therefore easier to bend. Rigid conduit generally comes with threaded ends; thin-wall conduit ends are not threaded, so connections are made with clamps or crimp-on connectors. Flexible conduit is usually used only for short turns, as to connect heavy-duty appliances to the main conduit run, to connect an outlet to the main conduit run, and the like.

OUTLET BOXES

Regardless of whether nonmetallic sheathed cable or conduit is used, outlet boxes are installed first. The outlet boxes are metal and are equipped with knockout holes for the placement of wire. The boxes may be equipped with a grounding screw for the ground wire; if not, the ground wire may be attached to the box edge with the appropriate grounding clip (this is for cable). If conduit is used, no grounding wire is used because the conduit may be used as a ground. (See Figures 15-3 and 15-4.)

The foregoing is very basic information. The electrical system is much too complex and poten-

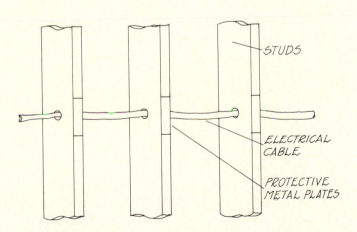

FIGURE 15-2 Electrical cable, strung through studs, protected by metal plates.

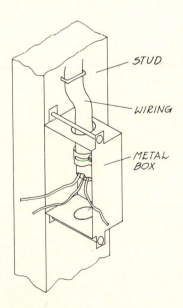

FIGURE 15-3 Wall outlet box.

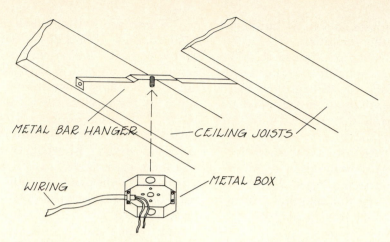

FIGURE 15-4 Ceiling outlet box.

tially dangerous to be designed or totally installed by an amateur, even if the amateur is a general contractor. It is recommended that the electrical designer be consulted early in the design stages of the house or houses; working closely with the electrical designer early can prevent costly reworking later. Also, the electrical designer is qualified to handle any negotiations with the utility company that may be necessary and to deal effectively with electrical inspectors.

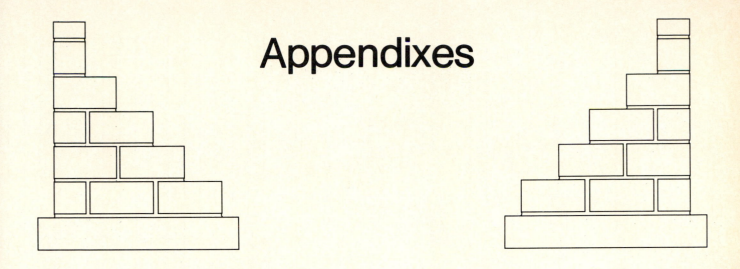

Appendixes

APPENDIX A

FHA FORM 2005: DESCRIPTION OF MATERIALS[1]

FHA Form 2005
VA Form 26-1852
Form FmHA 424-2
Rev. 4/77

U. S. DEPARTMENT OF HOUSING AND URBAN DEVELOPMENT
FEDERAL HOUSING ADMINISTRATION

For accurate register of carbon copies, form
may be separated along above fold. Staple
completed sheets together in original order.

Form Approved
OMB No. 63–R0055

DESCRIPTION OF MATERIALS

No. _____ (To be inserted by FHA, VA or FmHA)

☐ Proposed Construction   PLAN #1001, COUNTRY PLACE SUBDIVISION

☐ Under Construction

Property address SEE ATTACHED PLANS City COLE COUNTY State IOWA

Mortgagor or Sponsor _____
(Name)

Contractor or Builder QUALITY HOMES, INC. _____ SIOUX CITY, IOWA
(Name)   (Address)
                                                          (Address)

INSTRUCTIONS

1. For additional information on how this form is to be submitted, number
of copies, etc., see the instructions applicable to the FHA Application for
Mortgage Insurance, VA Request for Determination of Reasonable Value, or
FmHA Property Information and Appraisal Report, as the case may be.
2. Describe all materials and equipment to be used, whether or not shown
on the drawings, by marking an X in each appropriate check-box and entering
the information called for in each space. If space is inadequate, enter
"See misc.", and describe under item 27 or on an attached sheet. THE USE
OF PAINT CONTAINING MORE THAN THE PERCENTAGE OF LEAD
BY WEIGHT PERMITTED BY LAW IS PROHIBITED.
3. Work not specifically described or shown will not be considered unless

required, then the minimum acceptable will be assumed. Work exceeding
minimum requirements cannot be considered unless specifically described.
4. Include no alternates, "or equal" phrases, or contradictory items.
(Consideration of a request for acceptance of substitute materials or equipment is not thereby precluded.)
5. Include signatures required at the end of this form.
6. The construction shall be completed in compliance with the related
drawings and specifications, as amended during processing. The specifications include this Description of Materials and the applicable Minimum
Property Standards.

1. EXCAVATION:

Bearing soil, type FIRM CLAY

2. FOUNDATIONS:

Footings: concrete mix 1:3:5 ; strength psi 2,500   Reinforcing _____

Foundation wall: material 8" CONCRETE BLOCKS   Reinforcing _____

Interior foundation wall: material _____ Party foundation wall _____

Columns: material and sizes _____ Piers: material and reinforcing 8" CONC. FILLED POSTHOLE

Girders: material and sizes _____ Sills: material _____

Basement entrance areaway _____ Window areaways _____

Waterproofing _____ Footing drains _____

Termite protection BONDED SYSTEM – SOIL POISON

Basementless space: ground cover _____ ; insulation _____ ; foundation vents _____

Special foundations _____

Additional information: _____

3. CHIMNEYS:

Material BRICK   Prefabricated (make and size) _____

Flue lining: material T.C.   Heater flue size _____ Fireplace flue size 13"X13"

Vents (material and size): gas or oil heater _____ ; water heater _____

Additional information: _____

4. FIREPLACES:

Type: ☒ solid fuel; ☐ gas-burning; ☐ circulator *(make and size)* _____; Ash dump and clean-out _____

Fireplace: facing _FIRE BRICK_; lining _FIRE BRICK_; hearth _BRICK_; mantel _WOOD_

Additional information: _____

5. EXTERIOR WALLS:

Wood frame: wood grade, and species _GRADE MARK #2 SPRUCE_   ☒ Corner bracing. Building paper or felt _#15_

Sheathing _FIBERBOARD_; thickness _½"_; width _48"_; ☐ solid; ☐ spaced ____"o. c.; ☐ diagonal;

Siding _MASONITE_; grade _PRIMED_; type _1 X 8_; size _7"_; exposure ____"; fastening _GALV'N NAILS_

Shingles _____; grade _____; type _____; size _____; exposure ____"; fastening _____

Stucco _____; thickness ____"; Lath _____; weight _____ lb.

Masonry veneer _BRICK ALL WALLS_ Sills _M. BRICK_ _____ Lintels _STEEL_ Base flashing _____

Masonry: ☒ solid ☐ faced ☐ stuccoed; total-wall thickness _9½"_; facing thickness _4"_; facing material _BRICK_

Backup material _____; thickness _____; bonding _____

Door sills _CONC. OR BRICK_ Window sills _BRICK OR WOOD_ Lintels _STEEL_ Base flashing _____

Interior surfaces: dampproofing, _____ coats of _____; furring _____

Additional information: _____

Exterior painting: material _PAINT READY MIX EXTERIOR GRADE_; number of coats _3_

Gable wall construction: ☐ same as main walls; ☒ other construction _STUDS 15 # FELT & SIDING_

6. FLOOR FRAMING:

Joists: wood, grade, and species _____; other _____; bridging _____; anchors _____

Concrete slab: ☐ basement floor; ☒ first floor; ☒ ground supported; ☐ self-supporting; mix _1:3½:4½_; thickness _4"_;

reinforcing _6 X 6 10 ga. WIRE MESH_; insulation _1" A 24" PERIMETER_; membrane _- 000 POLY_

Fill under slab: material _WASHED GRAVEL_; thickness _4"_. Additional information: _____

8" CONC. FILLED PILASTERS AT 4' O.C. BOTH WAYS & UNDER LOAD BRG WALLS

7. SUBFLOORING: *(Describe underflooring for special floors under item 21.)*

Material: grade and species _N/A_

Laid: ☐ first floor; ☐ second floor; ☐ attic _____ sq. ft.; ☐ diagonal; ☐ right angles. Additional information: _____

8. FINISH FLOORING: *(Wood only.* Describe other finish flooring under item 21.)

LOCATION	ROOMS	GRADE	SPECIES	THICKNESS	WIDTH	BLDG. PAPER	FINISH
First floor	N/A						
Second floor							
Attic floor	___ sq. ft.						

Additional information: _____

FHA Form 2005
VA Form 26–1852
Form FmHA 424–2

DESCRIPTION OF MATERIALS

1

9. PARTITION FRAMING:
Studs: wood, grade, and species *#2 CEDAR OR SPRUCE* size and spacing *2"X4" @ 16" O.C.* Other _____
Additional information: *DOUBLE STUDS AT ALL PLUMBING WALLS*

10. CEILING FRAMING:
Joists: wood, grade, and species *#2 YP* Other _____ Bridging *SOLID*
Additional information: *16" O.C. (SEE PLANS FOR SIZE & SPACING)*

11. ROOF FRAMING:
Rafters: wood, grade, and species *#2 YP* Roof trusses (see detail): grade and species _____
Additional information: *SEE PLANS FOR SIZE & SPACING*

12. ROOFING:
Sheathing; wood, grade, and species *PLYWOOD ½" THICK*; ☒ solid: ☐ spaced _____ " o.c.
Roofing *ASPHALT SHINGLES*; grade *C*; size *12X36*; type *# 235*
Underlay *15# FELT WHERE PITCH LESS 7/12*; weight or thickness _____; size _____; fastening _____
Built-up roofing _____; number of plies _____; surfacing material _____
Flashing: material *BONDERIZED - IRON*; gage or weight *26 GAGE*; ☐ gravel stops; ☐ snow guards
Additional information: *SHINGLES TO HAVE 5" EXPOSURE EXCEPT 3/12 TO HAVE 4" EXPOSURE*
AND 2 LAYERS 15# FELT.

13. GUTTERS AND DOWNSPOUTS: *SEE PLANS*
Gutters: material *GALV. IRON*; gage or weight *26*; size *3X5*; shape *O.G.*
Downspouts: material *GALV. IRON*; gage or weight *26*; size *3X5*; shape *RECT.*; number *AS REQ.*
Downspouts connected to: ☐ Storm sewer; ☐ sanitary sewer; ☐ dry-well. ☒ Splash blocks: material and size *CONCRETE*
Additional information: *WATER DIVERTERS OVER ENTRANCES IF GUTTERS ARE NOT USED.*

14. LATH AND PLASTER:
Lath ☐ walls, ☐ ceilings: material _____; weight or thickness _____ Plaster: coats _____; finish _____
Dry-wall ☒ walls, ☒ ceilings: material *GYP. BOARD*; thickness *½"*; finish *SMOOTH*
Joint treatment *FILL, TAPE, FLOAT, AND SAND & SPRAY CEILINGS.*

15. DECORATING: (Paint, wallpaper, etc.)

Rooms	Wall Finish Material and Application	Ceiling Finish Material and Application
Kitchen	*WALLPAPER - 3 COATS ENAMEL*	*BLOWN*
Bath	*WALLPAPER*	"
Other	*3 COATS LATEX*	"
GREAT RM.	*PANELS*	"

Additional information: _____

16. INTERIOR DOORS AND TRIM:
Doors: type *6-PANEL*; material *CARADCO W.P.*; thickness *1⅜"*
Door trim: type *CLAMSHELL*; material *FIR*; Base: type *STOCK 2-PC*; material *FIR*; size *3½"*
Finish: doors *STAIN OR PAINT*; trim *STAIN OR PAINT*
Other trim (item, type and location) *¼" PREFINISHED PANELING*
Additional information: _____

17. WINDOWS:

Windows: type _SH_ ; make _AIR CONTROL_ ; material _ALUMINUM_ ; sash thickness _1½"_

Glass: grade _SSB_ ☐ sash weights; ☐ balances, type _____ ; head flashing _____

Trim: type _CLAMSHELL_ ; material _FIR_ Paint _PAINT OR STAIN_ ; number coats _2_

Weatherstripping: type _SPRING_ ; material _ALUMINUM_ Storm sash, number _____

Screens: ☐ full; ☒ half; type _____ ; number _____ ; screen cloth material _____

Basement windows: type _____ ; material _____ ; screens, number _____ ; Storm sash, number _____

Special windows _SEE PLANS_

Additional information:

18. ENTRANCES AND EXTERIOR DETAIL:

Main entrance door: material _FIR_ ; width _36"_ ; thickness _1¾"_. Frame: material _YP_ ; thickness _1½"_

Other entrance doors: material _FIR (SEE PLAN)_ ; width _32"_ ; thickness _1¾"_. Frame: material _YP_ ; thickness _1½"_

Head flashing _22 GAGE BOND IRON_ Weatherstripping: type _SPRING BRONZE_ ; saddles _INTERLOCKING_

Screen doors: thickness _____ "; number _____ ; screen cloth material _____ Storm doors: thickness _____ "; number _____

Combination storm and screen doors: thickness _1⅛"_ ; number _____ ; screen cloth material _SEE PLANS_

Shutters: ☐ hinged; ☒ fixed. Railings _____ Attic louvers _26 GA. GALV. IRON W/SCREEN_

Exterior millwork: grade and species _D GRADE FIR_ Paint _EXTERIOR GRADE_ ; number coats _2_

Additional information:

19. CABINETS AND INTERIOR DETAIL:

Kitchen cabinets, wall units: material _BIRCH MILL MADE_ ; lineal feet of shelves _PLAN_ ; shelf width _12"_

Base units: material _BIRCH_ ; counter top _POST FORM.FORMICA_ ; edging _FORMICA_

Back and end splash _FORMICA_ Finish of cabinets _STAIN OR LAQUER_ ; number coats _3_

Medicine cabinets: make _____ ; model _____

Other cabinets and built-in furniture _MILL-MADE VANITIES, PAINTED_

Additional information: _MAN-MADE MARBLE TOPS OR VANITIES_

20. STAIRS:

STAIR	TREADS		RISERS		STRINGS		HANDRAIL		BALUSTERS	
	Material	Thickness	Material	Thickness	Material	Size	Material	Size	Material	Size
Basement										
Main										
Attic										

Disappearing: make and model number _CENTURY STAIR — IF SHOWN ON PLAN OR 24"X30" ACCESS._

Additional information:

2

21. SPECIAL FLOORS AND WAINSCOT: (Describe Carpet as listed in Certified Products Directory)

FLOORS

Location	Material, Color, Border, Sizes, Gage, Etc.	Threshold Material	Wall Base Material	Underfloor Material
Kitchen	UTILITY VINYL	METAL	WOOD	CONCRETE
Bath	CERAMIC TILE	WOOD	CERAMIC	11
ENTRY	BROKEN MARBLE OR SLATE OR PARQUET	METAL	WOOD	11

WAINSCOT

Location	Material, Color, Border, Cap. Sizes, Gage, Etc.	Height	Height Over Tub	Height in Showers (From Floor)
Bath	CERAMIC TILE @ TUB, SEE PLANS	48"	72"	

Bathroom accessories: ☒ Recessed; material CHINA SEE PLANS ; number 6 ; ☐ Attached; material _____ ; number _____
Additional information: _____

22. PLUMBING:

Fixture	Number	Location	Make	Mfr's Fixture Identification No.	Size	Color
Sink	1	SEE PLANS	GERBER			
Lavatory	2	BATHS				
Water closet		BATHS				
Bathtub	SEE PLANS					
Shower over tub △	INCLUDED W/BATH	GERBER				
Stall shower △						
Laundry trays		ROUGH-IN FOR WASHING MACHINE CONNECTIONS				

△ ☐ Curtain rod △ ☐ Door ☒ Shower pan: material COMPOTITE ★
Water supply: ☒ public; ☐ community system; ☐ individual (private) system. ★
Sewage disposal: ☒ public; ☐ community system; ☐ individual (private) system. ★
★ Show and describe individual system in complete detail in separate drawings and specifications according to requirements.
House drain (inside): ☐ cast iron; ☐ tile; ☐ other PVC_____ House sewer (outside): ☐ cast iron; ☒ tile; ☐ other _____
Water piping: ☐ galvanized steel; ☒ copper tubing; ☐ other _____ ; Sill cocks, number 2
Domestic water heater: type _____ ; make and model RHEEM OR EQUAL; heating capacity 214,500 W _____
_____ gph. 100° rise. Storage tank: material GLASS LINED ; capacity 40 gallons.
Gas service: ☐ utility company; ☐ liq. pet. gas; ☐ other _____ Gas piping: ☐ cooking; ☐ house heating.
Footing drains connected to: ☐ storm sewer; ☐ sanitary sewer; ☐ dry well. Sump pump; make and model _____
_____ ; capacity _____ ; discharges into _____.

23. HEATING: SEE PLANS (HEATING), AIR CONDITIONING INCLUDED (HEAT PUMPS)

☐ Hot water. ☐ Steam. ☐ Vapor. ☐ One-pipe system. ☐ Two-pipe system.

☐ Radiators. ☐ Convectors. ☐ Baseboard radiation. Make and model _____

Radiant panel: ☐ floor; ☐ wall; ☐ ceiling. Panel coil: material _____

☐ Circulator. ☐ Return pump. Make and model _____

Boiler: make and model _____ Output _____ Btuh.; net rating _____ Btuh.

Additional information: _____

Warm air: ☐ Gravity. ☒ Forced. Type of system CENTRAL

Duct material: supply G.I. ; return G.I. Insulation _____, thickness _____

Furnace: make and model _____ Input _____ Btuh.; output _____ Btuh.

Additional information: _____

☐ Space heater; ☐ floor furnace; ☐ wall heater. Input _____ Btuh.; output _____ Btuh.; number units _____

Make, model _____ Additional information: _____

Controls: make and types AUTO WALL THERMOSTAT _____

Additional information: _____

Fuel: ☐ Coal; ☐ oil; ☐ gas; ☐ liq. pet. gas; ☒ electric; ☐ other _____ ; storage capacity _____

Additional information: _____

Firing equipment furnished separately: ☐ Gas burner, conversion type. ☐ Stoker: hopper feed ☐; bin feed ☐

Oil burner: ☐ pressure atomizing; ☐ vaporizing

Make and model _____ Control _____

Additional information: _____

Electric heating system: type _____ Input _____ volts; output _____ Btuh.

Additional information: _____

Ventilating equipment: attic fan, make and model SEE ITEM #26 _____ ; capacity _____ cfm.

kitchen exhaust fan, make and model 8" EXHAUST FAN IN BATHS WHERE INDICATED ON PLANS.

Other heating, ventilating, or cooling equipment _____

24. ELECTRIC WIRING:

Service: ☐ overhead; ☒ underground. Panel: ☐ fuse box; ☒ circuit-breaker; make 200 AMP. MIN AMP's As REQ'D No. circuits _____

Wiring: ☐ conduit; ☐ armored cable; ☐ nonmetallic cable; ☐ knob and tube; ☒ other COPPER WIRING

Special outlets: ☒ range; ☐ water heater; ☒ other DRYER DISHWASHER, DISPOSAL

☐ Doorbell. ☒ Chimes. Push-button locations FRONT DOOR Additional information: _____

25. LIGHTING FIXTURES: SEE PLAN

Total number of fixtures SEE PLAN Total allowance for fixtures, typical installation, $ 2,275.00

Nontypical installation _____

Additional information: _____

DESCRIPTION OF MATERIALS

3

DESCRIPTION OF MATERIALS

26. INSULATION:

Location	Thickness	Material, Type, and Method of Installation	Vapor Barrier
Roof	6"	FIBERGLASS R-19, VAULTED AREA	
Ceiling	8½"	BLOWN INSULATION R-19	
Wall	3½"	FIBERGLASS R-11	
Floor	1"x 24"	PERIMETER INSULATION	

27. MISCELLANEOUS: (Describe any main dwelling materials, equipment, or construction items not shown elsewhere; or use to provide additional information where the space provided was inadequate. Always reference by item number to correspond to numbering used on this form.)

1. WALLPAPER ALLOWANCE $275.00
2. CARPET LIVING, DINING HALL, ALL BEDROOMS & CLOSETS
   (ALLOW $8.00 PER YD.)
3. DEN CARPET ALLOW $3.00 PER YD.
4. PARQUET FLOOR PER PLANS, ALLOW $3.50 PER SQ. FT.

HARDWARE: (make, material, and finish.) $250.00

SPECIAL EQUIPMENT: (State material or make, model and quantity. Include only equipment and appliances which are acceptable by local law, custom and applicable FHA standards. Do not include items which, by established custom, are supplied by occupant and removed when he vacates premises or chattles prohibited by law from becoming realty.)

SDW4800 DISHWASHER
RDE 6300 DROP-IN RANGE
RDE 2300 OVEN
RDE 8400 COOKTOP
SYD60 DISPOSAL

PORCHES: SEE PLANS

TERRACES:

SEE PLANS

GARAGES:

SEE PLANS

SEE PLOT PLAN

WALKS AND DRIVEWAYS:
Driveway: width _L'_ ; base material _CLAY_ ; thickness _____ "; surfacing material _CONCRETE_ ; thickness _4 "_
Front walk: width _3"_ ; material _CONCRETE_; thickness _4 "_. Service walk: width _3'_ ; material _CONC._ ; thickness _4 "_
Steps: material _CONCRETE_ ; treads _12 "_; risers _7 "_. Check walls _CONCRETE_

OTHER ONSITE IMPROVEMENTS:
(Specify all exterior onsite improvements not described elsewhere, including items such as unusual grading, drainage structures, retaining walls, fence, railings, and accessory structures.)

FINE GRADE LOT BEFORE SODDING

LANDSCAPING, PLANTING, AND FINISH GRADING:
Topsoil _____ " thick: ☐ front yard; ☐ side yards; ☐ rear yard to _____ feet behind main building. _SOD_
Lawns (seeded, sodded, or sprigged): ☒ front yard _SOD_ ; ☒ side yards _SOD_ ; ☒ rear yard _SOD_
Planting: ☐ as specified and shown on drawings; ☐ as follows:

_L_ Shade trees, deciduous.	_2 "_ caliper.		
_____ Low flowering trees, deciduous,	_____ ' to _____ '		
_____ High-growing shrubs, deciduous,	_____ ' to _____ '		
_____ Medium-growing shrubs, deciduous,	_____ ' to _____ '		
_____ Low-growing shrubs, deciduous,	_____ ' to _____ '		
	Evergreen trees.	_____ ' to _____ ',	B & B.
	Evergreen shrubs.	_____ ' to _____ ',	B & B.
	Vines, 2-year		

$ 680.00 LANDSCAPING
18 NATIVE SHRUBS

IDENTIFICATION.—This exhibit shall be identified by the signature of the builder, or sponsor, and/or the proposed mortgagor if the latter is known at the time of application.

Date _____

_____
Signature

_____
Signature

4

FHA Form 2005
VA Form 26–1852
Form FmHA 424–2

GOVERNMENT PRINTING OFFICE : 1977 O – 236–421

B-1 UNIFIED SOIL CLASSIFICATION SYSTEM

Major Divisons			Group Symbols	Typical Descriptions
COARSE-GRAINED SOILS (more than 50% retained on no. 200 sieve[a])	GRAVELS (50% or more of coarse fraction retained on no. 4 sieve)	Clean Gravels	GW	Well-graded gravels and gravel-sand mixtures, little or no fines
			GP	Poorly graded gravels and gravel-sand mixtures, little or no fines
		Gravels with Fines	GM	Silty gravels, gravel-sand-silt mixtures
			GC	Clayey gravels, gravel-sand-clay mixtures
	SANDS (more than 50% of coarse fraction passes no. 4 sieve)	Clean Sands	SW	Well-graded sands and gravelly sands, little or no fines
			SP	Poorly graded sands and gravelly sands, little or no fines
		Sands with Fines	SM	Silty sands, sand-silt mixtures
			SC	Clayey sands, sand-clay mixtures
FINE-GRAINED SOILS (50% or more passes no. 200 sieve[a])	SILTS AND CLAYS (liquid limit 50% or less)		ML	Inorganic silts, very fine sands, rock flour, silty or clayey fine sands
			CL	Inorganic clays of low to medium plasticity, gravelly clays, sandy clays, silty clays, lean clays
			OL	Organic silts and organic silty clays of low plasticity
	SILTS AND CLAYS (liquid limit greater than 50%)		MH	Inorganic silts, micaceous or diatomaceous fine sands or silts, elastic silts
			CH	Inorganic clays of high plasticity, fat clays
			OH	Organic clays of medium to high plasticity
Highly Organic Soils			PT	Peat, muck, and other highly organic soils

[a]Based on the material passing the 3-in. (75-mm) sieve.

Source: Federal Housing Administration

B-2 SOIL TREATMENT CHEMICALS

Chemicals	Concentrations
Aldrin	0.5% applied in water emulsion
Chlordane	1.0% applied in water emulsion
Dieldrin	0.5% applied in water emulsion
Heptachlor	0.5% applied in water emulsion
Chlordane and Heptachlor	0.5% Chlordane plus 0.25% Heptachlor, a 0.75% solution applied in water emulsion

Note: Apply according to applicable regulations.
Source: Federal Housing Administration

B-3 SEISMIC RISK MAP

SEISMIC RISK MAP

Zone 0: No damage
Zone 1: Minor damage
Zone 2: Moderate damage
Zone 3: Major damage

C-1 LUMBER MEASUREMENTS AND SIZES

METRIC SIZES

Millimeters	Inches	Millimeters	Inches
16 × 75	⅝ × 3	44 × 150	1¾ × 6
16 × 100	⅝ × 4	44 × 175	1¾ × 7
16 × 125	⅝ × 5	44 × 200	1¾ × 8
16 × 150	⅝ × 6	44 × 225	1¾ × 9
19 × 75	¾ × 3	44 × 250	1¾ × 10
19 × 100	¾ × 4	44 × 300	1¾ × 12
19 × 125	¾ × 5	50 × 75	2 × 3
19 × 150	¾ × 6	50 × 100	2 × 4
22 × 75	⅞ × 3	50 × 125	2 × 5
22 × 100	⅞ × 4	50 × 150	2 × 6
22 × 125	⅞ × 5	50 × 175	2 × 7
22 × 150	⅞ × 6	50 × 200	2 × 8
25 × 75	1 × 3	50 × 225	2 × 9
25 × 100	1 × 4	50 × 250	2 × 10
25 × 125	1 × 5	50 × 300	2 × 12
25 × 150	1 × 6	63 × 100	2½ × 4
25 × 175	1 × 7	63 × 125	2½ × 5
25 × 200	1 × 8	63 × 150	2½ × 6
25 × 225	1 × 9	63 × 175	2½ × 7
25 × 250	1 × 10	63 × 200	2½ × 8
25 × 300	1 × 12	63 × 225	2½ × 9
32 × 75	1¼ × 3	75 × 100	3 × 4
32 × 100	1¼ × 4	75 × 125	3 × 5
32 × 125	1¼ × 5	75 × 150	3 × 6
32 × 150	1¼ × 6	75 × 175	3 × 7
32 × 175	1¼ × 7	75 × 200	3 × 8
32 × 200	1¼ × 8	75 × 225	3 × 9
32 × 225	1¼ × 9	75 × 250	3 × 10
32 × 250	1¼ × 10	75 × 300	3 × 12
32 × 300	1¼ × 12	100 × 100	4 × 4
38 × 75	1½ × 3	100 × 150	4 × 6
38 × 100	1½ × 4	100 × 200	4 × 8
38 × 125	1½ × 5	100 × 250	4 × 10
38 × 150	1½ × 6	100 × 300	4 × 12
38 × 175	1½ × 7	150 × 150	6 × 6
38 × 200	1½ × 8	150 × 200	6 × 8
38 × 225	1½ × 9	150 × 300	6 × 12
44 × 75	1¾ × 3	200 × 200	8 × 8
44 × 100	1¾ × 4	250 × 250	10 × 10
44 × 125	1¾ × 5	300 × 300	12 × 12

METRIC LENGTHS
(1 inch = 25 mm)

Length (meters)	Equivalent (feet and inches)
1.8	5–10⅞
2.1	6–10⅝
2.4	7–10½
2.7	8–10¼
3.0	9–10⅛
3.3	10–9⅞
3.6	11–9¾
3.9	12–9½
4.2	13–9⅜
4.5	14–9⅓
4.8	15–9
5.1	16–8¾
5.4	17–8⅝
5.7	18–8⅜
6.0	19–8¼
6.3	20–8
6.6	21–7⅞
6.9	22–7⅝
7.2	23–7½
7.5	24–7¼
7.8	25–7⅛

LUMBER DIMENSIONS

Nominal Size (feet)	Actual Size (feet)
1 × 2	¾ × 1½
1 × 3	¾ × 2½
1 × 4	¾ × 3½
1 × 5	¾ × 4½
1 × 6	¾ × 5½
1 × 8	¾ × 7¼
1 × 10	¾ × 9¼
1 × 12	¾ × 11¼
2 × 2	1½ × 1½
2 × 3	1½ × 2½
2 × 4	1½ × 3½
2 × 6	1½ × 5½
2 × 8	1½ × 7¼
2 × 10	1½ × 9¼
2 × 12	1½ × 11¼
4 × 4	3½ × 3½
4 × 6	3½ × 5½
4 × 10	3½ × 9¼
6 × 6	5½ × 5½

C-2 DECIMAL EQUIVALENTS OF A FOOT

	0″	1″	2″	3″	4″	5″	6″	7″	8″	9″	10″	11″	
0″	.00	.08	.17	.25	.33	.42	.50	.58	.67	.75	.83	.92	*0″*
1/8″	.01	.09	.18	.26	.34	.43	.51	.59	.68	.76	.84	.93	*1/8″*
1/4″	.02	.10	.19	.27	.35	.44	.52	.60	.69	.77	.85	.94	*1/4″*
3/8″	.03	.11	.20	.28	.36	.45	.53	.61	.70	.78	.86	.95	*3/8″*
1/2″	.04	.12	.21	.29	.37	.46	.54	.62	.71	.79	.87	.96	*1/2″*
5/8″	.05	.14	.22	.30	.39	.47	.55	.64	.72	.80	.89	.97	*5/8″*
3/4″	.06	.15	.23	.31	.40	.48	.56	.65	.73	.81	.90	.98	*3/4″*
7/8″	.07	.16	.24	.32	.41	.49	.57	.66	.74	.82	.91	.99	*7/8″*

C-3 NOMINAL AND MINIMUM DRESSED SIZES FOR SOFTWOOD LUMBER (PS-20)

American Lumber Standards for Softwood Lumber, PS-20, directs all grading rules to establish a minimum dressed thickness and width for lumber surfaced dry or green. The table summarizes some of the actual sizes permitted by PS-20. For sizes of other products, refer to the American Lumber Standard or to the applicable grading rule book.

NOMINAL AND MINIMUM DRESSED LUMBER SIZES
(inches)

Nominal Size	Surfaced Dry	Minimum Dressed Size[a]		
		Surfaced Green		
		Redwood, Western Red Cedar, Northern White Cedar	Other Species	
1	¾[b]	25/32[b]	25/32[b]	
2	1½	1 9/16	1 9/16	
4	3½	3 9/16	3 9/16	
6	5½	5 9/16	5⅝	
8	7¼	7⅜	7½	
10	9¼	9⅜	9½	
12	11¼	11⅜	11½	

[a]Shrinkage that may occur after dressing to a standard size may be recognized through the allowance of a tolerance below the above sizes on the basis of 1% shrinkage (0.7% for redwood, western red cedar, and northern white cedar) for each 4 points of moisture content reduction below the applicable maximum (30% for green sizes).

[b]Boards less than the minimum thickness for 1 inch nominal and ⅝ inch or greater thickness dry (11/16 inch green) may be used but should be marked to show size and condition of seasoning at time of dressing.

Source: Federal Housing Administration.

C-4 SPAN TABLES[1]

Explanation of Tables

The span tables for joists and rafters on pp. 214-49 are calculated on the basis of a series of modulus of elasticity (E) and fiber bending stress (F_b) values. The range of values in the tables provides allowable spans for all species and grades of nominal 2-inch framing lumber customarily used in construction.

Tables J-1 through J-6 list spans for floor and ceiling joists used over a single span with calculations based on E and the required F_b values shown.

Tables R-1 through R-15 list spans for rafters used over a single span with calculations based on F_b and the required E values shown.

Applicable design criteria for each condition of use appear at the top of each table. While these criteria are directed principally to residential construction, they are suitable for other occupancies having similar conditions of loading. Tabulated spans for rafters also apply to other types of occupancy, since the occupancy has little bearing on roof loading.

Lumber Sizes

Tabulated spans apply to surfaced (S4S) lumber having dimensions that conform to the American Softwood Lumber Standard PS 20-70. These sizes are as follows:

[1]Tables and information in Appendix C-4 courtesy of National Forest Products Assoc.

Nominal Reference	Dressed Size (inches)	
	Surfaced Dry	Surfaced Green
2 × 4	1½ × 3½	1⁹⁄₁₆ × 3⁹⁄₁₆
2 × 6	1½ × 5½	1⁹⁄₁₆ × 5⅝
2 × 8	1½ × 7¼	1⁹⁄₁₆ × 7½
2 × 10	1½ × 9¼	1⁹⁄₁₆ × 9½
2 × 12	1½ × 11¼	1⁹⁄₁₆ × 11½

Moisture Content

The listed dry and green sizes are related at 19% maximum moisture content. Tabulated spans are calculated on the basis of the dry sizes and are also applicable to the corresponding green sizes. The spans in these tables are intended for use in covered structures or where moisture content in use does not exceed 19%.

Span Measurement

Tabulated spans are the clear distance between supports. For sloping rafters, the span is measured along the horizontal projection. Appendix C-5 provides a chart by which horizontal distance can be converted to sloping distance, or vice versa.

Lumber Design Values

Use of these span tables requires reference to the applicable design values for the various species and grades of lumber. Modulus of elasticity (E) and fiber bending stress (F_b) values therein are based on the National Design Specification for Wood Construction (formerly National Design Specification for Stress Grade Lumber and Its Fastenings) and incorporate adjustments appropriate for repetitive-member use under various durations of load.

Repetitive-member use is that condition where framing members such as joists, rafters, studs, planks, decking, or similar members are spaced not more than 24 inches, are not less than three in number, and are joined by floor, roof, or other load-distributing elements adequate to support the design load. Design values in bending (F_b) for such use are 15% greater than for single-member use.

For rafters, design values in bending (F_b) may be greater than the design values for normal duration of load, by the following amounts:

15% for 2 months' duration, as for snow
25% for 7 days' duration, as for construction loading

Roof Loads

Rafter spans are tabulated for the most common roof loads. For roof loads intermediate between those tabulated, straight-line interpolation may be used. For roof live loads less than 20 pounds per square foot (psf), rafter spans and required E values tabulated for 20 psf may be adjusted in accordance with Table C-4-1.

Lumber Identification

When used with the tabulated spans in these tables, lumber should be identified by the grademark of an agency recognized as being competent by the Board of Review of the American Lumber Standards Committee or the Canadian Lumber Standards Administration Board.

Use of the Span Tables

Spans for floor and ceiling joists are calculated on the basis of the modulus of elasticity (E), with

TABLE C-4-1

Table number	For roof live loads of 12 psf or 16 psf			
	Multiply tabulated span by		Multiply tabulated E-value by	
	12 psf	16 psf	12 psf	16 psf
R-1	1.14	1.06	0.89	0.96
R-4	1.14	1.06	0.89	0.96
R-7	1.17	1.07	0.96	0.99
R-10	1.14	1.06	0.89	0.96
R-13	1.19	1.08	1.02	1.02

For intermediate values of roof live loads, use straight line interpolation.

the required fiber bending stress (F_b) listed below each span. Spans for rafters are calculated on the basis of fiber bending stress (F_b), with the required modulus of elasticity (E) listed below each span. Use of the tables is illustrated in the following examples.

> *Example 1. Floor Joists.* Assume a required span of 12'-9", a live load of 40 psf, and joists spaced 16 inches on centers. Table J-1 shows that a grade of two-by-eight having an E value of 1,600,000 psi and an F_b value of 1,250 psi would have a span of 12'-10", which satisfies the condition.
>
> *Example 2. Rafters.* Assume a horizontal projection span of 13'-0", a live load of 20 psf, dead load of 7 psf, no attached ceiling, and rafters spaced 16 inches on centers. Table R-13 shows that a two-by-six having an F_b value of 1,200 psi and an E value of 940,000 psi would have a span of 13'-0" of horizontal projection. Conversion of horizontal to sloping distance is illustrated in Appendix C-5.

Since many combinations of size, spacing, E, and F_b values are possible, it is recommended that the user examine the tables to determine which combination fits a particular case most effectively.

The spans for nominal two-by-five joists or rafters are 82% of the spans tabulated for the same spacing of nominal two-by-six joists or rafters. For each joist or rafter spacing, the required values of F_b or E for two-by-fives are the same as the tabulated values for two-by-sixes.

C-5   RAFTER SPAN CONVERSION DIAGRAM

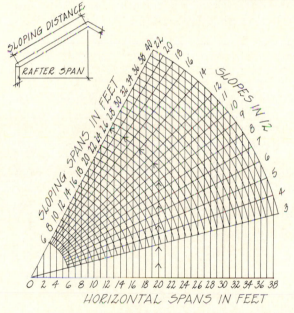

To find the rafter span when its horizontal span and slope are known, follow the vertical line from the horizontal span to its intersection with the radial line of the slope. From the intersection, follow the curve line to the sloping span. The diagram may also be used to determine the horizontal span when the sloping span and slope are known, as well as to determine the slope when the sloping and horizontal spans are known.

Example: For a horizontal span of 20 feet and a slope of 10 in 12, the sloping span of the rafter is read directly from the diagram as 26 feet.

C-6   SPANS FOR SOUTHERN PINE[1]

These abbreviated span tables are helpful for quick calculations. They are intended, however, for southern pine only.

Spans in the following tables are given in feet and inches of horizontal projection of the clear distance between the support of the member as follows:

For sloped roofs the actual length of the member is obtained by converting the horizontal span length to the sloped length.

End-jointed lumber may be used interchangeably with solid sawn lumber of the same grade.

[1] Reprinted with the permission of Southern Forest Products Assoc., New Orleans, LA.

DESIGN CRITERIA:
Deflection - For 40 lbs. per sq. ft. live load. Limited to span in inches divided by 360.
Strength - Live Load of 40 lbs. per sq. ft. plus dead load of 10 lbs. per sq. ft. determines the required fiber stress value.

TABLE J-1
FLOOR JOISTS

40 Lbs. Per Sq. Ft. Live Load
(All rooms except those used for sleeping areas and attic floors.)

JOIST SIZE (IN)	SPACING (IN)	Modulus of Elasticity, "E", in 1,000,000 psi																		
		0.4	0.5	0.6	0.7	0.8	0.9	1.0	1.1	1.2	1.3	1.4	1.5	1.6	1.7	1.8	1.9	2.0	2.2	2.4
2x6	12.0	6-9 450	7-3 520	7-9 590	8-2 660	8-6 720	8-10 780	9-2 830	9-6 890	9-9 940	10-0 990	10-3 1040	10-6 1090	10-9 1140	10-11 1190	11-2 1230	11-4 1280	11-7 1320	11-11 1410	12-3 1490
	13.7	6-6 470	7-0 550	7-5 620	7-9 690	8-2 750	8-6 810	8-9 870	9-1 930	9-4 980	9-7 1040	9-10 1090	10-0 1140	10-3 1190	10-6 1240	10-8 1290	10-10 1340	11-1 1380	11-5 1470	11-9 1560
	16.0	6-2 500	6-7 580	7-0 650	7-5 720	7-9 790	8-0 860	8-4 920	8-7 980	8-10 1040	9-1 1090	9-4 1150	9-6 1200	9-9 1250	9-11 1310	10-2 1360	10-4 1410	10-6 1460	10-10 1550	11-2 1640
	19.2	5-9 530	6-3 610	6-7 690	7-0 770	7-3 840	7-7 910	7-10 970	8-1 1040	8-4 1100	8-7 1160	8-9 1220	9-0 1280	9-2 1330	9-4 1390	9-6 1440	9-8 1500	9-10 1550	10-2 1650	10-6 1750
	24.0	5-4 570	5-9 660	6-2 750	6-6 830	6-9 900	7-0 980	7-3 1050	7-6 1120	7-9 1190	7-11 1250	8-2 1310	8-4 1380	8-6 1440	8-8 1500	8-10 1550	9-0 1610	9-2 1670	9-6 1780	9-9 1880
	32.0					6-2 1010	6-5 1090	6-7 1150	6-10 1230	7-0 1300	7-3 1390	7-5 1450	7-7 1520	7-9 1590	7-11 1660	8-0 1690	8-2 1760	8-4 1840	8-7 1950	8-10 2060
2x8	12.0	8-11 450	9-7 520	10-2 590	10-9 660	11-3 720	11-8 780	12-1 830	12-6 890	12-10 940	13-2 990	13-6 1040	13-10 1090	14-2 1140	14-5 1190	14-8 1230	15-0 1280	15-3 1320	15-9 1410	16-2 1490
	13.7	8-6 470	9-2 550	9-9 620	10-3 690	10-9 750	11-2 810	11-7 870	11-11 930	12-3 980	12-7 1040	12-11 1090	13-3 1140	13-6 1190	13-10 1240	14-1 1290	14-4 1340	14-7 1380	15-0 1470	15-6 1560
	16.0	8-1 500	8-9 580	9-3 650	9-9 720	10-2 790	10-7 850	11-0 920	11-4 980	11-8 1040	12-0 1090	12-3 1150	12-7 1200	12-10 1250	13-1 1310	13-4 1360	13-7 1410	13-10 1460	14-3 1550	14-8 1640
	19.2	7-7 530	8-2 610	8-9 690	9-2 770	9-7 840	10-0 910	10-4 970	10-8 1040	11-0 1100	11-3 1160	11-7 1220	11-10 1280	12-1 1330	12-4 1390	12-7 1440	12-10 1500	13-0 1550	13-5 1650	13-10 1750
	24.0	7-1 570	7-7 660	8-1 750	8-6 830	8-11 900	9-3 980	9-7 1050	9-11 1120	10-2 1190	10-6 1250	10-9 1310	11-0 1380	11-3 1440	11-5 1500	11-8 1550	11-11 1610	12-1 1670	12-6 1780	12-10 1880
	32.0					8-1 990	8-5 1080	8-9 1170	9-0 1230	9-3 1300	9-6 1370	9-9 1450	10-0 1520	10-2 1570	10-5 1650	10-7 1700	10-10 1790	11-0 1840	11-4 1950	11-8 2070
2x10	12.0	11-4 450	12-3 520	13-0 590	13-8 660	14-4 720	14-11 780	15-5 830	15-11 890	16-5 940	16-10 990	17-3 1040	17-8 1090	18-0 1140	18-5 1190	18-9 1230	19-1 1280	19-5 1320	20-1 1410	20-8 1490
	13.7	10-10 470	11-8 550	12-5 620	13-1 690	13-8 750	14-3 810	14-9 870	15-3 930	15-8 980	16-1 1040	16-6 1090	16-11 1140	17-3 1190	17-7 1240	17-11 1290	18-3 1340	18-7 1380	19-2 1470	19-9 1560
	16.0	10-4 500	11-1 580	11-10 650	12-5 720	13-0 790	13-6 850	14-0 920	14-6 980	14-11 1040	15-3 1090	15-8 1150	16-0 1200	16-5 1250	16-9 1310	17-0 1360	17-4 1410	17-8 1460	18-3 1550	18-9 1640
	19.2	9-9 530	10-6 610	11-1 690	11-8 770	12-3 840	12-9 910	13-2 970	13-7 1040	14-0 1100	14-5 1160	14-9 1220	15-1 1280	15-5 1330	15-9 1390	16-0 1440	16-4 1500	16-7 1550	17-2 1650	17-8 1750
	24.0	9-0 570	9-9 660	10-4 750	10-10 830	11-4 900	11-10 980	12-3 1050	12-8 1120	13-0 1190	13-4 1250	13-8 1310	14-0 1380	14-4 1440	14-7 1500	14-11 1550	15-2 1610	15-5 1670	15-11 1780	16-5 1880
	32.0					10-4 1000	10-9 1080	11-1 1150	11-6 1240	11-10 1310	12-2 1380	12-5 1440	12-9 1520	13-0 1580	13-3 1640	13-6 1700	13-9 1770	14-0 1830	14-6 1970	14-11 2080
2x12	12.0	13-10 450	14-11 520	15-10 590	16-8 660	17-5 720	18-1 780	18-9 830	19-4 890	19-11 940	20-6 990	21-0 1040	21-6 1090	21-11 1140	22-5 1190	22-10 1230	23-3 1280	23-7 1320	24-5 1410	25-1 1490
	13.7	13-3 470	14-3 550	15-2 620	15-11 690	16-8 750	17-4 810	17-11 870	18-6 930	19-1 980	19-7 1040	20-1 1090	20-6 1140	21-0 1190	21-5 1240	21-10 1290	22-3 1340	22-7 1380	23-4 1470	24-0 1560
	16.0	12-7 500	13-6 580	14-4 650	15-2 720	15-10 790	16-5 860	17-0 920	17-7 980	18-1 1040	18-7 1090	19-1 1150	19-6 1200	19-11 1250	20-4 1310	20-9 1360	21-1 1410	21-6 1460	22-2 1550	22-10 1640
	19.2	11-10 530	12-9 610	13-6 690	14-3 770	14-11 840	15-6 910	16-0 970	16-7 1040	17-0 1100	17-6 1160	17-11 1220	18-4 1280	18-9 1330	19-2 1390	19-6 1440	19-10 1500	20-2 1550	20-10 1650	21-6 1750
	24.0	11-0 570	11-10 660	12-7 750	13-3 830	13-10 900	14-4 980	14-11 1050	15-4 1120	15-10 1190	16-3 1250	16-8 1310	17-0 1380	17-5 1440	17-9 1500	18-1 1550	18-5 1610	18-9 1670	19-4 1780	19-11 1880
	32.0					12-7 1000	13-1 1080	13-6 1150	13-11 1220	14-4 1300	14-9 1380	15-2 1450	15-6 1520	15-10 1580	16-2 1650	16-5 1700	16-9 1770	17-0 1830	17-7 1950	18-1 2070

Note: The required extreme fiber stress in bending, "F_b", in pounds per square inch is shown below each span.

TABLE J-2
FLOOR JOISTS
30 lbs. Per Sq. Ft. Live Load
(All rooms used for sleeping areas and attic floors.)

DESIGN CRITERIA:
Deflection - For 30 lbs. per sq. ft. live load.
Limited to span in inches divided by 360.
Strength - Live Load of 30 lbs. per sq. ft. plus
dead load of 10 lbs. per sq. ft. determines
the required fiber stress value.

Modulus of Elasticity, "E", in 1,000,000 psi

Joist Size	Spacing (IN)	0.4	0.5	0.6	0.7	0.8	0.9	1.0	1.1	1.2	1.3	1.4	1.5	1.6	1.7	1.8	1.9	2.0	2.2	2.4
2×6	12.0	7-5 / 440	8-0 / 510	8-6 / 570	8-11 / 640	9-4 / 700	9-9 / 750	10-1 / 810	10-5 / 860	10-9 / 910	11-0 / 960	11-3 / 1010	11-7 / 1060	11-10 / 1100	12-0 / 1150	12-3 / 1200	12-6 / 1240	12-9 / 1280	13-1 / 1370	13-6 / 1450
	13.7	7-1 / 460	7-8 / 530	8-2 / 600	8-7 / 670	8-11 / 730	9-4 / 790	9-8 / 840	10-0 / 900	10-3 / 950	10-6 / 1010	10-10 / 1060	11-1 / 1110	11-3 / 1160	11-6 / 1200	11-9 / 1250	11-11 / 1300	12-2 / 1340	12-7 / 1430	12-11 / 1510
	16.0	6-9 / 480	7-3 / 560	7-9 / 630	8-2 / 700	8-6 / 770	8-10 / 830	9-2 / 890	9-6 / 950	9-9 / 1000	10-0 / 1060	10-3 / 1110	10-6 / 1160	10-9 / 1220	10-11 / 1270	11-2 / 1320	11-4 / 1360	11-7 / 1410	11-11 / 1500	12-3 / 1590
	19.2	6-4 / 510	6-10 / 600	7-3 / 670	7-8 / 740	8-0 / 810	8-4 / 880	8-8 / 940	8-11 / 1010	9-2 / 1070	9-5 / 1130	9-8 / 1180	9-10 / 1240	10-1 / 1290	10-4 / 1350	10-6 / 1400	10-8 / 1450	10-10 / 1500	11-3 / 1600	11-7 / 1690
	24.0	5-11 / 550	6-4 / 640	6-9 / 720	7-1 / 800	7-5 / 880	7-9 / 950	8-0 / 1020	8-3 / 1080	8-6 / 1150	8-9 / 1210	8-11 / 1270	9-2 / 1330	9-4 / 1390	9-7 / 1450	9-9 / 1510	9-11 / 1560	10-1 / 1620	10-5 / 1720	10-9 / 1820
	32.0					6-9 / 960	7-0 / 1040	7-3 / 1110	7-6 / 1190	7-9 / 1270	7-11 / 1330	8-2 / 1410	8-4 / 1470	8-6 / 1530	8-8 / 1590	8-10 / 1650	9-0 / 1710	9-2 / 1780	9-6 / 1910	9-9 / 2010
2×8	12.0	9-10 / 440	10-7 / 510	11-3 / 570	11-10 / 640	12-4 / 700	12-10 / 750	13-4 / 810	13-9 / 860	14-2 / 910	14-6 / 960	14-11 / 1010	15-3 / 1060	15-7 / 1100	15-10 / 1150	16-2 / 1200	16-6 / 1240	16-9 / 1280	17-4 / 1370	17-10 / 1450
	13.7	9-4 / 460	10-1 / 530	10-9 / 600	11-4 / 670	11-10 / 730	12-3 / 790	12-9 / 840	13-2 / 900	13-6 / 950	13-11 / 1010	14-3 / 1060	14-7 / 1110	14-11 / 1160	15-2 / 1200	15-6 / 1250	15-9 / 1300	16-0 / 1340	16-7 / 1430	17-0 / 1510
	16.0	8-11 / 480	9-7 / 560	10-2 / 630	10-9 / 700	11-3 / 770	11-8 / 830	12-1 / 890	12-6 / 950	12-10 / 1000	13-2 / 1060	13-6 / 1110	13-10 / 1160	14-2 / 1220	14-5 / 1270	14-8 / 1320	15-0 / 1360	15-3 / 1410	15-9 / 1500	16-2 / 1590
	19.2	8-5 / 510	9-0 / 600	9-7 / 670	10-1 / 740	10-7 / 810	11-0 / 880	11-4 / 940	11-9 / 1010	12-1 / 1070	12-5 / 1130	12-9 / 1180	13-0 / 1240	13-4 / 1290	13-7 / 1350	13-10 / 1400	14-1 / 1450	14-4 / 1500	14-9 / 1600	15-3 / 1690
	24.0	7-9 / 550	8-5 / 640	8-11 / 720	9-4 / 800	9-10 / 880	10-2 / 950	10-7 / 1020	10-11 / 1080	11-3 / 1150	11-6 / 1210	11-10 / 1270	12-1 / 1330	12-4 / 1390	12-7 / 1450	12-10 / 1510	13-1 / 1560	13-4 / 1620	13-9 / 1720	14-2 / 1820
	32.0					8-11 / 970	9-3 / 1040	9-7 / 1120	9-11 / 1200	10-2 / 1260	10-6 / 1340	10-9 / 1410	11-0 / 1470	11-3 / 1540	11-5 / 1590	11-8 / 1660	11-11 / 1730	12-1 / 1780	12-6 / 1900	12-10 / 2010
2×10	12.0	12-6 / 440	13-6 / 510	14-4 / 570	15-1 / 640	15-9 / 700	16-5 / 750	17-0 / 810	17-6 / 860	18-0 / 910	18-6 / 960	19-0 / 1010	19-5 / 1060	19-10 / 1100	20-3 / 1150	20-8 / 1200	21-0 / 1240	21-5 / 1280	22-1 / 1370	22-9 / 1450
	13.7	11-11 / 460	12-11 / 530	13-8 / 600	14-5 / 670	15-1 / 730	15-8 / 790	16-3 / 840	16-9 / 900	17-3 / 950	17-9 / 1010	18-2 / 1060	18-7 / 1110	19-0 / 1160	19-4 / 1200	19-9 / 1250	20-1 / 1300	20-5 / 1340	21-1 / 1430	21-9 / 1510
	16.0	11-4 / 480	12-3 / 560	13-0 / 630	13-8 / 700	14-4 / 770	14-11 / 830	15-5 / 890	15-11 / 950	16-5 / 1000	16-10 / 1060	17-3 / 1110	17-8 / 1160	18-0 / 1220	18-5 / 1270	18-9 / 1320	19-1 / 1360	19-5 / 1410	20-1 / 1500	20-8 / 1590
	19.2	10-8 / 510	11-6 / 600	12-3 / 670	12-11 / 740	13-6 / 810	14-0 / 880	14-6 / 940	15-0 / 1010	15-5 / 1070	15-10 / 1130	16-3 / 1180	16-7 / 1240	17-0 / 1290	17-4 / 1350	17-8 / 1400	18-0 / 1450	18-3 / 1500	18-10 / 1600	19-5 / 1690
	24.0	9-11 / 550	10-8 / 640	11-4 / 720	11-11 / 800	12-6 / 880	13-0 / 950	13-6 / 1020	13-11 / 1080	14-4 / 1150	14-8 / 1210	15-1 / 1270	15-5 / 1330	15-9 / 1390	16-1 / 1450	16-5 / 1510	16-8 / 1560	17-0 / 1620	17-6 / 1720	18-0 / 1820
	32.0					11-4 / 960	11-10 / 1050	12-3 / 1120	12-8 / 1200	13-0 / 1260	13-4 / 1330	13-8 / 1400	14-0 / 1470	14-4 / 1540	14-7 / 1590	14-11 / 1660	15-2 / 1720	15-5 / 1780	15-11 / 1890	16-5 / 2020
2×12	12.0	15-2 / 440	16-5 / 510	17-5 / 570	18-4 / 640	19-2 / 700	19-11 / 750	20-8 / 810	21-4 / 860	21-11 / 910	22-6 / 960	23-1 / 1010	23-7 / 1060	24-2 / 1100	24-8 / 1150	25-1 / 1200	25-7 / 1240	26-0 / 1280	26-10 / 1370	27-8 / 1450
	13.7	14-7 / 460	15-8 / 530	16-8 / 600	17-6 / 670	18-4 / 730	19-1 / 790	19-9 / 840	20-5 / 900	21-0 / 950	21-7 / 1010	22-1 / 1060	22-7 / 1110	23-1 / 1160	23-7 / 1200	24-0 / 1250	24-5 / 1300	24-10 / 1340	25-8 / 1430	26-5 / 1510
	16.0	13-10 / 480	14-11 / 560	15-10 / 630	16-8 / 700	17-5 / 770	18-1 / 830	18-9 / 890	19-4 / 950	19-11 / 1000	20-6 / 1060	21-0 / 1110	21-6 / 1160	21-11 / 1220	22-5 / 1270	22-10 / 1320	23-3 / 1360	23-7 / 1410	24-5 / 1500	25-1 / 1590
	19.2	13-0 / 510	14-0 / 600	14-11 / 670	15-8 / 740	16-5 / 810	17-0 / 880	17-8 / 940	18-3 / 1010	18-9 / 1070	19-3 / 1130	19-9 / 1180	20-2 / 1240	20-8 / 1290	21-1 / 1350	21-6 / 1400	21-10 / 1450	22-3 / 1500	22-11 / 1600	23-7 / 1690
	24.0	12-1 / 550	13-0 / 640	13-10 / 720	14-7 / 800	15-2 / 880	15-10 / 950	16-5 / 1020	16-11 / 1080	17-5 / 1150	17-11 / 1210	18-4 / 1270	18-9 / 1330	19-2 / 1390	19-7 / 1450	19-11 / 1510	20-3 / 1560	20-8 / 1620	21-4 / 1720	21-11 / 1820
	32.0					13-10 / 970	14-4 / 1040	14-11 / 1130	15-4 / 1190	15-10 / 1270	16-3 / 1340	16-8 / 1400	17-0 / 1460	17-5 / 1530	17-9 / 1590	18-1 / 1650	18-5 / 1720	18-9 / 1780	19-4 / 1890	19-11 / 2010

Note: The required extreme fiber stress in bending, "Fb", in pounds per square inch is shown below each span.

TABLE J-3
CEILING JOISTS
20 Lbs. Per Sq. Ft. Live Load
(Limited attic storage where development of future rooms is not possible)
(Plaster Ceiling)

DESIGN CRITERIA:
Deflection - For 20 lbs. per sq. ft. live load.
Limited to span in inches divided by 360.
Strength - Live load of 20 lbs. per sq. ft. plus dead load of 10 lbs. per sq. ft. determines required fiber stress value.

Modulus of Elasticity, "E", in 1,000,000 psi

JOIST SIZE	SPACING (IN)	0.4	0.5	0.6	0.7	0.8	0.9	1.0	1.1	1.2	1.3	1.4	1.5	1.6	1.7	1.8	1.9	2.0	2.2	2.4
2x4	12.0	5-5 / 430	5-10 / 500	6-2 / 560	6-6 / 630	6-10 / 680	7-1 / 740	7-4 / 790	7-7 / 850	7-10 / 900	8-0 / 950	8-3 / 990	8-5 / 1040	8-7 / 1090	8-9 / 1130	8-11 / 1170	9-1 / 1220	9-3 / 1260	9-7 / 1340	9-10 / 1420
	13.7	5-2 / 450	5-7 / 520	5-11 / 590	6-3 / 650	6-6 / 720	6-9 / 770	7-0 / 830	7-3 / 880	7-6 / 940	7-8 / 990	7-10 / 1040	8-1 / 1090	8-3 / 1140	8-5 / 1180	8-7 / 1230	8-8 / 1270	8-10 / 1320	9-2 / 1400	9-5 / 1490
	16.0	4-11 / 470	5-4 / 550	5-8 / 620	5-11 / 690	6-2 / 750	6-5 / 810	6-8 / 870	6-11 / 930	7-1 / 990	7-3 / 1040	7-6 / 1090	7-8 / 1140	7-10 / 1200	8-0 / 1240	8-1 / 1290	8-3 / 1340	8-5 / 1390	8-8 / 1480	8-11 / 1570
	19.2	4-8 / 500	5-0 / 580	5-4 / 660	5-7 / 730	5-10 / 800	6-1 / 870	6-3 / 930	6-6 / 990	6-8 / 1050	6-10 / 1110	7-0 / 1160	7-2 / 1220	7-4 / 1270	7-6 / 1320	7-8 / 1370	7-9 / 1420	7-11 / 1470	8-2 / 1570	8-5 / 1660
	24.0	4-4 / 540	4-8 / 630	4-11 / 710	5-2 / 790	5-5 / 860	5-8 / 930	5-10 / 1000	6-0 / 1070	6-2 / 1130	6-4 / 1190	6-6 / 1250	6-8 / 1310	6-10 / 1370	7-0 / 1420	7-1 / 1480	7-3 / 1530	7-4 / 1590	7-7 / 1690	7-10 / 1790
2x6	12.0	8-6 / 430	9-2 / 500	9-9 / 560	10-3 / 630	10-9 / 680	11-2 / 740	11-7 / 790	11-11 / 850	12-3 / 900	12-7 / 950	12-11 / 990	13-3 / 1040	13-6 / 1090	13-9 / 1130	14-1 / 1170	14-4 / 1220	14-7 / 1260	15-0 / 1340	15-6 / 1420
	13.7	8-2 / 450	8-9 / 520	9-4 / 590	9-10 / 650	10-3 / 720	10-8 / 770	11-1 / 830	11-5 / 880	11-9 / 940	12-1 / 990	12-4 / 1040	12-8 / 1090	12-11 / 1140	13-2 / 1180	13-5 / 1230	13-8 / 1270	13-11 / 1320	14-4 / 1400	14-9 / 1490
	16.0	7-9 / 470	8-4 / 550	8-10 / 620	9-4 / 690	9-9 / 750	10-2 / 810	10-6 / 870	10-10 / 930	11-2 / 990	11-5 / 1040	11-9 / 1090	12-0 / 1140	12-3 / 1200	12-6 / 1240	12-9 / 1290	13-0 / 1340	13-3 / 1390	13-8 / 1480	14-1 / 1570
	19.2	7-3 / 500	7-10 / 580	8-4 / 660	8-9 / 730	9-2 / 800	9-6 / 870	9-10 / 930	10-2 / 990	10-6 / 1050	10-9 / 1110	11-1 / 1160	11-4 / 1220	11-7 / 1270	11-9 / 1320	12-0 / 1370	12-3 / 1420	12-5 / 1470	12-10 / 1570	13-3 / 1660
	24.0	6-9 / 540	7-3 / 630	7-9 / 710	8-2 / 790	8-6 / 860	8-10 / 930	9-2 / 1000	9-6 / 1070	9-9 / 1130	10-0 / 1190	10-3 / 1250	10-6 / 1310	10-9 / 1370	10-11 / 1420	11-2 / 1480	11-4 / 1530	11-7 / 1590	11-11 / 1690	12-3 / 1790
2x8	12.0	11-3 / 430	12-1 / 500	12-10 / 560	13-6 / 630	14-2 / 680	14-8 / 740	15-3 / 790	15-9 / 850	16-2 / 900	16-7 / 950	17-0 / 990	17-5 / 1040	17-10 / 1090	18-2 / 1130	18-6 / 1170	18-10 / 1220	19-2 / 1260	19-10 / 1340	20-5 / 1420
	13.7	10-9 / 450	11-7 / 520	12-3 / 590	12-11 / 650	13-6 / 720	14-1 / 770	14-7 / 830	15-0 / 880	15-6 / 940	15-11 / 990	16-3 / 1040	16-8 / 1090	17-0 / 1140	17-5 / 1180	17-9 / 1230	18-0 / 1270	18-4 / 1320	18-11 / 1400	19-6 / 1490
	16.0	10-2 / 470	11-0 / 550	11-8 / 620	12-3 / 690	12-10 / 750	13-4 / 810	13-10 / 870	14-3 / 930	14-8 / 990	15-1 / 1040	15-6 / 1090	15-10 / 1140	16-2 / 1200	16-6 / 1240	16-10 / 1290	17-2 / 1340	17-5 / 1390	18-0 / 1480	18-6 / 1570
	19.2	9-7 / 500	10-4 / 580	11-0 / 660	11-7 / 730	12-1 / 800	12-7 / 870	13-0 / 930	13-5 / 990	13-10 / 1050	14-2 / 1110	14-7 / 1160	14-11 / 1220	15-3 / 1270	15-6 / 1320	15-10 / 1370	16-1 / 1420	16-5 / 1470	16-11 / 1570	17-5 / 1660
	24.0	8-11 / 540	9-7 / 630	10-2 / 710	10-9 / 790	11-3 / 860	11-8 / 930	12-1 / 1000	12-6 / 1070	12-10 / 1130	13-2 / 1190	13-6 / 1250	13-10 / 1310	14-2 / 1370	14-5 / 1420	14-8 / 1480	15-0 / 1530	15-3 / 1590	15-9 / 1690	16-2 / 1790
2x10	12.0	14-4 / 430	15-5 / 500	16-5 / 560	17-3 / 630	18-0 / 680	18-9 / 740	19-5 / 790	20-1 / 850	20-8 / 900	21-2 / 950	21-9 / 990	22-3 / 1040	22-9 / 1090	23-2 / 1130	23-8 / 1170	24-1 / 1220	24-6 / 1260	25-3 / 1340	26-0 / 1420
	13.7	13-8 / 450	14-9 / 520	15-8 / 590	16-6 / 650	17-3 / 720	17-11 / 770	18-7 / 830	19-2 / 880	19-9 / 940	20-3 / 990	20-9 / 1040	21-3 / 1090	21-9 / 1140	22-2 / 1180	22-7 / 1230	23-0 / 1270	23-5 / 1320	24-2 / 1400	24-10 / 1490
	16.0	13-0 / 470	14-0 / 550	14-11 / 620	15-8 / 690	16-5 / 750	17-0 / 810	17-8 / 870	18-3 / 930	18-9 / 990	19-3 / 1040	19-9 / 1090	20-2 / 1140	20-8 / 1200	21-1 / 1240	21-6 / 1290	21-10 / 1340	22-3 / 1390	22-11 / 1480	23-8 / 1570
	19.2	12-3 / 500	13-2 / 580	14-0 / 660	14-9 / 730	15-5 / 800	16-0 / 870	16-7 / 930	17-2 / 990	17-8 / 1050	18-1 / 1110	18-7 / 1160	19-0 / 1220	19-5 / 1270	19-10 / 1320	20-2 / 1370	20-7 / 1420	20-11 / 1470	21-7 / 1570	22-3 / 1660
	24.0	11-4 / 540	12-3 / 630	13-0 / 710	13-8 / 790	14-4 / 860	14-11 / 930	15-5 / 1000	15-11 / 1070	16-5 / 1130	16-10 / 1190	17-3 / 1250	17-8 / 1310	18-0 / 1370	18-5 / 1420	18-9 / 1480	19-1 / 1530	19-5 / 1590	20-1 / 1690	20-8 / 1790

Note: The required extreme fiber stress in bending, "F_b", in pounds per square inch is shown below each span.

TABLE J-4
CEILING JOISTS

20 Lbs. Per Sq. Ft. Live Load

(Limited attic storage where development of future rooms is not possible)
(Drywall Ceiling)

DESIGN CRITERIA:
Deflection - For 20 lbs. per sq. ft. live load.
Limited to span in inches divided by 240.
Strength - live load of 20 lbs. per sq. ft. plus dead load of 10 lbs. per sq. ft. determines required fiber stress value.

JOIST SIZE (IN)	SPACING (IN)	0.4	0.5	0.6	0.7	0.8	0.9	1.0	1.1	1.2	1.3	1.4	1.5	1.6	1.7	1.8	1.9	2.0	2.2	2.4
										Modulus of Elasticity, "E", in 1,000,000 psi									
2x4	12.0	6-2/560	6-8/660	7-1/740	7-6/820	7-10/900	8-1/970	8-5/1040	8-8/1110	8-11/1170	9-2/1240	9-5/1300	9-8/1360	9-10/1420	10-0/1480	10-3/1540	10-5/1600	10-7/1650	10-11/1760	11-3/1860
	13.7	5-11/590	6-5/690	6-9/770	7-2/860	7-6/940	7-9/1010	8-1/1090	8-4/1160	8-7/1230	8-9/1300	9-0/1360	9-3/1420	9-5/1490	9-7/1550	9-9/1610	10-0/1670	10-2/1730	10-6/1840	10-9/1950
	16.0	5-8/620	6-1/720	6-5/810	6-9/900	7-1/990	7-5/1070	7-8/1140	7-11/1220	8-1/1290	8-4/1360	8-7/1430	8-9/1500	8-11/1570	9-1/1630	9-4/1690	9-6/1760	9-8/1820	9-11/1940	10-3/2050
	19.2	5-4/660	5-9/770	6-1/870	6-5/960	6-8/1050	6-11/1130	7-2/1220	7-5/1300	7-8/1370	7-10/1450	8-1/1520	8-3/1590	8-5/1660	8-7/1730	8-9/1800	8-11/1870	9-1/1930	9-4/2060	9-8/2180
	24.0	4-11/710	5-4/830	5-8/930	5-11/1030	6-2/1130	6-5/1220	6-8/1310	6-11/1400	7-1/1480	7-3/1560	7-6/1640	7-8/1720	7-10/1790	8-0/1870	8-1/1940	8-3/2010	8-5/2080	8-8/2220	8-11/2350
2x6	12.0	9-9/560	10-6/660	11-2/740	11-9/820	12-3/900	12-9/970	13-3/1040	13-8/1110	14-1/1170	14-5/1240	14-9/1300	15-2/1360	15-6/1420	15-9/1480	16-1/1540	16-4/1600	16-8/1650	17-2/1760	17-8/1860
	13.7	9-4/590	10-0/690	10-8/770	11-3/860	11-9/940	12-3/1010	12-8/1090	13-1/1160	13-5/1230	13-10/1300	14-2/1360	14-6/1420	14-9/1490	15-1/1550	15-5/1610	15-8/1670	15-11/1730	16-5/1840	16-11/1950
	16.0	8-10/620	9-6/720	10-2/810	10-8/900	11-2/990	11-7/1070	12-0/1140	12-5/1220	12-9/1290	13-1/1360	13-5/1430	13-9/1500	14-1/1570	14-4/1630	14-7/1690	14-11/1760	15-2/1820	15-7/1940	16-1/2050
	19.2	8-4/660	9-0/770	9-6/870	10-0/960	10-6/1050	10-11/1130	11-4/1220	11-8/1300	12-0/1370	12-4/1450	12-8/1520	12-11/1590	13-3/1660	13-6/1730	13-9/1800	14-0/1870	14-3/1930	14-8/2060	15-2/2180
	24.0	7-9/710	8-4/830	8-10/930	9-4/1030	9-9/1130	10-2/1220	10-6/1310	10-10/1400	11-2/1480	11-5/1560	11-9/1640	12-0/1720	12-3/1790	12-6/1870	12-9/1940	13-0/2010	13-3/2080	13-8/2220	14-1/2350
2x8	12.0	12-10/560	13-10/660	14-8/740	15-6/820	16-2/900	16-10/970	17-5/1040	18-0/1110	18-6/1170	19-0/1240	19-6/1300	19-11/1360	20-5/1420	20-10/1480	21-2/1540	21-7/1600	21-11/1650	22-8/1760	23-4/1860
	13.7	12-3/590	13-3/690	14-1/770	14-10/860	15-6/940	16-1/1010	16-8/1090	17-2/1160	17-9/1230	18-2/1300	18-8/1360	19-1/1420	19-6/1490	19-11/1550	20-3/1610	20-8/1670	21-0/1730	21-8/1840	22-4/1950
	16.0	11-8/620	12-7/720	13-4/810	14-1/900	14-8/990	15-3/1070	15-10/1140	16-4/1220	16-10/1290	17-3/1360	17-9/1430	18-2/1500	18-6/1570	18-11/1630	19-3/1690	19-7/1760	19-11/1820	20-7/1940	21-2/2050
	19.2	11-0/660	11-10/770	12-7/870	13-3/960	13-10/1050	14-5/1130	14-11/1220	15-5/1300	15-10/1370	16-3/1450	16-8/1520	17-1/1590	17-5/1660	17-9/1730	18-2/1800	18-5/1870	18-9/1930	19-5/2060	19-11/2180
	24.0	10-2/710	11-0/830	11-8/930	12-3/1030	12-10/1130	13-4/1220	13-10/1310	14-3/1400	14-8/1480	15-1/1560	15-6/1640	15-10/1720	16-2/1790	16-6/1870	16-10/1940	17-2/2010	17-5/2080	18-0/2220	18-6/2350
2x10	12.0	16-5/560	17-8/660	18-9/740	19-9/820	20-8/900	21-6/970	22-3/1040	22-11/1110	23-8/1170	24-3/1240	24-10/1300	25-5/1360	26-0/1420	26-6/1480	27-1/1540	27-6/1600	28-0/1650	28-11/1760	29-9/1860
	13.7	15-8/590	16-11/690	17-11/770	18-11/860	19-9/940	20-6/1010	21-3/1090	21-11/1160	22-7/1230	23-3/1300	23-9/1360	24-4/1420	24-10/1490	25-5/1550	25-10/1610	26-4/1670	26-10/1730	27-8/1840	28-6/1950
	16.0	14-11/620	16-0/720	17-0/810	17-11/900	18-9/990	19-6/1070	20-2/1140	20-10/1220	21-6/1290	22-1/1360	22-7/1430	23-2/1500	23-8/1570	24-1/1630	24-7/1690	25-0/1760	25-5/1820	26-3/1940	27-1/2050
	19.2	14-0/660	15-1/770	16-0/870	16-11/960	17-8/1050	18-4/1130	19-0/1220	19-7/1300	20-2/1370	20-9/1450	21-3/1520	21-9/1590	22-3/1660	22-8/1730	23-2/1800	23-7/1870	23-11/1930	24-9/2060	25-5/2180
	24.0	13-0/710	14-0/830	14-11/930	15-8/1030	16-5/1130	17-0/1220	17-8/1310	18-3/1400	18-9/1480	19-3/1560	19-9/1640	20-2/1720	20-8/1790	21-1/1870	21-6/1940	21-10/2010	22-3/2080	22-11/2220	23-8/2350

Note: The required extreme fiber stress in bending, "F_b", in pounds per square inch is shown below each span.

217

TABLE J-5
CEILING JOISTS
10 Lbs. Per Sq. Ft. Live Load
(No attic storage and roof slope not steeper than 3 in 12)
(Plaster Ceiling)

DESIGN CRITERIA:
Deflection - For 10 lbs. per sq. ft. live load.
Limited to span in inches divided by 360.
Strength - live load of 10 lbs. per sq. ft. plus
dead load of 5 lbs. per sq. ft. determines
required fiber stress value.

Modulus of Elasticity, "E", in 1,000,000 psi

JOIST SIZE	SPACING (IN)	0.4	0.5	0.6	0.7	0.8	0.9	1.0	1.1	1.2	1.3	1.4	1.5	1.6	1.7	1.8	1.9	2.0	2.2	2.4
2x4	12.0	6-10 / 340	7-4 / 400	7-10 / 450	8-3 / 500	8-7 / 540	8-11 / 590	9-3 / 630	9-7 / 670	9-10 / 710	10-1 / 750	10-4 / 790	10-7 / 830	10-10 / 860	11-1 / 900	11-3 / 930	11-6 / 970	11-8 / 1000	12-1 / 1070	12-5 / 1130
	13.7	6-6 / 360	7-0 / 410	7-6 / 470	7-10 / 520	8-3 / 570	8-7 / 610	8-10 / 660	9-2 / 700	9-5 / 740	9-8 / 780	9-11 / 820	10-2 / 860	10-4 / 900	10-7 / 940	10-9 / 970	11-0 / 1010	11-2 / 1050	11-6 / 1110	11-10 / 1180
	16.0	6-2 / 380	6-8 / 440	7-1 / 490	7-6 / 550	7-10 / 600	8-1 / 650	8-5 / 690	8-8 / 740	8-11 / 780	9-2 / 830	9-5 / 870	9-8 / 910	9-10 / 950	10-0 / 990	10-3 / 1030	10-5 / 1060	10-7 / 1100	10-11 / 1170	11-3 / 1240
	19.2	5-10 / 400	6-3 / 460	6-8 / 520	7-0 / 580	7-4 / 630	7-8 / 690	7-11 / 740	8-2 / 790	8-5 / 830	8-8 / 880	8-10 / 920	9-1 / 970	9-3 / 1010	9-5 / 1050	9-8 / 1090	9-10 / 1130	10-0 / 1170	10-4 / 1250	10-7 / 1320
	24.0	5-5 / 430	5-10 / 500	6-2 / 560	6-6 / 630	6-10 / 680	7-1 / 740	7-4 / 790	7-7 / 850	7-10 / 900	8-0 / 950	8-3 / 990	8-5 / 1040	8-7 / 1090	8-9 / 1130	8-11 / 1170	9-1 / 1220	9-3 / 1260	9-7 / 1340	9-10 / 1420
2x6	12.0	10-9 / 340	11-7 / 400	12-3 / 450	12-11 / 500	13-6 / 540	14-1 / 590	14-7 / 630	15-0 / 670	15-6 / 710	15-11 / 750	16-3 / 790	16-8 / 830	17-0 / 860	17-4 / 900	17-8 / 930	18-0 / 970	18-4 / 1000	18-11 / 1070	19-6 / 1130
	13.7	10-3 / 360	11-1 / 410	11-9 / 470	12-4 / 520	12-11 / 570	13-5 / 610	13-11 / 660	14-4 / 700	14-9 / 740	15-2 / 780	15-7 / 820	15-11 / 860	16-3 / 900	16-7 / 940	16-11 / 970	17-3 / 1010	17-6 / 1050	18-1 / 1110	18-8 / 1180
	16.0	9-9 / 380	10-6 / 440	11-2 / 490	11-9 / 550	12-3 / 600	12-9 / 650	13-3 / 690	13-8 / 740	14-1 / 780	14-5 / 830	14-9 / 870	15-2 / 910	15-6 / 950	15-9 / 990	16-1 / 1030	16-4 / 1060	16-8 / 1100	17-2 / 1170	17-8 / 1240
	19.2	9-2 / 400	9-10 / 460	10-6 / 520	11-1 / 580	11-7 / 630	12-0 / 690	12-5 / 740	12-10 / 790	13-3 / 830	13-7 / 880	13-11 / 920	14-3 / 970	14-7 / 1010	14-10 / 1050	15-2 / 1090	15-5 / 1130	15-8 / 1170	16-2 / 1250	16-8 / 1320
	24.0	8-6 / 430	9-2 / 500	9-9 / 560	10-3 / 630	10-9 / 680	11-2 / 740	11-7 / 790	11-11 / 850	12-3 / 900	12-7 / 950	12-11 / 990	13-3 / 1040	13-6 / 1090	13-9 / 1130	14-1 / 1170	14-4 / 1220	14-7 / 1260	15-0 / 1340	15-6 / 1420
2x8	12.0	14-2 / 340	15-3 / 400	16-2 / 450	17-0 / 500	17-10 / 540	18-6 / 590	19-2 / 630	19-10 / 670	20-5 / 710	20-11 / 750	21-5 / 790	21-11 / 830	22-5 / 860	22-11 / 900	23-4 / 930	23-9 / 970	24-2 / 1000	24-11 / 1070	25-8 / 1130
	13.7	13-6 / 360	14-7 / 410	15-6 / 470	16-3 / 520	17-0 / 570	17-9 / 610	18-4 / 660	18-11 / 700	19-6 / 740	20-0 / 780	20-6 / 820	21-0 / 860	21-5 / 900	21-11 / 940	22-4 / 970	22-9 / 1010	23-1 / 1050	23-10 / 1110	24-7 / 1180
	16.0	12-10 / 380	13-10 / 440	14-8 / 490	15-6 / 550	16-2 / 600	16-10 / 650	17-5 / 690	18-0 / 740	18-6 / 780	19-0 / 830	19-6 / 870	19-11 / 910	20-5 / 950	20-10 / 990	21-2 / 1030	21-7 / 1060	21-11 / 1100	22-8 / 1170	23-4 / 1240
	19.2	12-1 / 400	13-0 / 460	13-10 / 520	14-7 / 580	15-3 / 630	15-10 / 690	16-5 / 740	16-11 / 790	17-5 / 830	17-11 / 880	18-4 / 920	18-9 / 970	19-2 / 1010	19-7 / 1050	19-11 / 1090	20-4 / 1130	20-8 / 1170	21-4 / 1250	21-11 / 1320
	24.0	11-3 / 430	12-1 / 500	12-10 / 560	13-6 / 630	14-2 / 680	14-8 / 740	15-3 / 790	15-9 / 850	16-2 / 900	16-7 / 950	17-0 / 990	17-5 / 1040	17-10 / 1090	18-2 / 1130	18-6 / 1170	18-10 / 1220	19-2 / 1260	19-10 / 1340	20-5 / 1420
2x10	12.0	18-0 / 340	19-5 / 400	20-8 / 450	21-9 / 500	22-9 / 540	23-8 / 590	24-6 / 630	25-3 / 670	26-0 / 710	26-9 / 750	27-5 / 790	28-0 / 830	28-7 / 860	29-2 / 900	29-9 / 930	30-4 / 970	30-10 / 1000	31-10 / 1070	32-9 / 1130
	13.7	17-3 / 360	18-7 / 410	19-9 / 470	20-9 / 520	21-9 / 570	22-7 / 610	23-5 / 660	24-2 / 700	24-10 / 740	25-7 / 780	26-2 / 820	26-10 / 860	27-5 / 900	27-11 / 940	28-6 / 970	29-0 / 1010	29-6 / 1050	30-5 / 1110	31-4 / 1180
	16.0	16-5 / 380	17-8 / 440	18-9 / 490	19-9 / 550	20-8 / 600	21-6 / 650	22-3 / 690	22-11 / 740	23-8 / 780	24-3 / 830	24-10 / 870	25-5 / 910	26-0 / 950	26-6 / 990	27-1 / 1030	27-6 / 1060	28-0 / 1100	28-11 / 1170	29-9 / 1240
	19.2	15-5 / 400	16-7 / 460	17-8 / 520	18-7 / 580	19-5 / 630	20-2 / 690	20-11 / 740	21-7 / 790	22-3 / 830	22-10 / 880	23-5 / 920	23-11 / 970	24-6 / 1010	25-0 / 1050	25-5 / 1090	25-11 / 1130	26-4 / 1170	27-3 / 1250	28-0 / 1320
	24.0	14-4 / 430	15-5 / 500	16-5 / 560	17-3 / 630	18-0 / 680	18-9 / 740	19-5 / 790	20-1 / 850	20-8 / 900	21-2 / 950	21-9 / 990	22-3 / 1040	22-9 / 1090	23-2 / 1130	23-8 / 1170	24-1 / 1220	24-6 / 1260	25-3 / 1340	26-0 / 1420

Note: The required extreme fiber stress in bending, "F_b" in pounds per square inch is shown below each span.

TABLE J-6
CEILING JOISTS
10 Lbs. Per Sq. Ft. Live Load
(No attic storage and roof slope not steeper than 3 in 12)
(Drywall Ceiling)

DESIGN CRITERIA:
Deflection - For 10 lbs. per sq. ft. live load.
Limited to span in inches divided by 240.
Strength - live load of 10 lbs. per sq. ft. plus
dead load of 5 lbs. per sq. ft. determines
required fiber stress value.

Each cell shows the allowable span (feet-inches) with the required extreme fiber stress in bending, F_b, in pounds per square inch shown below each span.

Modulus of Elasticity, "E", in 1,000,000 psi

Joist Size	Spacing (IN)	0.4	0.5	0.6	0.7	0.8	0.9	1.0	1.1	1.2	1.3	1.4	1.5	1.6	1.7	1.8	1.9	2.0	2.2	2.4
2x4	12.0	7-10 / 450	8-5 / 520	8-11 / 590	9-5 / 650	9-10 / 710	10-3 / 770	10-7 / 830	10-11 / 880	11-3 / 930	11-7 / 980	11-10 / 1030	12-2 / 1080	12-5 / 1130	12-8 / 1180	12-11 / 1220	13-2 / 1270	13-4 / 1310	13-9 / 1400	14-2 / 1480
2x4	13.7	7-6 / 470	8-1 / 540	8-7 / 610	9-0 / 680	9-5 / 740	9-9 / 800	10-2 / 860	10-6 / 920	10-9 / 970	11-1 / 1030	11-4 / 1080	11-7 / 1130	11-10 / 1180	12-1 / 1230	12-4 / 1280	12-7 / 1320	12-9 / 1370	13-2 / 1460	13-7 / 1550
2x4	16.0	7-1 / 490	7-8 / 570	8-1 / 650	8-7 / 720	8-11 / 780	9-4 / 850	9-8 / 910	9-11 / 970	10-3 / 1030	10-6 / 1080	10-9 / 1140	11-0 / 1190	11-3 / 1240	11-6 / 1290	11-9 / 1340	11-11 / 1390	12-2 / 1440	12-6 / 1540	12-11 / 1630
2x4	19.2	6-8 / 520	7-2 / 610	7-8 / 690	8-1 / 760	8-5 / 830	8-9 / 900	9-1 / 970	9-4 / 1030	9-8 / 1090	9-11 / 1150	10-2 / 1210	10-4 / 1270	10-7 / 1320	10-10 / 1380	11-0 / 1430	11-3 / 1480	11-5 / 1530	11-9 / 1630	12-2 / 1730
2x4	24.0	6-2 / 560	6-8 / 660	7-1 / 740	7-6 / 820	7-10 / 900	8-1 / 970	8-5 / 1040	8-8 / 1110	8-11 / 1170	9-2 / 1240	9-5 / 1300	9-8 / 1360	9-10 / 1420	10-0 / 1480	10-3 / 1540	10-5 / 1600	10-7 / 1650	10-11 / 1760	11-3 / 1860
2x6	12.0	12-3 / 450	13-3 / 520	14-1 / 590	14-9 / 650	15-6 / 710	16-1 / 770	16-8 / 830	17-2 / 880	17-8 / 930	18-2 / 980	18-8 / 1030	19-1 / 1080	19-6 / 1130	19-11 / 1180	20-3 / 1220	20-8 / 1270	21-0 / 1310	21-8 / 1400	22-4 / 1480
2x6	13.7	11-9 / 470	12-8 / 540	13-5 / 610	14-2 / 680	14-9 / 740	15-5 / 800	15-11 / 860	16-5 / 920	16-11 / 970	17-5 / 1030	17-10 / 1080	18-3 / 1130	18-8 / 1180	19-0 / 1230	19-5 / 1280	19-9 / 1320	20-1 / 1370	20-9 / 1460	21-4 / 1550
2x6	16.0	11-2 / 490	12-0 / 570	12-9 / 650	13-5 / 720	14-1 / 780	14-7 / 850	15-2 / 910	15-7 / 970	16-1 / 1030	16-6 / 1080	16-11 / 1140	17-4 / 1190	17-8 / 1240	18-1 / 1290	18-5 / 1340	18-9 / 1390	19-1 / 1440	19-8 / 1540	20-3 / 1630
2x6	19.2	10-6 / 520	11-4 / 610	12-0 / 690	12-8 / 760	13-3 / 830	13-9 / 900	14-3 / 970	14-8 / 1030	15-2 / 1090	15-7 / 1150	15-11 / 1210	16-4 / 1270	16-8 / 1320	17-0 / 1380	17-4 / 1430	17-8 / 1480	17-11 / 1530	18-6 / 1630	19-1 / 1730
2x6	24.0	9-9 / 560	10-6 / 660	11-2 / 740	11-9 / 820	12-3 / 900	12-9 / 970	13-3 / 1040	13-8 / 1110	14-1 / 1170	14-5 / 1240	14-9 / 1300	15-2 / 1360	15-6 / 1420	15-9 / 1480	16-1 / 1540	16-4 / 1600	16-8 / 1650	17-2 / 1760	17-8 / 1860
2x8	12.0	16-2 / 450	17-5 / 520	18-6 / 590	19-6 / 650	20-5 / 710	21-2 / 770	21-11 / 830	22-8 / 880	23-4 / 930	24-0 / 980	24-7 / 1030	25-2 / 1080	25-8 / 1130	26-2 / 1180	26-9 / 1220	27-2 / 1270	27-8 / 1310	28-7 / 1400	29-5 / 1480
2x8	13.7	15-6 / 470	16-8 / 540	17-9 / 610	18-8 / 680	19-6 / 740	20-3 / 800	21-0 / 860	21-8 / 920	22-4 / 970	22-11 / 1030	23-6 / 1080	24-0 / 1130	24-7 / 1180	25-1 / 1230	25-7 / 1280	26-0 / 1320	26-6 / 1370	27-4 / 1460	28-1 / 1550
2x8	16.0	14-8 / 490	15-10 / 570	16-10 / 650	17-9 / 720	18-6 / 780	19-3 / 850	19-11 / 910	20-7 / 970	21-2 / 1030	21-9 / 1080	22-4 / 1140	22-10 / 1190	23-4 / 1240	23-10 / 1290	24-3 / 1340	24-8 / 1390	25-2 / 1440	25-11 / 1540	26-9 / 1630
2x8	19.2	13-10 / 520	14-11 / 610	15-10 / 690	16-8 / 760	17-5 / 830	18-2 / 900	18-9 / 970	19-5 / 1030	19-11 / 1090	20-6 / 1150	21-0 / 1210	21-6 / 1270	21-11 / 1320	22-5 / 1380	22-10 / 1430	23-3 / 1480	23-8 / 1530	24-5 / 1630	25-2 / 1730
2x8	24.0	12-10 / 560	13-10 / 660	14-8 / 740	15-6 / 820	16-2 / 900	16-10 / 970	17-5 / 1040	18-0 / 1110	18-6 / 1170	19-0 / 1240	19-6 / 1300	19-11 / 1360	20-5 / 1420	20-10 / 1480	21-2 / 1540	21-7 / 1600	21-11 / 1650	22-8 / 1760	23-4 / 1860
2x10	12.0	20-8 / 450	22-3 / 520	23-8 / 590	24-10 / 650	26-0 / 710	27-1 / 770	28-0 / 830	28-11 / 880	29-9 / 930	30-7 / 980	31-4 / 1030	32-1 / 1080	32-9 / 1130	33-5 / 1180	34-1 / 1220	34-8 / 1270	35-4 / 1310	36-5 / 1400	37-6 / 1480
2x10	13.7	19-9 / 470	21-3 / 540	22-7 / 610	23-9 / 680	24-10 / 740	25-10 / 800	26-10 / 860	27-8 / 920	28-6 / 970	29-3 / 1030	30-0 / 1080	30-8 / 1130	31-4 / 1180	32-0 / 1230	32-7 / 1280	33-2 / 1320	33-9 / 1370	34-10 / 1460	35-10 / 1550
2x10	16.0	18-9 / 490	20-2 / 570	21-6 / 650	22-7 / 720	23-8 / 780	24-7 / 850	25-5 / 910	26-3 / 970	27-1 / 1030	27-9 / 1080	28-6 / 1140	29-2 / 1190	29-9 / 1240	30-5 / 1290	31-0 / 1340	31-6 / 1390	32-1 / 1440	33-1 / 1540	34-1 / 1630
2x10	19.2	17-8 / 520	19-0 / 610	20-2 / 690	21-3 / 760	22-3 / 830	23-2 / 900	23-11 / 970	24-9 / 1030	25-5 / 1090	26-2 / 1150	26-10 / 1210	27-5 / 1270	28-0 / 1320	28-7 / 1380	29-2 / 1430	29-8 / 1480	30-2 / 1530	31-2 / 1630	32-1 / 1730
2x10	24.0	16-5 / 560	17-8 / 660	18-9 / 740	19-9 / 820	20-8 / 900	21-6 / 970	22-3 / 1040	22-11 / 1110	23-8 / 1170	24-3 / 1240	24-10 / 1300	25-5 / 1360	26-0 / 1420	26-6 / 1480	27-1 / 1540	27-6 / 1600	28-0 / 1650	28-11 / 1760	29-9 / 1860

Note: The required extreme fiber stress in bending, "F_b", in pounds per square inch is shown below each span.

TABLE R-1
FLAT OR SLOPED RAFTERS
Supporting Drywall Ceiling
(Flat roof or cathedral ceiling with no attic space)
Live Load - 20 lb. per sq. ft.

DESIGN CRITERIA:
Strength - 15 lbs. per sq. ft. dead load plus 20 lbs. per sq. ft. live load determines required fiber stress.
Deflection - For 20 lbs. per sq. ft. live load. Limited to span in inches divided by 240.

RAFTER SIZE (IN)	SPACING (IN)	Extreme Fiber Stress in Bending, "F_b" (psi).										
		300	400	500	600	700	800	900	1000	1100	1200	1300
2x6	12.0	6-7 0.12	7-7 0.19	8-6 0.26	9-4 0.35	10-0 0.44	10-9 0.54	11-5 0.64	12-0 0.75	12-7 0.86	13-2 0.98	13-8 1.11
	13.7	6-2 0.12	7-1 0.18	7-11 0.25	8-8 0.33	9-5 0.41	10-0 0.50	10-8 0.60	11-3 0.70	11-9 0.81	12-4 0.92	12-10 1.04
	16.0	5-8 0.11	6-7 0.16	7-4 0.23	8-1 0.30	8-8 0.38	9-4 0.46	9-10 0.55	10-5 0.65	10-11 0.75	11-5 0.85	11-10 0.96
	19.2	5-2 0.10	6-0 0.15	6-9 0.21	7-4 0.27	7-11 0.35	8-6 0.42	9-0 0.51	9-6 0.59	9-11 0.68	10-5 0.78	10-10 0.88
	24.0	4-8 0.09	5-4 0.13	6-0 0.19	6-7 0.25	7-1 0.31	7-7 0.38	8-1 0.45	8-6 0.53	8-11 0.61	9-4 0.70	9-8 0.78
2x8	12.0	8-8 0.12	10-0 0.19	11-2 0.26	12-3 0.35	13-3 0.44	14-2 0.54	15-0 0.64	15-10 0.75	16-7 0.86	17-4 0.98	18-0 1.11
	13.7	8-1 0.12	9-4 0.18	10-6 0.25	11-6 0.33	12-5 0.41	13-3 0.50	14-0 0.60	14-10 0.70	15-6 0.81	16-3 0.92	16-10 1.04
	16.0	7-6 0.11	8-8 0.16	9-8 0.23	10-7 0.30	11-6 0.38	12-3 0.46	13-0 0.55	13-8 0.65	14-4 0.75	15-0 0.85	15-7 0.96
	19.2	6-10 0.10	7-11 0.15	8-10 0.21	9-8 0.27	10-6 0.35	11-2 0.42	11-10 0.51	12-6 0.59	13-1 0.68	13-8 0.78	14-3 0.88
	24.0	6-2 0.09	7-1 0.13	7-11 0.19	8-8 0.25	9-4 0.31	10-0 0.38	10-7 0.45	11-2 0.53	11-9 0.61	12-3 0.70	12-9 0.78
2x10	12.0	11-1 0.12	12-9 0.19	14-3 0.26	15-8 0.35	16-11 0.44	18-1 0.54	19-2 0.64	20-2 0.75	21-2 0.86	22-1 0.98	23-0 1.11
	13.7	10-4 0.12	11-11 0.18	13-4 0.25	14-8 0.33	15-10 0.41	16-11 0.50	17-11 0.60	18-11 0.70	19-10 0.81	20-8 0.92	21-6 1.04
	16.0	9-7 0.11	11-1 0.16	12-4 0.23	13-6 0.30	14-8 0.38	15-8 0.46	16-7 0.55	17-6 0.65	18-4 0.75	19-2 0.85	19-11 0.96
	19.2	8-9 0.10	10-1 0.15	11-3 0.21	12-4 0.27	13-4 0.35	14-3 0.42	15-2 0.51	15-11 0.59	16-9 0.68	17-6 0.78	18-2 0.88
	24.0	7-10 0.09	9-0 0.13	10-1 0.19	11-1 0.25	11-11 0.31	12-9 0.38	13-6 0.45	14-3 0.53	15-0 0.61	15-8 0.70	16-3 0.78
2x12	12.0	13-5 0.12	15-6 0.19	17-4 0.26	19-0 0.35	20-6 0.44	21-11 0.54	23-3 0.64	24-7 0.75	25-9 0.86	26-11 0.98	28-0 1.11
	13.7	12-7 0.12	14-6 0.18	16-3 0.25	17-9 0.33	19-3 0.41	20-6 0.50	21-9 0.60	23-0 0.70	24-1 0.81	25-2 0.92	26-2 1.04
	16.0	11-8 0.11	13-5 0.16	15-0 0.23	16-6 0.30	17-9 0.38	19-0 0.46	20-2 0.55	21-3 0.65	22-4 0.75	23-3 0.85	24-3 0.96
	19.2	10-8 0.10	12-3 0.15	13-9 0.21	15-0 0.27	16-3 0.35	17-4 0.42	18-5 0.51	19-5 0.59	20-4 0.68	21-3 0.78	22-2 0.88
	24.0	9-6 0.09	11-0 0.13	12-3 0.19	13-5 0.25	14-6 0.31	15-6 0.38	16-6 0.45	17-4 0.53	18-2 0.61	19-0 0.70	19-10 0.78

Note: The required modulus of elasticity, "E", in 1,000,000 pounds per square inch is shown below each span.

TABLE R-1 (cont.)

RAFTERS: Spans are measured along the
horizontal projection and loads are
considered as applied on the horizontal
projection.

Extreme Fiber Stress in Bending, "F_b" (psi).											RAFTER SPACING (IN)	SIZE (IN)
1400	1500	1600	1700	1800	1900	2000	2100	2200	2400	2700		
14-2 1.24	14-8 1.37	15-2 1.51	15-8 1.66	16-1 1.81	16-7 1.96	17-0 2.12	17-5 2.28	17-10 2.44			12.0	
13-3 1.16	13-9 1.29	14-2 1.42	14-8 1.55	15-1 1.69	15-6 1.83	15-11 1.98	16-3 2.13	16-8 2.28	17-5 2.60		13.7	
12-4 1.07	12-9 1.19	13-2 1.31	13-7 1.44	13-11 1.56	14-4 1.70	14-8 1.83	15-1 1.97	15-5 2.11	16-1 2.41		16.0	2x6
11-3 0.98	11-7 1.09	12-0 1.20	12-4 1.31	12-9 1.43	13-1 1.55	13-5 1.67	13-9 1.80	14-1 1.93	14-8 2.20		19.2	
10-0 0.88	10-5 0.97	10-9 1.07	11-1 1.17	11-5 1.28	11-8 1.39	12-0 1.50	12-4 1.61	12-7 1.73	13-2 1.97	13-11 2.35	24.0	
18-9 1.24	19-5 1.37	20-0 1.51	20-8 1.66	21-3 1.81	21-10 1.96	22-4 2.12	22-11 2.28	23-6 2.44			12.0	
17-6 1.16	18-2 1.29	18-9 1.42	19-4 1.55	19-10 1.69	20-5 1.83	20-11 1.98	21-5 2.13	21-11 2.28	22-11 2.60		13.7	
16-3 1.07	16-9 1.19	17-4 1.31	17-10 1.44	18-5 1.56	18-11 1.70	19-5 1.83	19-10 1.97	20-4 2.11	21-3 2.41†		16.0	2x8
14-10 0.98	15-4 1.09	15-10 1.20	16-4 1.31	16-9 1.43	17-3 1.55	17-8 1.67	18-2 1.80	18-7 1.93	19-5 2.20		19.2	
13-3 0.88	13-8 0.97	14-2 1.07	14-7 1.17	15-0 1.28	15-5 1.39	15-10 1.50	16-3 1.61	16-7 1.73	17-4 1.97	18-5 2.35	24.0	
23-11 1.24	24-9 1.37	25-6 1.51	26-4 1.66	27-1 1.81	27-10 1.96	28-7 2.12	29-3 2.28	29-11 2.44			12.0	
22-4 1.16	23-2 1.29	23-11 1.42	24-7 1.55	25-4 1.69	26-0 1.83	26-8 1.98	27-4 2.13	28-0 2.28	29-3 2.60		13.7	
20-8 1.07	21-5 1.19	22-1 1.31	22-10 1.44	23-5 1.56	24-1 1.70	24-9 1.83	25-4 1.97	25-11 2.11	27-1 2.41		16.0	2x10
18-11 0.98	19-7 1.09	20-2 1.20	20-10 1.31	21-5 1.43	22-0 1.55	22-7 1.67	23-2 1.80	23-8 1.93	24-9 2.20		19.2	
16-11 0.88	17-6 0.97	18-1 1.07	18-7 1.17	19-2 1.28	19-8 1.39	20-2 1.50	20-8 1.61	21-2 1.73	22-1 1.97	23-5 2.35	24.0	
29-1 1.24	30-1 1.37	31-1 1.51	32-0 1.66	32-11 1.81	33-10 1.96	34-9 2.12	35-7 2.28	36-5 2.44			12.0	
27-2 1.16	28-2 1.29	29-1 1.42	29-11 1.55	30-10 1.69	31-8 1.83	32-6 1.98	33-3 2.13	34-1 2.28	35-7 2.60		13.7	
25-2 1.07	26-0 1.19	26-11 1.31	27-9 1.44	28-6 1.56	29-4 1.70	30-1 1.83	30-10 1.97	31-6 2.11	32-11 2.41		16.0	2x12
23-0 0.98	23-9 1.09	24-7 1.20	25-4 1.31	26-0 1.43	26-9 1.55	27-5 1.67	28-2 1.80	28-9 1.93	30-1 2.20		19.2	
20-6 0.88	21 3 0.97	21-11 1.07	22-8 1.17	23-3 1.28	23-11 1.39	24-7 1.50	25-2 1.61	25-9 1.73	26-11 1.97	28-6 2.35	24.0	

Note: The required modulus of elasticity, "E", in 1,000,000 pounds per square inch is shown below each span.

TABLE R-2
FLAT OR SLOPED RAFTERS
Supporting Drywall Ceiling
(Flat roof or cathedral ceiling with no attic space)
Live Load - 30 lb. per sq. ft.

DESIGN CRITERIA:
Strength - 15 lbs. per sq. ft. dead load plus 30
  lbs. per sq. ft. live load determines required
  fiber stress.
Deflection - For 30 lbs. per sq. ft. live load.
  Limited to span in inches divided by 240.

RAFTER SIZE (IN)	SPACING (IN)	Extreme Fiber Stress in Bending, "F_b" (psi).										
		300	400	500	600	700	800	900	1000	1100	1200	1300
2x6	12.0	5-10 0.13	6-8 0.19	7-6 0.27	8-2 0.36	8-10 0.45	9-6 0.55	10-0 0.66	10-7 0.77	11-1 0.89	11-7 1.01	12-1 1.14
	13.7	5-5 0.12	6-3 0.18	7-0 0.25	7-8 0.33	8-3 0.42	8-10 0.52	9-5 0.61	9-11 0.72	10-5 0.83	10-10 0.95	11-3 1.07
	16.0	5-0 0.11	5-10 0.17	6-6 0.24	7-1 0.31	7-8 0.39	8-2 0.48	8-8 0.57	9-2 0.67	9-7 0.77	10-0 0.88	10-5 0.99
	19.2	4-7 0.10	5-4 0.15	5-11 0.22	6-6 0.28	7-0 0.36	7-6 0.44	7-11 0.52	8-4 0.61	8-9 0.70	9-2 0.80	9-6 0.90
	24.0	4-1 0.09	4-9 0.14	5-4 0.19	5-10 0.25	6-3 0.32	6-8 0.39	7-1 0.46	7-6 0.54	7-10 0.63	8-2 0.72	8-6 0.81
2x8	12.0	7-8 0.13	8-10 0.19	9-10 0.27	10-10 0.36	11-8 0.45	12-6 0.55	13-3 0.66	13-11 0.77	14-8 0.89	15-3 1.01	15-11 1.14
	13.7	7-2 0.12	8-3 0.18	9-3 0.25	10-1 0.33	10-11 0.42	11-8 0.52	12-5 0.61	13-1 0.72	13-8 0.83	14-4 0.95	14-11 1.07
	16.0	6-7 0.11	7-8 0.17	8-7 0.24	9-4 0.31	10-1 0.39	10-10 0.48	11-6 0.57	12-1 0.67	12-8 0.77	13-3 0.88	13-9 0.99
	19.2	6-1 0.10	7-0 0.15	7-10 0.22	8-7 0.28	9-3 0.36	9-10 0.44	10-6 0.52	11-0 0.61	11-7 0.70	12-1 0.80	12-7 0.90
	24.0	5-5 0.09	6-3 0.14	7-0 0.19	7-8 0.25	8-3 0.32	8-10 0.39	9-4 0.46	9-10 0.54	10-4 0.63	10-10 0.72	11-3 0.81
2x10	12.0	9-9 0.13	11-3 0.19	12-7 0.27	13-9 0.36	14-11 0.45	15-11 0.55	16-11 0.66	17-10 0.77	18-8 0.89	19-6 1.01	20-4 1.14
	13.7	9-1 0.12	10-6 0.18	11-9 0.25	12-11 0.33	13-11 0.42	14-11 0.52	15-10 0.61	16-8 0.72	17-6 0.83	18-3 0.95	19-0 1.07
	16.0	8-5 0.11	9-9 0.17	10-11 0.24	11-11 0.31	12-11 0.39	13-9 0.48	14-8 0.57	15-5 0.67	16-2 0.77	16-11 0.88	17-7 0.99
	19.2	7-8 0.10	8-11 0.15	9-11 0.22	10-11 0.28	11-9 0.36	12-7 0.44	13-4 0.52	14-1 0.61	14-9 0.70	15-5 0.80	16-1 0.90
	24.0	6-11 0.09	8-0 0.14	8-11 0.19	9-9 0.25	10-6 0.32	11-3 0.39	11-11 0.46	12-7 0.54	13-2 0.63	13-9 0.72	14-4 0.81
2x12	12.0	11-10 0.13	13-8 0.19	15-4 0.27	16-9 0.36	18-1 0.45	19-4 0.55	20-6 0.66	21-8 0.77	22-8 0.89	23-9 1.01	24-8 1.14
	13.7	11-1 0.12	12-10 0.18	14-4 0.25	15-8 0.33	16-11 0.42	18-1 0.52	19-3 0.61	20-3 0.72	21-3 0.83	22-2 0.95	23-1 1.07
	16.0	10-3 0.11	11-10 0.17	13-3 0.24	14-6 0.31	15-8 0.39	16-9 0.48	17-9 0.57	18-9 0.67	19-8 0.77	20-6 0.88	21-5 0.99
	19.2	9-5 0.10	10-10 0.15	12-1 0.22	13-3 0.28	14-4 0.36	15-4 0.44	16-3 0.52	17-1 0.61	17-11 0.70	18-9 0.80	19-6 0.90
	24.0	8-5 0.09	9-8 0.14	10-10 0.19	11-10 0.25	12-10 0.32	13-8 0.39	14-6 0.46	15-4 0.54	16-1 0.63	16-9 0.72	17-5 0.81

Note: The required modulus of elasticity, "E", in 1,000,000 pounds per square inch is shown below each span.

TABLE R-2 (cont.)

RAFTERS: Spans are measured along the horizontal projection and loads are considered as applied on the horizontal projection.

Extreme Fiber Stress in Bending, "F_b" (psi).											RAFTER SPACING (IN)	SIZE (IN)
1400	1500	1600	1700	1800	1900	2000	2100	2200	2400	2700		
12-6 / 1.28	13-0 / 1.41	13-5 / 1.56	13-10 / 1.71	14-2 / 1.86	14-7 / 2.02	15-0 / 2.18	15-4 / 2.34	15-8 / 2.51			12.0	
11-9 / 1.19	12-2 / 1.32	12-6 / 1.46	12-11 / 1.60	13-3 / 1.74	13-8 / 1.89	14-0 / 2.04	14-4 / 2.19	14-8 / 2.35			13.7	
10-10 / 1.10	11-3 / 1.22	11-7 / 1.35	11-11 / 1.48	12-4 / 1.61	12-8 / 1.75	13-0 / 1.89	13-3 / 2.03	13-7 / 2.18	14-2 / 2.48		16.0	2x6
9-11 / 1.01	10-3 / 1.12	10-7 / 1.23	10-11 / 1.35	11-3 / 1.47	11-6 / 1.59	11-10 / 1.72	12-2 / 1.85	12-5 / 1.99	13-0 / 2.26		19.2	
8-10 / 0.90	9-2 / 1.00	9-6 / 1.10	9-9 / 1.21	10-0 / 1.31	10-4 / 1.43	10-7 / 1.54	10-10 / 1.66	11-1 / 1.78	11-7 / 2.02	12-4 / 2.41	24.0	
16-6 / 1.28	17-1 / 1.41	17-8 / 1.56	18-2 / 1.71	18-9 / 1.86	19-3 / 2.02	19-9 / 2.18	20-3 / 2.34	20-8 / 2.51			12.0	
15-5 / 1.19	16-0 / 1.32	16-6 / 1.46	17-0 / 1.60	17-6 / 1.74	18-0 / 1.89	18-5 / 2.04	18-11 / 2.19	19-4 / 2.35			13.7	
14-4 / 1.10	14-10 / 1.22	15-3 / 1.35	15-9 / 1.48	16-3 / 1.61	16-8 / 1.75	17-1 / 1.89	17-6 / 2.03	17-11 / 2.18	18-9 / 2.48		16.0	2x8
13-1 / 1.01	13-6 / 1.12	13-11 / 1.23	14-5 / 1.35	14-10 / 1.47	15-2 / 1.59	15-7 / 1.72	16-0 / 1.85	16-4 / 1.99	17-1 / 2.26		19.2	
11-8 / 0.90	12-1 / 1.00	12-6 / 1.10	12-10 / 1.21	13-3 / 1.31	13-7 / 1.43	13-11 / 1.54	14-4 / 1.66	14-8 / 1.78	15-3 / 2.02	16-3 / 2.41	24.0	
21-1 / 1.28	21-10 / 1.41	22-6 / 1.56	23-3 / 1.71	23-11 / 1.86	24-6 / 2.02	25-2 / 2.18	25-10 / 2.34	26-5 / 2.51			12.0	
19-8 / 1.19	20-5 / 1.32	21-1 / 1.46	21-9 / 1.60	22-4 / 1.74	22-11 / 1.89	23-7 / 2.04	24-2 / 2.19	24-8 / 2.35			13.7	
18-3 / 1.10	18-11 / 1.22	19-6 / 1.35	20-1 / 1.48	20-8 / 1.61	21-3 / 1.75	21-10 / 1.89	22-4 / 2.03	22-10 / 2.18	23-11 / 2.48		16.0	2x10
16-8 / 1.01	17-3 / 1.12	17-10 / 1.23	18-4 / 1.35	18-11 / 1.47	19-5 / 1.59	19-11 / 1.72	20-5 / 1.85	20-10 / 1.99	21-10 / 2.26		19.2	
14-11 / 0.90	15-5 / 1.00	15-11 / 1.10	16-5 / 1.21	16-11 / 1.31	17-4 / 1.43	17-10 / 1.54	18-3 / 1.66	18-8 / 1.78	19-6 / 2.02	20-8 / 2.41	24.0	
25-7 / 1.28	26-6 / 1.41	27-5 / 1.56	28-3 / 1.71	29-1 / 1.86	29-10 / 2.02	30-7 / 2.18	31-4 / 2.34	32-1 / 2.51			12.0	
24-0 / 1.19	24-10 / 1.32	25-7 / 1.46	26-5 / 1.60	27-2 / 1.74	27-11 / 1.89	28-8 / 2.04	29-4 / 2.19	30-0 / 2.35			13.7	
22-2 / 1.10	23-0 / 1.22	23-9 / 1.35	24-5 / 1.48	25-2 / 1.61	25-10 / 1.75	26-6 / 1.89	27-2 / 2.03	27-10 / 2.18	29-1 / 2.48		16.0	2x12
20-3 / 1.01	21-0 / 1.12	21-8 / 1.23	22-4 / 1.35	23-0 / 1.47	23-7 / 1.59	24-2 / 1.72	24-10 / 1.85	25-5 / 1.99	26-6 / 2.26		19.2	
18-1 / 0.90	18-9 / 1.00	19-4 / 1.10	20-0 / 1.21	20-6 / 1.31	21-1 / 1.43	21-8 / 1.54	22-2 / 1.66	22-8 / 1.78	23-9 / 2.02	25-2 / 2.41	24.0	

Note: The required modulus of elasticity, "E", in 1,000,000 pounds per square inch is shown below each span.

TABLE R-3
FLAT OR SLOPED RAFTERS
Supporting Drywall Ceiling
(Flat roof or cathedral ceiling with no attic space)
Live Load - 40 lb. per sq. ft.

DESIGN CRITERIA:
Strength - 15 lbs. per sq. ft. dead load plus 40
  lbs. per sq. ft. live load determines required
  fiber stress.
Deflection - For 40 lbs. per sq. ft. live load.
  Limited to span in inches divided by 240.

RAFTER SIZE (IN)	SPACING (IN)	Extreme Fiber Stress in Bending, "F_b" (psi).										
		300	400	500	600	700	800	900	1000	1100	1200	1300
2x6	12.0	5-3 0.12	6-1 0.19	6-9 0.27	7-5 0.35	8-0 0.44	8-7 0.54	9-1 0.65	9-7 0.76	10-0 0.88	10-6 1.00	10-11 1.13
	13.7	4-11 0.12	5-8 0.18	6-4 0.25	6-11 0.33	7-6 0.42	8-0 0.51	8-6 0.61	8-11 0.71	9-5 0.82	9-10 0.93	10-3 1.05
	16.0	4-6 0.11	5-3 0.17	5-10 0.23	6-5 0.31	6-11 0.39	7-5 0.47	7-10 0.56	8-3 0.66	8-8 0.76	9-1 0.86	9-5 0.98
	19.2	4-2 0.10	4-9 0.15	5-4 0.21	5-10 0.28	6-4 0.35	6-9 0.43	7-2 0.51	7-7 0.60	7-11 0.69	8-3 0.79	8-8 0.89
	24.0	3-8 0.09	4-3 0.14	4-9 0.19	5-3 0.25	5-8 0.31	6-1 0.38	6-5 0.46	6-9 0.54	7-1 0.62	7-5 0.71	7-9 0.80
2x8	12.0	6-11 0.12	8-0 0.19	8-11 0.27	9-9 0.35	10-7 0.44	11-3 0.54	12-0 0.65	12-7 0.76	13-3 0.88	13-10 1.00	14-5 1.13
	13.7	6-6 0.12	7-6 0.18	8-4 0.25	9-2 0.33	9-11 0.42	10-7 0.51	11-2 0.61	11-10 0.71	12-5 0.82	12-11 0.93	13-6 1.05
	16.0	6-0 0.11	6-11 0.17	7-9 0.23	8-6 0.31	9-2 0.39	9-9 0.47	10-4 0.56	10-11 0.66	11-6 0.76	12-0 0.86	12-6 0.98
	19.2	5-6 0.10	6-4 0.15	7-1 0.21	7-9 0.28	8-4 0.35	8-11 0.43	9-6 0.51	10-0 0.60	10-6 0.69	10-11 0.79	11-5 0.89
	24.0	4-11 0.09	5-8 0.14	6-4 0.19	6-11 0.25	7-6 0.31	8-0 0.38	8-6 0.46	8-11 0.54	9-4 0.62	9-9 0.71	10-2 0.80
2x10	12.0	8-10 0.12	10-2 0.19	11-5 0.27	12-6 0.35	13-6 0.44	14-5 0.54	15-3 0.65	16-1 0.76	16-11 0.88	17-8 1.00	18-4 1.13
	13.7	8-3 0.12	9-6 0.18	10-8 0.25	11-8 0.33	12-7 0.42	13-6 0.51	14-3 0.61	15-1 0.71	15-10 0.82	16-6 0.93	17-2 1.05
	16.0	7-8 0.11	8-10 0.17	9-10 0.23	10-10 0.31	11-8 0.39	12-6 0.47	13-3 0.56	13-11 0.66	14-8 0.76	15-3 0.86	15-11 0.98
	19.2	7-0 0.10	8-1 0.15	9-0 0.21	9-10 0.28	10-8 0.35	11-5 0.43	12-1 0.51	12-9 0.60	13-4 0.69	13-11 0.79	14-6 0.89
	24.0	6-3 0.09	7-2 0.14	8-1 0.19	8-10 0.25	9-6 0.31	10-2 0.38	10-10 0.46	11-5 0.54	11-11 0.62	12-6 0.71	13-0 0.80
2x12	12.0	10-9 0.12	12-5 0.19	13-10 0.27	15-2 0.35	16-5 0.44	17-6 0.54	18-7 0.65	19-7 0.76	20-6 0.88	21-5 1.00	22-4 1.13
	13.7	10-0 0.12	11-7 0.18	12-11 0.25	14-2 0.33	15-4 0.42	16-5 0.51	17-5 0.61	18-4 0.71	19-3 0.82	20-1 0.93	20-11 1.05
	16.0	9-3 0.11	10-9 0.17	12-0 0.23	13-2 0.31	14-2 0.39	15-2 0.47	16-1 0.56	17-0 0.66	17-9 0.76	18-7 0.86	19-4 0.98
	19.2	8-6 0.10	9-10 0.15	10-11 0.21	12-0 0.28	12-11 0.35	13-10 0.43	14-8 0.51	15-6 0.60	16-3 0.69	17-0 0.79	17-8 0.89
	24.0	7-7 0.09	8-9 0.14	9-10 0.19	10-9 0.25	11-7 0.31	12-5 0.38	13-2 0.46	13-10 0.54	14-6 0.62	15-2 0.71	15-9 0.80

Note: The required modulus of elasticity, "E", in 1,000,000 pounds per square inch is shown below each span.

TABLE R-3 (cont.)

RAFTERS: Spans are measured along the horizontal projection and loads are considered as applied on the horizontal projection.

Extreme Fiber Stress in Bending, "F_b" (psi).											RAFTER SPACING (IN)	SIZE (IN)
1400	1500	1600	1700	1800	1900	2000	2100	2200	2400	2700		
11-4 1.26	11-9 1.40	12-1 1.54	12-6 1.68	12-10 1.83	13-2 1.99	13-6 2.15	13-10 2.31	14-2 2.48			12.0	
10-7 1.18	11-0 1.31	11-4 1.44	11-8 1.57	12-0 1.72	12-4 1.86	12-8 2.01	13-0 2.16	13-3 2.32			13.7	
9-10 1.09	10-2 1.21	10-6 1.33	10-10 1.46	11-1 1.59	11-5 1.72	11-9 1.86	12-0 2.00	12-4 2.15	12-10 2.45		16.0	2x6
8-11 0.99	9-3 1.10	9-7 1.22	9-10 1.33	10-2 1.45	10-5 1.57	10-8 1.70	11-0 1.83	11-3 1.96	11-9 2.23		19.2	
8-0 0.89	8-3 0.99	8-7 1.09	8-10 1.19	9-1 1.30	9-4 1.41	9-7 1.52	9-10 1.63	10-0 1.75	10-6 2.00	11-1 2.38	24.0	
14-11 1.26	15-5 1.40	16-0 1.54	16-5 1.68	16-11 1.83	17-5 1.99	17-10 2.15	18-3 2.31	18-9 2.48			12.0	
14-0 1.18	14-6 1.31	14-11 1.44	15-5 1.57	15-10 1.72	16-3 1.86	16-8 2.01	17-1 2.16	17-6 2.32			13.7	
12-11 1.09	13-5 1.21	13-10 1.33	14-3 1.46	14-8 1.59	15-1 1.72	15-5 1.86	15-10 2.00	16-3 2.15	16-11 2.45		16.0	2x8
11-10 0.99	12-3 1.10	12-7 1.22	13-0 1.33	13-5 1.45	13-9 1.57	14-1 1.70	14-6 1.83	14-10 1.96	15-5 2.23		19.2	
10-7 0.89	10-11 0.99	11-3 1.09	11-8 1.19	12-0 1.30	12-4 1.41	12-7 1.52	12-11 1.63	13-3 1.75	13-10 2.00	14-8 2.38	24.0	
19-1 1.26	19-9 1.40	20-4 1.54	21-0 1.68	21-7 1.83	22-2 1.99	22-9 2.15	23-4 2.31	23-11 2.48			12.0	
17-10 1.18	18-5 1.31	19-1 1.44	19-8 1.57	20-2 1.72	20-9 1.86	21-4 2.01	21-10 2.16	22-4 2.32			13.7	
16-6 1.09	17-1 1.21	17-8 1.33	18-2 1.46	18-9 1.59	19-3 1.72	19-9 1.86	20-2 2.00	20-8 2.15	21-7 2.45		16.0	2x10
15-1 0.99	15-7 1.10	16-1 1.22	16-7 1.33	17-1 1.45	17-7 1.57	18-0 1.70	18-5 1.83	18-11 1.96	19-9 2.23		19.2	
13-6 0.89	13-11 0.99	14-5 1.09	14-10 1.19	15-3 1.30	15-8 1.41	16-1 1.52	16-6 1.63	16-11 1.75	17-8 2.00	18-9 2.38	24.0	
23-2 1.26	24-0 1.40	24-9 1.54	25-6 1.68	26-3 1.83	27-0 1.99	27-8 2.15	28-5 2.31	29-1 2.48			12.0	
21-8 1.18	22-5 1.31	23-2 1.44	23-11 1.57	24-7 1.72	25-3 1.86	25-11 2.01	26-7 2.16	27-2 2.32			13.7	
20-1 1.09	20-9 1.21	21-5 1.33	22-1 1.46	22-9 1.59	23-5 1.72	24-0 1.86	24-7 2.00	25-2 2.15	26-3 2.45		16.0	2x12
18-4 0.99	19-0 1.10	19-7 1.22	20-2 1.33	20-9 1.45	21-4 1.57	21-11 1.70	22-5 1.83	23-0 1.96	24-0 2.23		19.2	
16-5 0.89	17-0 0.99	17-6 1.09	18-1 1.19	18-7 1.30	19-1 1.41	19-7 1.52	20-1 1.63	20-6 1.75	21-5 2.00	22-9 2.38	24.0	

Note: The required modulus of elasticity, "E", in 1,000,000 pounds per square inch is shown below each span.

TABLE R-4
FLAT OR SLOPED RAFTERS
Supporting Plaster Ceiling
(Flat roof or cathedral ceiling with no attic space)
Live Load - 20 lb. per sq. ft.

DESIGN CRITERIA:
Strength - 15 lbs. per sq. ft. dead load plus 20
lbs. per sq. ft. live load determines required
fiber stress.
Deflection - For 20 lbs. per sq. ft. live load.
Limited to span in inches divided by 360.

RAFTER SIZE (IN)	SPACING (IN)	Extreme Fiber Stress in Bending, "F$_b$" (psi).									
		300	400	500	600	700	800	900	1000	1100	1200
2x6	12.0	6-7 0.18	7-7 0.28	8-6 0.40	9-4 0.52	10-0 0.66	10-9 0.80	11-5 0.96	12-0 1.12	12-7 1.29	13-2 1.48
	13.7	6-2 0.17	7-1 0.27	7-11 0.37	8-8 0.49	9-5 0.61	10-0 0.75	10-8 0.90	11-3 1.05	11-9 1.21	12-4 1.38
	16.0	5-8 0.16	6-7 0.25	7-4 0.34	8-1 0.45	8-8 0.57	9-4 0.70	9-10 0.83	10-5 0.97	10-11 1.12	11-5 1.28
	19.2	5-2 0.15	6-0 0.22	6-9 0.31	7-4 0.41	7-11 0.52	8-6 0.63	9-0 0.76	9-6 0.89	9-11 1.02	10-5 1.17
	24.0	4-8 0.13	5-4 0.20	6-0 0.28	6-7 0.37	7-1 0.46	7-7 0.57	8-1 0.68	8-6 0.79	8-11 0.92	9-4 1.04
2x8	12.0	8-8 0.18	10-0 0.28	11-2 0.40	12-3 0.52	13-3 0.66	14-2 0.80	15-0 0.96	15-10 1.12	16-7 1.29	17-4 1.48
	13.7	8-1 0.17	9-4 0.27	10-6 0.37	11-6 0.49	12-5 0.61	13-3 0.75	14-0 0.90	14-10 1.05	15-6 1.21	16-3 1.38
	16.0	7-6 0.16	8-8 0.25	9-8 0.34	10-7 0.45	11-6 0.57	12-3 0.70	13-0 0.83	13-8 0.97	14-4 1.12	15-0 1.28
	19.2	6-10 0.15	7-11 0.22	8-10 0.31	9-8 0.41	10-6 0.52	11-2 0.63	11-10 0.76	12-6 0.89	13-1 1.02	13-8 1.17
	24.0	6-2 0.13	7-1 0.20	7-11 0.28	8-8 0.37	9-4 0.46	10-0 0.57	10-7 0.68	11-2 0.79	11-9 0.92	12-3 1.04
2x10	12.0	11-1 0.18	12-9 0.28	14-3 0.40	15-8 0.52	16-11 0.66	18-1 0.80	19-2 0.96	20-2 1.12	21-2 1.29	22-1 1.48
	13.7	10-4 0.17	11-11 0.27	13-4 0.37	14-8 0.49	15-10 0.61	16-11 0.75	17-11 0.90	18-11 1.05	19-10 1.21	20-8 1.38
	16.0	9-7 0.16	11-1 0.25	12-4 0.34	13-6 0.45	14-8 0.57	15-8 0.70	16-7 0.83	17-6 0.97	18-4 1.12	19-2 1.28
	19.2	8-9 0.15	10-1 0.22	11-3 0.31	12-4 0.41	13-4 0.52	14-3 0.63	15-2 0.76	15-11 0.89	16-9 1.02	17-6 1.17
	24.0	7-10 0.13	9-0 0.20	10-1 0.28	11-1 0.37	11-11 0.46	12-9 0.57	13-6 0.68	14-3 0.79	15-0 0.92	15-8 1.04
2x12	12.0	13-5 0.18	15-6 0.28	17-4 0.40	19-0 0.52	20-6 0.66	21-11 0.80	23-3 0.96	24-7 1.12	25-9 1.29	26-11 1.48
	13.7	12-7 0.17	14-6 0.27	16-3 0.37	17-9 0.49	19-3 0.61	20-6 0.75	21-9 0.90	23-0 1.05	24-1 1.21	25-2 1.38
	16.0	11-8 0.16	13-5 0.25	15-0 0.34	16-6 0.45	17-9 0.57	19-0 0.70	20-2 0.83	21-3 0.97	22-4 1.12	23-3 1.28
	19.2	10-8 0.15	12-3 0.22	13-9 0.31	15-0 0.41	16-3 0.52	17-4 0.63	18-5 0.76	19-5 0.89	20-4 1.02	21-3 1.17
	24.0	9-6 0.13	11-0 0.20	12-3 0.28	13-5 0.37	14-6 0.46	15-6 0.57	16-6 0.68	17-4 0.79	18-2 0.92	19-0 1.04

Note: The required modulus of elasticity, "E", in 1,000,000 pounds per square inch is shown
below each span.

TABLE R-4 (cont.)

RAFTERS: Spans are measured along the
horizontal projection and loads are
considered as applied on the horizontal
projection.

Extreme Fiber Stress in Bending, "F$_b$" (psi).									RAFTER	
1300	1400	1500	1600	1700	1800	1900	2000	2100	SPACING (IN)	SIZE (IN)
13-8 1.66	14-2 1.86	14-8 2.06	15-2 2.27	15-8 2.49					12.0	
12-10 1.56	13-3 1.74	13-9 1.93	14-2 2.12	14-8 2.33	15-1 2.54				13.7	
11-10 1.44	12-4 1.61	12-9 1.79	13-2 1.97	13-7 2.15	13-11 2.35	14-4 2.55			16.0	2x6
10-10 1.32	11-3 1.47	11-7 1.63	12-0 1.80	12-4 1.97	12-9 2.14	13-1 2.32	13-5 2.51		19.2	
9-8 1.18	10-0 1.31	10-5 1.46	10-9 1.61	11-1 1.76	11-5 1.92	11-8 2.08	12-0 2.24	12-4 2.41	24.0	
18-0 1.66	18-9 1.86	19-5 2.06	20-0 2.27	20-8 2.49					12.0	
16-10 1.56	17-6 1.74	18-2 1.93	18-9 2.12	19-4 2.33	19-10 2.54				13.7	
15-7 1.44	16-3 1.61	16-9 1.79	17-4 1.97	17-10 2.15	18-5 2.35	18-11 2.55			16.0	2x8
14-3 1.32	14-10 1.47	15-4 1.63	15-10 1.80	16-4 1.97	16-9 2.14	17-3 2.32	17-8 2.51		19.2	
12-9 1.18	13-3 1.31	13-8 1.46	14-2 1.61	14-7 1.76	15-0 1.92	15-5 2.08	15-10 2.24	16-3 2.41	24.0	
23-0 1.66	23-11 1.86	24-9 2.06	25-6 2.27	26-4 2.49					12.0	
21-6 1.56	22-4 1.74	23-2 1.93	23-11 2.12	24-7 2.33	25-4 2.54				13.7	
19-11 1.44	20-8 1.61	21-5 1.79	22-1 1.97	22-10 2.15	23-5 2.35	24-1 2.55			16.0	2x10
18-2 1.32	18-11 1.47	19-7 1.63	20-2 1.80	20-10 1.97	21-5 2.14	22-0 2.32	22-7 2.51		19.2	
16-3 1.18	16-11 1.31	17-6 1.46	18-1 1.61	18-7 1.76	19-2 1.92	19-8 2.08	20-2 2.24	20-8 2.41	24.0	
28-0 1.66	29-1 1.86	30-1 2.06	31-1 2.27	32-0 2.49					12.0	
26-2 1.56	27-2 1.74	28-2 1.93	29-1 2.12	29-11 2.33	30-10 2.54				13.7	
24-3 1.44	25-2 1.61	26-0 1.79	26-11 1.97	27-9 2.15	28-6 2.35	29-4 2.55			16.0	2x12
22-2 1.32	23-0 1.47	23-9 1.63	24-7 1.80	25-4 1.97	26-0 2.14	26-9 2.32	27-5 2.51		19.2	
19-10 1.18	20-6 1.31	21-3 1.46	21-11 1.61	22-8 1.76	23-3 1.92	23-11 2.08	24-7 2.24	25-2 2.41	24.0	

Note: The required modulus of elasticity, "E", in 1,000,000 pounds per square inch is shown
below each span.

TABLE R-5
FLAT OR SLOPED RAFTERS
Supporting Plaster Ceiling
(Flat roof or cathedral ceiling with no attic space)
Live Load - 30 lb. per sq. ft.

DESIGN CRITERIA:

Strength - 15 lbs. per sq. ft. dead load plus 30 lbs. per sq. ft. live load determines required fiber stress.

Deflection - For 30 lbs. per sq. ft. live load. Limited to Span in inches divided by 360.

RAFTER SIZE (IN)	SPACING (IN)	Extreme Fiber Stress in Bending, "F_b" (psi).									
		300	400	500	600	700	890	900	1000	1100	1200
2x6	12.0	5-10 0.19	6-8 0.29	7-6 0.41	8-2 0.54	8-10 0.68	9-6 0.83	10-0 0.99	10-7 1.15	11-1 1.33	11-7 1.52
	13.7	5-5 0.18	6-3 0.27	7-0 0.38	7-8 0.50	8-3 0.63	8-10 0.77	9-5 0.92	9-11 1.08	10-5 1.25	10-10 1.42
	16.0	5-0 0.16	5-10 0.25	6-6 0.35	7-1 0.46	7-8 0.59	8-2 0.72	8-8 0.85	9-2 1.00	9-7 1.15	10-0 1.31
	19.2	4-7 0.15	5-4 0.23	5-11 0.32	6-6 0.42	7-0 0.53	7-6 0.65	7-11 0.78	8-4 0.91	8-9 1.05	9-2 1.20
	24.0	4-1 0.13	4-9 0.21	5-4 0.29	5-10 0.38	6-3 0.48	6-8 0.58	7-1 0.70	7-6 0.82	7-10 0.94	8-2 1.07
2x8	12.0	7-8 0.19	8-10 0.29	9-10 0.41	10-10 0.54	11-8 0.68	12-6 0.83	13-3 0.99	13-11 1.15	14-8 1.33	15-3 1.52
	13.7	7-2 0.18	8-3 0.27	9-3 0.38	10-1 0.50	10-11 0.63	11-8 0.77	12-5 0.92	13-1 1.08	13-8 1.25	14-4 1.42
	16.0	6-7 0.16	7-8 0.25	8-7 0.35	9-4 0.46	10-1 0.59	10-10 0.72	11-6 0.85	12-1 1.00	12-8 1.15	13-3 1.31
	19.2	6-1 0.15	7-0 0.23	7-10 0.32	8-7 0.42	9-3 0.53	9-10 0.65	10-6 0.78	11-0 0.91	11-7 1.05	12-1 1.20
	24.0	5-5 0.13	6-3 0.21	7-0 0.29	7-8 0.38	8-3 0.48	8-10 0.58	9-4 0.70	9-10 0.82	10-4 0.94	10-10 1.07
2x10	12.0	9-9 0.19	11-3 0.29	12-7 0.41	13-9 0.54	14-11 0.68	15-11 0.83	16-11 0.99	17-10 1.15	18-8 1.33	19-6 1.52
	13.7	9-1 0.18	10-6 0.27	11-9 0.38	12-11 0.50	13-11 0.63	14-11 0.77	15-10 0.92	16-8 1.08	17-6 1.25	18-3 1.42
	16.0	8-5 0.16	9-9 0.25	10-11 0.35	11-11 0.46	12-11 0.59	13-9 0.72	14-8 0.85	15-5 1.00	16-2 1.15	16-11 1.31
	19.2	7-8 0.15	8-11 0.23	9-11 0.32	10-11 0.42	11-9 0.53	12-7 0.65	13-4 0.78	14-1 0.91	14-9 1.05	15-5 1.20
	24.0	6-11 0.13	8-0 0.21	8-11 0.29	9-9 0.38	10-6 0.48	11-3 0.58	11-11 0.70	12-7 0.82	13-2 0.94	13-9 1.07
2x12	12.0	11-10 0.19	13-8 0.29	15-4 0.41	16-9 0.54	18-1 0.68	19-4 0.83	20-6 0.99	21-8 1.15	22-8 1.33	23-9 1.52
	13.7	11-1 0.18	12-10 0.27	14-4 0.38	15-8 0.50	16-11 0.63	18-1 0.77	19-3 0.92	20-3 1.08	21-3 1.25	22-2 1.42
	16.0	10-3 0.16	11-10 0.25	13-3 0.35	14-6 0.46	15-8 0.59	16-9 0.72	17-9 0.85	18-9 1.00	19-8 1.15	20-6 1.31
	19.2	9-5 0.15	10-10 0.23	12-1 0.32	13-3 0.42	14-4 0.53	15-4 0.65	16-3 0.78	17-1 0.91	17-11 1.05	18-9 1.20
	24.0	8-5 0.13	9-8 0.21	10-10 0.29	11-10 0.38	12-10 0.48	13-8 0.58	14-6 0.70	15-4 0.82	16-1 0.94	16-9 1.07

Note: The required modulus of elasticity, "E", in 1,000,000 pounds per square inch is shown below each span.

TABLE R-5 (cont.)

RAFTERS: Spans are measured along the
horizontal projection and loads are
considered as applied on the horizontal
projection.

\multicolumn{9}{c}{Extreme Fiber Stress in Bending, "F_b" (psi).}	RAFTER SPACING (IN)	SIZE (IN)
1300	1400	1500	1600	1700	1800	1900	2000	2100		
12-1 1.71	12-6 1.91	13-0 2.12	13-5 2.34	13-10 2.56					12.0	
11-3 1.60	11-9 1.79	12-2 1.98	12-6 2.19	12-11 2.39					13.7	
10-5 1.48	10-10 1.66	11-3 1.84	11-7 2.02	11-11 2.22	12-4 2.41				16.0	2x6
9-6 1.35	9-11 1.51	10-3 1.68	10-7 1.85	10-11 2.02	11-3 2.20	11-6 2.39	11-10 2.58		19.2	
8-6 1.21	8-10 1.35	9-2 1.50	9-6 1.65	9-9 1.81	10-0 1.97	10-4 2.14	10-7 2.31	10-10 2.48	24.0	
15-11 1.71	16-6 1.91	17-1 2.12	17-8 2.34	18-2 2.56					12.0	
14-11 1.60	15-5 1.79	16-0 1.98	16-6 2.19	17-0 2.39					13.7	
13-9 1.48	14-4 1.66	14-10 1.84	15-3 2.02	15-9 2.22	16-3 2.41				16.0	2x8
12-7 1.35	13-1 1.51	13-6 1.68	13-11 1.85	14-5 2.02	14-10 2.20	15-2 2.39	15-7 2.58		19.2	
11-3 1.21	11-8 1.35	12-1 1.50	12-6 1.65	12-10 1.81	13-3 1.97	13-7 2.14	13-11 2.31	14-4 2.48	24.0	
20-4 1.71	21-1 1.91	21-10 2.12	22-6 2.34	23-3 2.56					12.0	
19-0 1.60	19-8 1.79	20-5 1.98	21-1 2.19	21-9 2.39					13.7	
17-7 1.48	18-3 1.66	18-11 1.84	19-6 2.02	20-1 2.22	20-8 2.41				16.0	2x10
16-1 1.35	16-8 1.51	17-3 1.68	17-10 1.85	18-4 2.02	18-11 2.20	19-5 2.39	19-11 2.58		19.2	
14-4 1.21	14-11 1.35	15-5 1.50	15-11 1.65	16-5 1.81	16-11 1.97	17-4 2.14	17-10 2.31	18-3 2.48	24.0	
24-8 1.71	25-7 1.91	26-6 2.12	27-5 2.34	28-3 2.56					12.0	
23-1 1.60	24-0 1.79	24-10 1.98	25-7 2.19	26-5 2.39					13.7	
21-5 1.48	22-2 1.66	23-0 1.84	23-9 2.02	24-5 2.22	25-2 2 41				16.0	2x12
19-6 1.35	20-3 1.51	21-0 1.68	21-8 1.85	22-4 2.02	23-0 2.20	23-7 2.39	24-2 2.58		19.2	
17-5 1.21	18-1 1.35	18-9 1.50	19-4 1.65	20-0 1.81	20-6 1.97	21-1 2.14	21-8 2.31	22-2 2.48	24.0	

Note: The required modulus of elasticity, "E", in 1,000,000 pounds per square inch is shown
below each span.

TABLE R-6
FLAT OR SLOPED RAFTERS
Supporting Plaster Ceiling
(Flat roof or cathedral ceiling with no attic space)
Live Load - 40 lb. per sq. ft.

DESIGN CRITERIA:
Strength - 15 lbs. per sq. ft. dead load plus 40
lbs. per sq. ft. live load determines required
fiber stress.
Deflection - For 40 lbs. per sq. ft. live load.
Limited to span in inches divided by 360.

RAFTER SIZE (IN)	SPACING (IN)	Extreme Fiber Stress in Bending, "F_b" (psi).									
		300	400	500	600	700	800	900	1000	1100	1200
2x6	12.0	5-3 0.19	6-1 0.29	6-9 0.40	7-5 0.53	8-0 0.67	8-7 0.82	9-1 0.97	9-7 1.14	10-0 1.31	10-6 1.50
	13.7	4-11 0.18	5-8 0.27	6-4 0.38	6-11 0.50	7-6 0.62	8-0 0.76	8-6 0.91	8-11 1.07	9-5 1.23	9-10 1.40
	16.0	4-6 0.16	5-3 0.25	5-10 0.35	6-5 0.46	6-11 0.58	7-5 0.71	7-10 0.84	8-3 0.99	8-8 1.14	9-1 1.30
	19.2	4-2 0.15	4-9 0.23	5-4 0.32	5-10 0.42	6-4 0.53	6-9 0.64	7-2 0.77	7-7 0.90	7-11 1.04	8-3 1.18
	24.0	3-8 0.13	4-3 0.20	4-9 0.28	5-3 0.37	5-8 0.47	6-1 0.58	6-5 0.69	6-9 0.81	7-1 0.93	7-5 1.06
2x8	12.0	6-11 0.19	8-0 0.29	8-11 0.40	9-9 0.53	10-7 0.67	11-3 0.82	12-0 0.97	12-7 1.14	13-3 1.31	13-10 1.50
	13.7	6-6 0.18	7-6 0.27	8-4 0.38	9-2 0.50	9-11 0.62	10-7 0.76	11-2 0.91	11-10 1.07	12-5 1.23	12-11 1.40
	16.0	6-0 0.16	6-11 0.25	7-9 0.35	8-6 0.46	9-2 0.58	9-9 0.71	10-4 0.84	10-11 0.99	11-6 1.14	12-0 1.30
	19.2	5-6 0.15	6-4 0.23	7-1 0.32	7-9 0.42	8-4 0.53	8-11 0.64	9-6 0.77	10-0 0.90	10-6 1.04	10-11 1.18
	24.0	4-11 0.13	5-8 0.20	6-4 0.28	6-11 0.37	7-6 0.47	8-0 0.58	8-6 0.69	8-11 0.81	9-4 0.93	9-9 1.06
2x10	12.0	8-10 0.19	10-2 0.29	11-5 0.40	12-6 0.53	13-6 0.67	14-5 0.82	15-3 0.97	16-1 1.14	16-11 1.31	17-8 1.50
	13.7	8-3 0.18	9-6 0.27	10-8 0.38	11-8 0.50	12-7 0.62	13-6 0.76	14-3 0.91	15-1 1.07	15-10 1.23	16-6 1.40
	16.0	7-8 0.16	8-10 0.25	9-10 0.35	10-10 0.46	11-8 0.58	12-6 0.71	13-3 0.84	13-11 0.99	14-8 1.14	15-3 1.30
	19.2	7-0 0.15	8-1 0.23	9-0 0.32	9-10 0.42	10-8 0.53	11-5 0.64	12-1 0.77	12-9 0.90	13-4 1.04	13-11 1.18
	24.0	6-3 0.13	7-2 0.20	8-1 0.28	8-10 0.37	9-6 0.47	10-2 0.58	10-10 0.69	11-5 0.81	11-11 0.93	12-6 1.06
2x12	12.0	10-9 0.19	12-5 0.29	13-10 0.40	15-2 0.53	16-5 0.67	17-6 0.82	18-7 0.97	19-7 1.14	20-6 1.31	21-5 1.50
	13.7	10-0 0.18	11-7 0.27	12-11 0.38	14-2 0.50	15-4 0.62	16-5 0.76	17-5 0.91	18-4 1.07	19-3 1.23	20-1 1.40
	16.0	9-3 0.16	10-9 0.25	12-0 0.35	13-2 0.46	14-2 0.58	15-2 0.71	16-1 0.84	17-0 0.99	17-9 1.14	18-7 1.30
	19.2	8-6 0.15	9-10 0.23	10-11 0.32	12-0 0.42	12-11 0.53	13-10 0.64	14-8 0.77	15-6 0.90	16-3 1.04	17-0 1.18
	24.0	7-7 0.13	8-9 0.20	9-10 0.28	10-9 0.37	11-7 0.47	12-5 0.58	13-2 0.69	13-10 0.81	14-6 0.93	15-2 1.06

Note: The required modulus of elasticity, "E", in 1,000,000 pounds per square inch is shown below
each span.

TABLE R-6 (cont.)

RAFTERS: Spans are measured along the
horizontal projection and loads are
considered as applied on the horizontal
projection.

Extreme Fiber Stress in Bending, "F_b" (psi).									RAFTER	SPACING SIZE
1300	1400	1500	1600	1700	1800	1900	2000	2100	(IN)	(IN)
10-11 1.69	11-4 1.89	11-9 2.09	12-1 2.31	12-6 2.53					12.0	
10-3 1.58	10-7 1.77	11-0 1.96	11-4 2.16	11-8 2.36	12-0 2.57				13.7	
9-5 1.46	9-10 1.63	10-2 1.81	10-6 2.00	10-10 2.19	11-1 2.38	11-5 2.58			16.0	2x6
8-8 1.34	8-11 1.49	9-3 1.65	9-7 1.82	9-10 2.00	10-2 2.18	10-5 2.36	10-8 2.55		19.2	
7-9 1.19	8-0 1.33	8-3 1.48	8-7 1.63	8-10 1.79	9-1 1.95	9-4 2.11	9-7 2.28	9-10 2.45	24.0	
14-5 1.69	14-11 1.89	15-5 2.09	16-0 2.31	16-5 2.53					12.0	
13-6 1.58	14-0 1.77	14-6 1.96	14-11 2.16	15-5 2.36	15-10 2.57				13.7	
12-6 1.46	12-11 1.63	13-5 1.81	13-10 2.00	14-3 2.19	14-8 2.38	15-1 2.58			16.0	2x8
11-5 1.34	11-10 1.49	12-3 1.65	12-7 1.82	13-0 2.00	13-5 2.18	13-9 2.36	14-1 2.55		19.2	
10-2 1.19	10-7 1.33	10-11 1.48	11-3 1.63	11-8 1.79	12-0 1.95	12-4 2.11	12-7 2.28	12-11 2.45	24.0	
18-4 1.69	19-1 1.89	19-9 2.09	20-4 2.31	21-0 2.53					12.0	
17-2 1.58	17-10 1.77	18-5 1.96	19-1 2.16	19-8 2.36	20-2 2.57				13.7	
15-11 1.46	16-6 1.63	17-1 1.81	17-8 2.00	18-2 2.19	18-9 2.38	19-3 2.58			16.0	2x10
14-6 1.34	15-1 1.49	15-7 1.65	16-1 1.82	16-7 2.00	17-1 2.18	17-7 2.36	18-0 2.55		19.2	
13-0 1.19	13-6 1.33	13-11 1.48	14-5 1.63	14-10 1.79	15-3 1.95	15-8 2.11	16-1 2.28	16-6 2.45	24.0	
22-4 1.69	23-2 1.89	24-0 2.09	24-9 2.31	25-6 2.53					12.0	
20-11 1.58	21-8 1.77	22-5 1.96	23-2 2.16	23-11 2.36	24-7 2.57				13.7	
19-4 1.46	20-1 1.63	20-9 1.81	21-5 2.00	22-1 2.19	22-9 2.38	23-5 2.58			16.0	2x12
17-8 1.34	18-4 1.49	19-0 1.65	19-7 1.82	20-2 2.00	20-9 2.18	21-4 2.36	21-11 2.55		19.2	
15-9 1.19	16-5 1.33	17-0 1.48	17-6 1.63	18-1 1.79	18-7 1.95	19-1 2.11	19-7 2.28	20-1 2.45	24.0	

Note: The required modulus of elasticity, "E", in 1,000,000 pounds per square inch is shown
below each span.

TABLE R-7
FLAT OR LOW SLOPE RAFTERS
No Ceiling Load
Slope 3 in 12 or less
Live Load - 20 lb. per. sq. ft.

DESIGN CRITERIA:
Strength - 10 lbs. per sq. ft. dead load plus 20
  lbs. per sq. ft. live load determines required
  fiber stress.
Deflection - For 20 lbs. per sq. ft. live load.
  Limited to span in inches divided by 240.

RAFTER SIZE (IN)	SPACING (IN)	Extreme Fiber Stress in Bending, "Fb" (psi).										
		300	400	500	600	700	800	900	1000	1100	1200	1300
2x6	12.0	7-1 0.15	8-2 0.24	9-2 0.33	10-0 0.44	10-10 0.55	11-7 0.67	12-4 0.80	13-0 0.94	13-7 1.09	14-2 1.24	14-9 1.40
	13.7	6-8 0.14	7-8 0.22	8-7 0.31	9-5 0.41	10-2 0.52	10-10 0.63	11-6 0.75	12-2 0.88	12-9 1.02	13-3 1.16	13-10 1.31
	16.0	6-2 0.13	7-1 0.21	7-11 0.29	8-8 0.38	9-5 0.48	10-0 0.58	10-8 0.70	11-3 0.82	11-9 0.94	12-4 1.07	12-10 1.21
	19.2	5-7 0.12	6-6 0.19	7-3 0.26	7-11 0.35	8-7 0.44	9-2 0.53	9-9 0.64	10-3 0.75	10-9 0.86	11-3 0.98	11-8 1.10
	24.0	5-0 0.11	5-10 0.17	6-6 0.24	7-1 0.31	7-8 0.39	8-2 0.48	8-8 0.57	9-2 0.67	9-7 0.77	10-0 0.88	10-5 0.99
2x8	12.0	9-4 0.15	10-10 0.24	12-1 0.33	13-3 0.44	14-4 0.55	15-3 0.67	16-3 0.80	17-1 0.94	17-11 1.09	18-9 1.24	19-6 1.40
	13.7	8-9 0.14	10-1 0.22	11-4 0.31	12-5 0.41	13-4 0.52	14-4 0.63	15-2 0.75	16-0 0.88	16-9 1.02	17-6 1.16	18-3 1.31
	16.0	8-1 0.13	9-4 0.21	10-6 0.29	11-6 0.38	12-5 0.48	13-3 0.58	14-0 0.70	14-10 0.82	15-6 0.94	16-3 1.07	16-10 1.21
	19.2	7-5 0.12	8-7 0.19	9-7 0.26	10-6 0.35	11-4 0.44	12-1 0.53	12-10 0.64	13-6 0.75	14-2 0.86	14-10 0.98	15-5 1.10
	24.0	6-7 0.11	7-8 0.17	8-7 0.24	9-4 0.31	10-1 0.39	10-10 0.48	11-6 0.57	12-1 0.67	12-8 0.77	13-3 0.88	13-9 0.99
2x10	12.0	11-11 0.15	13-9 0.24	15-5 0.33	16-11 0.44	18-3 0.55	19-6 0.67	20-8 0.80	21-10 0.94	22-10 1.09	23-11 1.24	24-10 1.40
	13.7	11-2 0.14	12-11 0.22	14-5 0.31	15-10 0.41	17-1 0.52	18-3 0.63	19-4 0.75	20-5 0.88	21-5 1.02	22-4 1.16	23-3 1.31
	16.0	10-4 0.13	11-11 0.21	13-4 0.29	14-8 0.38	15-10 0.48	16-11 0.58	17-11 0.70	18-11 0.82	19-10 0.94	20-8 1.07	21-6 1.21
	19.2	9-5 0.12	10-11 0.19	12-2 0.26	13-4 0.35	14-5 0.44	15-5 0.53	16-4 0.64	17-3 0.75	18-1 0.86	18-11 0.98	19-8 1.10
	24.0	8-5 0.11	9-9 0.17	10-11 0.24	11-11 0.31	12-11 0.39	13-9 0.48	14-8 0.57	15-5 0.67	16-2 0.77	16-11 0.88	17-7 0.99
2x12	12.0	14-6 0.15	16-9 0.24	18-9 0.33	20-6 0.44	22-2 0.55	23-9 0.67	25-2 0.80	26-6 0.94	27-10 1.09	29-1 1.24	30-3 1.40
	13.7	13-7 0.14	15-8 0.22	17-6 0.31	19-3 0.41	20-9 0.52	22-2 0.63	23-6 0.75	24-10 0.88	26-0 1.02	27-2 1.16	28-3 1.31
	16.0	12-7 0.13	14-6 0.21	16-3 0.29	17-9 0.38	19-3 0.48	20-6 0.58	21-9 0.70	23-0 0.82	24-1 0.94	25-2 1.07	26-2 1.21
	19.2	11-6 0.12	13-3 0.19	14-10 0.26	16-3 0.35	17-6 0.44	18-9 0.53	19-11 0.64	21-0 0.75	22-0 0.86	23-0 0.98	23-11 1.10
	24.0	10-3 0.11	11-10 0.17	13-3 0.24	14-6 0.31	15-8 0.39	16-9 0.48	17-9 0.57	18-9 0.67	19-8 0.77	20-6 0.88	21-5 0.99

Note: The required modulus of elasticity, "E", in 1,000,000 pounds per square inch is shown below each span.

TABLE R-7 (cont.)

RAFTERS: Spans are measured along the horizontal projection and loads are considered as applied on the horizontal projection.

Extreme Fiber Stress in Bending, "F_b" (psi).										RAFTER SPACING (IN)	SIZE (IN)
1400	1500	1600	1700	1800	1900	2000	2100	2200	2400		
15-4 1.56	15-11 1.73	16-5 1.91	16-11 2.09	17-5 2.28	17-10 2.47					12.0	
14-4 1.46	14-10 1.62	15-4 1.78	15-10 1.95	16-3 2.13	16-9 2.31	17-2 2.49				13.7	
13-3 1.35	13-9 1.50	14-2 1.65	14-8 1.81	15-1 1.97	15-6 2.14	15-11 2.31	16-3 2.48			16.0	2x6
12-2 1.23	12-7 1.37	13-0 1.51	13-4 1.65	13-9 1.80	14-2 1.95	14-6 2.11	14-10 2.27	15-2 2.43		19.2	
10-10 1.10	11-3 1.22	11-7 1.35	11-11 1.48	12-4 1.61	12-8 1.75	13-0 1.89	13-3 2.03	13-7 2.18	14-2 2.48	24.0	
20-3 1.56	20-11 1.73	21-7 1.91	22-3 2.09	22-11 2.28	23-7 2.47					12.0	
18-11 1.46	19-7 1.62	20-3 1.78	20-10 1.95	21-5 2.13	22-0 2.31	22-7 2.49				13.7	
17-6 1.35	18-2 1.50	18-9 1.65	19-4 1.81	19-10 1.97	20-5 2.14	20-11 2.31	21-5 2.48			16.0	2x8
16-0 1.23	16-7 1.37	17-1 1.51	17-7 1.65	18-2 1.80	18-7 1.95	19-1 2.11	19-7 2.27	20-0 2.43		19.2	
14-4 1.10	14-10 1.22	15-3 1.35	15-9 1.48	16-3 1.61	16-8 1.75	17-1 1.89	17-6 2.03	17-11 2.18	18-9 2.48	24.0	
25-10 1.56	26-8 1.73	27-7 1.91	28-5 2.09	29-3 2.28	30-1 2.47					12.0	
24-2 1.46	25-0 1.62	25-10 1.78	26-7 1.95	27-4 2.13	28-1 2.31	28-10 2.49				13.7	
22-4 1.35	23-2 1.50	23-11 1.65	24-7 1.81	25-4 1.97	26-0 2.14	26-8 2.31	27-4 2.48			16.0	2x10
20-5 1.23	21-1 1.37	21-10 1.51	22-6 1.65	23-2 1.80	23-9 1.95	24-5 2.11	25-0 2.27	25-7 2.43		19.2	
18-3 1.10	18-11 1.22	19-6 1.35	20-1 1.48	20-8 1.61	21-3 1.75	21-10 1.89	22-4 2.03	22-10 2.18	23-11 2.48	24.0	
31-4 1.56	32-6 1.73	33-6 1.91	34-7 2.09	35-7 2.28	36-7 2.47					12.0	
29-4 1.46	30-5 1.62	31-4 1.78	32-4 1.95	33-3 2.13	34-2 2.31	35-1 2.49				13.7	
27-2 1.35	28-2 1.50	29-1 1.65	29-11 1.81	30-10 1.97	31-8 2.14	32-6 2.31	33-3 2.48			16.0	2x12
24-10 1.23	25-8 1.37	26-6 1.51	27-4 1.65	28-2 1.80	28-11 1.95	29-8 2.11	30-5 2.27	31-1 2.43		19.2	
22-2 1.10	23-0 1.22	23-9 1.35	24-5 1.48	25-2 1.61	25-10 1.75	26-6 1.89	27-2 2.03	27-10 2.18	29-1 2.48	24.0	

Note: The required modulus of elasticity, "E", in 1,000,000 pounds per inch is shown below each span.

TABLE R-8
FLAT OR LOW SLOPE RAFTERS
No Ceiling Load
Slope 3 in 12 or less
Live Load - 30 lb. per. sq. ft.

DESIGN CRITERIA:
Strength - 10 lbs. per sq. ft. dead load plus 30
 lbs. per sq. ft. live load determines required
 fiber stress.
Deflection - For 30 lbs. per sq. ft. live load.
 Limited to span in inches divided by 240.

RAFTER SIZE (IN)	SPACING (IN)	Extreme Fiber Stress In Bending, "F_b" (psi).										
		300	400	500	600	700	800	900	1000	1100	1200	1300
2x6	12.0	6-2 / 0.15	7-1 / 0.23	7-11 / 0.32	8-8 / 0.43	9-5 / 0.54	10-0 / 0.66	10-8 / 0.78	11-3 / 0.92	11-9 / 1.06	12-4 / 1.21	12-10 / 1.36
	13.7	5-9 / 0.14	6-8 / 0.22	7-5 / 0.30	8-2 / 0.40	8-9 / 0.50	9-5 / 0.61	10-0 / 0.73	10-6 / 0.86	11-0 / 0.99	11-6 / 1.13	12-0 / 1.27
	16.0	5-4 / 0.13	6-2 / 0.20	6-11 / 0.28	7-6 / 0.37	8-2 / 0.47	8-8 / 0.57	9-3 / 0.68	9-9 / 0.80	10-2 / 0.92	10-8 / 1.05	11-1 / 1.18
	19.2	4-10 / 0.12	5-7 / 0.18	6-3 / 0.26	6-11 / 0.34	7-5 / 0.43	7-11 / 0.52	8-5 / 0.62	8-11 / 0.73	9-4 / 0.84	9-9 / 0.95	10-1 / 1.08
	24.0	4-4 / 0.11	5-0 / 0.16	5-7 / 0.23	6-2 / 0.30	6-8 / 0.38	7-1 / 0.46	7-6 / 0.55	7-11 / 0.65	8-4 / 0.75	8-8 / 0.85	9-1 / 0.96
2x8	12.0	8-1 / 0.15	9-4 / 0.23	10-6 / 0.32	11-6 / 0.43	12-5 / 0.54	13-3 / 0.66	14-0 / 0.78	14-10 / 0.92	15-6 / 1.06	16-3 / 1.21	16-10 / 1.36
	13.7	7-7 / 0.14	8-9 / 0.22	9-9 / 0.30	10-9 / 0.40	11-7 / 0.50	12-5 / 0.61	13-2 / 0.73	13-10 / 0.86	14-6 / 0.99	15-2 / 1.13	15-9 / 1.27
	16.0	7-0 / 0.13	8-1 / 0.20	9-1 / 0.28	9-11 / 0.37	10-9 / 0.47	11-6 / 0.57	12-2 / 0.68	12-10 / 0.80	13-5 / 0.92	14-0 / 1.05	14-7 / 1.18
	19.2	6-5 / 0.12	7-5 / 0.18	8-3 / 0.26	9-1 / 0.34	9-9 / 0.43	10-6 / 0.52	11-1 / 0.62	11-8 / 0.73	12-3 / 0.84	12-10 / 0.95	13-4 / 1.08
	24.0	5-9 / 0.11	6-7 / 0.16	7-5 / 0.23	8-1 / 0.30	8-9 / 0.38	9-4 / 0.46	9-11 / 0.55	10-6 / 0.65	11-0 / 0.75	11-6 / 0.85	11-11 / 0.96
2x10	12.0	10-4 / 0.15	11-11 / 0.23	13-4 / 0.32	14-8 / 0.43	15-10 / 0.54	16-11 / 0.66	17-11 / 0.78	18-11 / 0.92	19-10 / 1.06	20-8 / 1.21	21-6 / 1.36
	13.7	9-8 / 0.14	11-2 / 0.22	12-6 / 0.30	13-8 / 0.40	14-9 / 0.50	15-10 / 0.61	16-9 / 0.73	17-8 / 0.86	18-6 / 0.99	19-4 / 1.13	20-2 / 1.27
	16.0	8-11 / 0.13	10-4 / 0.20	11-7 / 0.28	12-8 / 0.37	13-8 / 0.47	14-8 / 0.57	15-6 / 0.68	16-4 / 0.80	17-2 / 0.92	17-11 / 1.05	18-8 / 1.18
	19.2	8-2 / 0.12	9-5 / 0.18	10-7 / 0.26	11-7 / 0.34	12-6 / 0.43	13-4 / 0.52	14-2 / 0.62	14-11 / 0.73	15-8 / 0.84	16-4 / 0.95	17-0 / 1.08
	24.0	7-4 / 0.11	8-5 / 0.16	9-5 / 0.23	10-4 / 0.30	11-2 / 0.38	11-11 / 0.46	12-8 / 0.55	13-4 / 0.65	14-0 / 0.75	14-8 / 0.85	15-3 / 0.96
2x12	12.0	12-7 / 0.15	14-6 / 0.23	16-3 / 0.32	17-9 / 0.43	19-3 / 0.54	20-6 / 0.66	21-9 / 0.78	23-0 / 0.92	24-1 / 1.06	25-2 / 1.21	26-2 / 1.36
	13.7	11-9 / 0.14	13-7 / 0.22	15-2 / 0.30	16-8 / 0.40	18-0 / 0.50	19-3 / 0.61	20-5 / 0.73	21-6 / 0.86	22-6 / 0.99	23-6 / 1.13	24-6 / 1.27
	16.0	10-11 / 0.13	12-7 / 0.20	14-1 / 0.28	15-5 / 0.37	16-8 / 0.47	17-9 / 0.57	18-10 / 0.68	19-11 / 0.80	20-10 / 0.92	21-9 / 1.05	22-8 / 1.18
	19.2	9-11 / 0.12	11-6 / 0.18	12-10 / 0.26	14-1 / 0.34	15-2 / 0.43	16-3 / 0.52	17-3 / 0.62	18-2 / 0.73	19-0 / 0.84	19-11 / 0.95	20-8 / 1.08
	24.0	8-11 / 0.11	10-3 / 0.16	11-6 / 0.23	12-7 / 0.30	13-7 / 0.38	14-6 / 0.46	15-5 / 0.55	16-3 / 0.65	17-0 / 0.75	17-9 / 0.85	18-6 / 0.96

Note: The required modulus of elasticity, "E", in 1,000,000 pounds per square inch is shown below each span.

TABLE R-8 (cont.)

RAFTERS: Spans are measured along the
horizontal projection and loads are
considered as applied on the horizontal
projection.

Extreme Fiber Stress in Bending, "F_b" (psi).										RAFTER SPACING (IN)	SIZE (IN)
1400	1500	1600	1700	1800	1900	2000	2100	2200	2400		
13-3 1.52	13-9 1 69	14-2 1.86	14-8 2.04	15-1 2.22	15-6 2.41	15-11 2.60				12.0	
12-5 1.42	12-10 1.58	13-3 1.74	13-8 1.90	14-1 2.08	14-6 2.25	14-10 2.43				13.7	
11-6 1.32	11-11 1.46	12-4 1.61	12-8 1.76	13-1 1.92	13-5 2.08	13-9 2.25	14-1 2.42	14-5 2.60		16.0	2x6
10-6 1.20	10-10 1.33	11-3 1.47	11-7 1.61	11-11 1.75	12-3 1.90	12-7 2.05	12-10 2.21	13-2 2.37		19.2	
9-5 1.08	9-9 1.19	10-0 1.31	10-4 1.44	10-8 1.57	10-11 1.70	11-3 1.84	11-6 1.98	11-9 2.12	12-4 2.41	24.0	
17-6 1.52	18-2 1.69	18-9 1.86	19-4 2.04	19-10 2.22	20-5 2.41	20-11 2.60				12.0	
16-5 1.42	16-11 1.58	17-6 1.74	18-1 1.90	18-7 2.08	19-1 2.25	19-7 2.43				13.7	
15-2 1.32	15-8 1.46	16-3 1.61	16-9 1.76	17-2 1.92	17-8 2.08	18-2 2.25	18-7 2.42	19-0 2.60		16.0	2x8
13-10 1.20	14-4 1.33	14-10 1.47	15-3 1.61	15-8 1.75	16-2 1.90	16-7 2.05	16-11 2.21	17-4 2.37		19.2	
12-5 1.08	12-10 1.19	13-3 1.31	13-8 1.44	14-0 1.57	14-5 1.70	14-10 1.84	15-2 1.98	15-6 2.12	16-3 2.41	24.0	
22-4 1.52	23-2 1.69	23-11 1.86	24-7 2.04	25-4 2.22	26-0 2.41	26-8 2.60				12.0	
20-11 1.42	21-8 1.58	22-4 1.74	23-0 1.90	23-8 2.08	24-4 2.25	25-0 2.43				13.7	
19-4 1.32	20-0 1.46	20-8 1.61	21-4 1.76	21-11 1.92	22-6 2.08	23-2 2.25	23-8 2.42	24-3 2.60		16.0	2x10
17-8 1.20	18-3 1.33	18-11 1.47	19-6 1.61	20-0 1.75	20-7 1.90	21-1 2.05	21-8 2.21	22-2 2.37		19.2	
15-10 1.08	16-4 1.19	16-11 1.31	17-5 1.44	17-11 1.57	18-5 1.70	18-11 1.84	19-4 1.98	19-10 2.12	20-8 2.41	24.0	
27-2 1.52	28-2 1.69	29-1 1.86	29-11 2.04	30-10 2.22	31-8 2.41	32-6 2.60				12.0	
25-5 1.42	26-4 1.58	27-2 1.74	28-0 1.90	28-10 2.08	29-7 2.25	30-5 2.43				13.7	
23-6 1.32	24-4 1.46	25-2 1.61	25-11 1.76	26-8 1.92	27-5 2.08	28-2 2.25	28-10 2.42	29-6 2.60		16.0	2x12
21-6 1.20	22-3 1.33	23-0 1.47	23-8 1.61	24-4 1.75	25-0 1.90	25-8 2.05	26-4 2.21	26-11 2.37		19.2	
19-3 1.08	19-11 1.19	20-6 1.31	21-2 1.44	21-9 1.57	22-5 1.70	23-0 1.84	23-6 1.98	24-1 2.12	25-2 2.41	24.0	

Note: The required modulus of elasticity, "E", in 1,000,000 pounds per square inch is shown below
each span.

TABLE R-9
FLAT OR LOW SLOPE RAFTERS
No Ceiling Load
Slope 3 in 12 or less
Live Load - 40 lb. per. sq. ft.

DESIGN CRITERIA:
Strength - 10 lbs. per sq. ft. dead load plus 40
  lbs. per sq. ft. live load determines required
  fiber stress.
Deflection - For 40 lbs. per sq. ft. live load.
  Limited to span in inches divided by 240.

RAFTER SIZE (IN)	SPACING (IN)	Extreme Fiber Stress in Bending, "F_b" (psi).										
		300	400	500	600	700	800	900	1000	1100	1200	1300
2x6	12.0	5-6 0.14	6-4 0.22	7-1 0.31	7-9 0.41	8-5 0.51	9-0 0.63	9-6 0.75	10-0 0.88	10-6 1.01	11-0 1.15	11-5 1.30
	13.7	5-2 0.13	5-11 0.21	6-8 0.29	7-3 0.38	7-10 0.48	8-5 0.59	8-11 0.70	9-5 0.82	9-10 0.95	10-3 1.08	10-9 1.22
	16.0	4-9 0.12	5-6 0.19	6-2 0.27	6-9 0.35	7-3 0.44	7-9 0.54	8-3 0.65	8-8 0.76	9-1 0.88	9-6 1.00	9-11 1.12
	19.2	4-4 0.11	5-0 0.18	5-7 0.24	6-2 0.32	6-8 0.41	7-1 0.50	7-6 0.59	7-11 0.69	8-4 0.80	8-8 0.91	9-1 1.03
	24.0	3-11 0.10	4-6 0.16	5-0 0.22	5-6 0.29	5-11 0.36	6-4 0.44	6-9 0.53	7-1 0.62	7-5 0.71	7-9 0.81	8-1 0.92
2x8	12.0	7-3 0.14	8-4 0.22	9-4 0.31	10-3 0.41	11-1 0.51	11-10 0.63	12-7 0.75	13-3 0.88	13-11 1.01	14-6 1.15	15-1 1.30
	13.7	6-9 0.13	7-10 0.21	8-9 0.29	9-7 0.38	10-4 0.48	11-1 0.59	11-9 0.70	12-5 0.82	13-0 0.95	13-7 1.08	14-1 1.22
	16.0	6-3 0.12	7-3 0.19	8-1 0.27	8-11 0.35	9-7 0.44	10-3 0.54	10-11 0.65	11-6 0.76	12-0 0.88	12-7 1.00	13-1 1.12
	19.2	5-9 0.11	6-7 0.18	7-5 0.24	8-1 0.32	8-9 0.41	9-4 0.50	9-11 0.59	10-6 0.69	11-0 0.80	11-6 0.91	11-11 1.03
	24.0	5-2 0.10	5-11 0.16	6-7 0.22	7-3 0.29	7-10 0.36	8-4 0.44	8-11 0.53	9-4 0.62	9-10 0.71	10-3 0.81	10-8 0.92
2x10	12.0	9-3 0.14	10-8 0.22	11-11 0.31	13-1 0.41	14-2 0.51	15-1 0.63	16-0 0.75	16-11 0.88	17-9 1.01	18-6 1.15	19-3 1.30
	13.7	8-8 0.13	10-0 0.21	11-2 0.29	12-3 0.38	13-3 0.48	14-2 0.59	15-0 0.70	15-10 0.82	16-7 0.95	17-4 1.08	18-0 1.22
	16.0	8-0 0.12	9-3 0.19	10-4 0.27	11-4 0.35	12-3 0.44	13-1 0.54	13-11 0.65	14-8 0.76	15-4 0.88	16-0 1.00	16-8 1.12
	19.2	7-4 0.11	8-5 0.18	9-5 0.24	10-4 0.32	11-2 0.41	11-11 0.50	12-8 0.59	13-4 0.69	14-0 0.80	14-8 0.91	15-3 1.03
	24.0	6-6 0.10	7-7 0.16	8-5 0.22	9-3 0.29	10-0 0.36	10-8 0.44	11-4 0.53	11-11 0.62	12-6 0.71	13-1 0.81	13-7 0.92
2x12	12.0	11-3 0.14	13-0 0.22	14-6 0.31	15-11 0.41	17-2 0.51	18-4 0.63	19-6 0.75	20-6 0.88	21-7 1.01	22-6 1.15	23-5 1.30
	13.7	10-6 0.13	12-2 0.21	13-7 0.29	14-11 0.38	16-1 0.48	17-2 0.59	18-3 0.70	19-3 0.82	20-2 0.95	21-1 1.08	21-11 1.22
	16.0	9-9 0.12	11-3 0.19	12-7 0.27	13-9 0.35	14-11 0.44	15-11 0.54	16-11 0.65	17-9 0.76	18-8 0.88	19-6 1.00	20-3 1.12
	19.2	8-11 0.11	10-3 0.18	11-6 0.24	12-7 0.32	13-7 0.41	14-6 0.50	15-5 0.59	16-3 0.69	17-0 0.80	17-9 0.91	18-6 1.03
	24.0	7-11 0.10	9-2 0.16	10-3 0.22	11-3 0.29	12-2 0.36	13-0 0.44	13-9 0.53	14-6 0.62	15-3 0.71	15-11 0.81	16-7 0.92

Note: The required modulus of elasticity, "E", in 1,000,000 pounds per square inch is shown below each span.

TABLE R-9 (cont.)

RAFTERS: Spans are measured along the horizontal projection and loads are considered as applied on the horizontal projection.

Extreme Fiber Stress in Bending, "F_b" (psi).										RAFTER SPACING (IN)	SIZE (IN)
1400	1500	1600	1700	1800	1900	2000	2100	2200	2400		
11-11 / 1.45	12-4 / 1.61	12-8 / 1.77	13-1 / 1.94	13-6 / 2.12	13-10 / 2.30	14-2 / 2.48				12.0	
11-1 / 1.36	11-6 / 1.51	11-11 / 1.66	12-3 / 1.82	12-7 / 1.98	12-11 / 2.15	13-3 / 2.32	13-7 / 2.49			13.7	
10-3 / 1.26	10-8 / 1.39	11-0 / 1.54	11-4 / 1.68	11-8 / 1.83	12-0 / 1.99	12-4 / 2.15	12-7 / 2.31	12-11 / 2.48		16.0	2x6
9-5 / 1.15	9-9 / 1.27	10-0 / 1.40	10-4 / 1.54	10-8 / 1.67	10-11 / 1.81	11-3 / 1.96	11-6 / 2.11	11-9 / 2.26	12-4 / 2.58	19.2	
8-5 / 1.03	8-8 / 1.14	9-0 / 1.25	9-3 / 1.37	9-6 / 1.50	9-9 / 1.62	10-0 / 1.75	10-3 / 1.89	10-6 / 2.02	11-0 / 2.30	24.0	
15-8 / 1.45	16-3 / 1.61	16-9 / 1.77	17-3 / 1.94	17-9 / 2.12	18-3 / 2.30	18-9 / 2.48				12.0	
14-8 / 1.36	15-2 / 1.51	15-8 / 1.66	16-2 / 1.82	16-7 / 1.98	17-1 / 2.15	17-6 / 2.32	17-11 / 2.49			13.7	
13-7 / 1.26	14-0 / 1.39	14-6 / 1.54	14-11 / 1.68	15-5 / 1.83	15-10 / 1.99	16-3 / 2.15	16-7 / 2.31	17-0 / 2.48		16.0	2x8
12-5 / 1.15	12-10 / 1.27	13-3 / 1.40	13-8 / 1.54	14-0 / 1.67	14-5 / 1.81	14-10 / 1.96	15-2 / 2.11	15-6 / 2.26	16-3 / 2.58	19.2	
11-1 / 1.03	11-6 / 1.14	11-10 / 1.25	12-2 / 1.37	12-7 / 1.50	12-11 / 1.62	13-3 / 1.75	13-7 / 1.89	13-11 / 2.02	14-6 / 2.30	24.0	
20-0 / 1.45	20-8 / 1.61	21-4 / 1.77	22-0 / 1.94	22-8 / 2.12	23-3 / 2.30	23-11 / 2.48				12.0	
18-8 / 1.36	19-4 / 1.51	20-0 / 1.66	20-7 / 1.82	21-2 / 1.98	21-9 / 2.15	22-4 / 2.32	22-11 / 2.49			13.7	
17-4 / 1.26	17-11 / 1.39	18-6 / 1.54	19-1 / 1.68	19-7 / 1.83	20-2 / 1.99	20-8 / 2.15	21-2 / 2.31	21-8 / 2.48		16.0	2x10
15-10 / 1.15	16-4 / 1.27	16-11 / 1.40	17-5 / 1.54	17-11 / 1.67	18-5 / 1.81	18-11 / 1.96	19-4 / 2.11	19-10 / 2.26	20-8 / 2.58	19.2	
14-2 / 1.03	14-8 / 1.14	15-1 / 1.25	15-7 / 1.37	16-0 / 1.50	16-6 / 1.62	16-11 / 1.75	17-4 / 1.89	17-9 / 2.02	18-6 / 2.30	24.0	
24-4 / 1.45	25-2 / 1.61	26-0 / 1.77	26-9 / 1.94	27-7 / 2.12	28-4 / 2.30	29-1 / 2.48				12.0	
22-9 / 1.36	23-6 / 1.51	24-4 / 1.66	25-1 / 1.82	25-9 / 1.98	26-6 / 2.15	27-2 / 2.32	27-10 / 2.49			13.7	
21-1 / 1.26	21-9 / 1.39	22-6 / 1.54	23-2 / 1.68	23-10 / 1.83	24-6 / 1.99	25-2 / 2.15	25-9 / 2.31	26-5 / 2.48		16.0	2x12
19-3 / 1.15	19-11 / 1.27	20-6 / 1.40	21-2 / 1.54	21-9 / 1.67	22-5 / 1.81	23-0 / 1.96	23-6 / 2.11	24-1 / 2.26	25-2 / 2.58	19.2	
17-2 / 1.03	17-9 / 1.14	18-4 / 1.25	18-11 / 1.37	19-6 / 1.50	20-0 / 1.62	20-6 / 1.75	21-1 / 1.89	21-7 / 2.02	22-6 / 2.30	24.0	

Note: The required modulus of elasticity, "E", in 1,000,000 pounds per inch is shown below each span.

TABLE R-10
MEDIUM OR HIGH SLOPE RAFTERS
No Ceiling Load
Slope over 3 in 12
Live Load - 20 lb. per. sq. ft.
(Heavy roof covering)

DESIGN CRITERIA:
Strength - 15 lbs per sq. ft. dead load plus 20 lbs. per sq. ft. live load determines required fiber stress.
Deflection - For 20 lbs. per sq. ft. live load. Limited to span in inches divided by 180.

RAFTER SIZE (IN)	SPACING (IN)	Extreme Fiber Stress in Bending, "F_b" (psi).											
		200	300	400	500	600	700	800	900	1000	1100	1200	1300
2x4	12.0	3-5 0.05	4-2 0.09	4-10 0.14	5-5 0.20	5-11 0.26	6-5 0.33	6-10 0.40	7-3 0.48	7-8 0.56	8-0 0.65	8-4 0.74	8-8 0.83
	13.7	3-2 0.05	3-11 0.09	4-6 0.13	5-1 0.19	5-6 0.24	6-0 0.31	6-5 0.38	6-9 0.45	7-2 0.52	7-6 0.61	7-10 0.69	8-2 0.78
	16.0	2-11 0.04	3-7 0.08	4-2 0.12	4-8 0.17	5-1 0.23	5-6 0.28	5-11 0.35	6-3 0.41	6-7 0.49	6-11 0.56	7-3 0.64	7-6 0.72
	19.2	2-8 0.04	3-4 0.07	3-10 0.11	4-3 0.16	4-8 0.21	5-1 0.26	5-5 0.32	5-9 0.38	6-0 0.44	6-4 0.51	6-7 0.58	6-11 0.66
	24.0	2-5 0.04	2-11 0.07	3-5 0.10	3-10 0.14	4-2 0.18	4-6 0.23	4-10 0.28	5-1 0.34	5-5 0.40	5-8 0.46	5-11 0.52	6-2 0.59
2x6	12.0	5-4 0.05	6-7 0.09	7-7 0.14	8-6 0.20	9-4 0.26	10-0 0.33	10-9 0.40	11-5 0.48	12-0 0.56	12-7 0.65	13-2 0.74	13-8 0.83
	13.7	5-0 0.05	6-2 0.09	7-1 0.13	7-11 0.19	8-8 0.24	9-5 0.31	10-0 0.38	10-8 0.45	11-3 0.52	11-9 0.61	12-4 0.69	12-10 0.78
	16.0	4-8 0.04	5-8 0.08	6-7 0.12	7-4 0.17	8-1 0.23	8-8 0.28	9-4 0.35	9-10 0.41	10-5 0.49	10-11 0.56	11-5 0.64	11-10 0.72
	19.2	4-3 0.04	5-2 0.07	6-0 0.11	6-9 0.16	7-4 0.21	7-11 0.26	8-6 0.32	9-0 0.38	9-6 0.44	9-11 0.51	10-5 0.58	10-10 0.66
	24.0	3-10 0.04	4-8 0.07	5-4 0.10	6-0 0.14	6-7 0.18	7-1 0.23	7-7 0.28	8-1 0.34	8-6 0.40	8-11 0.46	9-4 0.52	9-8 0.59
2x8	12.0	7-1 0.05	8-8 0.09	10-0 0.14	11-2 0.20	12-3 0.26	13-3 0.33	14-2 0.40	15-0 0.48	15-10 0.56	16-7 0.65	17-4 0.74	18-0 0.83
	13.7	6-7 0.05	8-1 0.09	9-4 0.13	10-6 0.19	11-6 0.24	12-5 0.31	13-3 0.38	14-0 0.45	14-10 0.52	15-6 0.61	16-3 0.69	16-10 0.78
	16.0	6-2 0.04	7-6 0.08	8-8 0.12	9-8 0.17	10-7 0.23	11-6 0.28	12-3 0.35	13-0 0.41	13-8 0.49	14-4 0.56	15-0 0.64	15-7 0.72
	19.2	5-7 0.04	6-10 0.07	7-11 0.11	8-10 0.16	9-8 0.21	10-6 0.26	11-2 0.32	11-10 0.38	12-6 0.44	13-1 0.51	13-8 0.58	14-3 0.66
	24.0	5-0 0.04	6-2 0.07	7-1 0.10	7-11 0.14	8-8 0.18	9-4 0.23	10-0 0.28	10-7 0.34	11-2 0.40	11-9 0.46	12-3 0.52	12-9 0.59
2x10	12.0	9-0 0.05	11-1 0.09	12-9 0.14	14-3 0.20	15-8 0.26	16-11 0.33	18-1 0.40	19-2 0.48	20-2 0.56	21-2 0.65	22-1 0.74	23-0 0.83
	13.7	8-5 0.05	10-4 0.09	11-11 0.13	13-4 0.19	14-8 0.24	15-10 0.31	16-11 0.38	17-11 0.45	18-11 0.52	19-10 0.61	20-8 0.69	21-6 0.78
	16.0	7-10 0.04	9-7 0.08	11-1 0.12	12-4 0.17	13-6 0.23	14-8 0.28	15-8 0.35	16-7 0.41	17-6 0.49	18-4 0.56	19-2 0.64	19-11 0.72
	19.2	7-2 0.04	8-9 0.07	10-1 0.11	11-3 0.16	12-4 0.21	13-4 0.26	14-3 0.32	15-2 0.38	15-11 0.44	16-9 0.51	17-6 0.58	18-2 0.66
	24.0	6-5 0.04	7-10 0.07	9-0 0.10	10-1 0.14	11-1 0.18	11-11 0.23	12-9 0.28	13-6 0.34	14-3 0.40	15-0 0.46	15-8 0.52	16-3 0.59

Note: The required modulus of elasticity, "E", in 1,000,000 pounds per square inch is shown below each span.

TABLE R-10 (cont.)

RAFTERS: Spans are measured along the horizontal projection and loads are considered as applied on the horizontal projection.

| Extreme Fiber Stress in Bending, "F_b" (psi). | | | | | | | | | | | | RAFTER SPACING (IN) | SIZE (IN) |
1400	1500	1600	1700	1800	1900	2000	2100	2200	2400	2700	3000		
9-0 / 0.93	9-4 / 1.03	9-8 / 1.14	9-11 / 1.24	10-3 / 1.36	10-6 / 1.47	10-10 / 1.59	11-1 / 1.71	11-4 / 1.83	11-10 / 2.09	12-7 / 2.49		12.0	
8-5 / 0.87	8-9 / 0.96	9-0 / 1.06	9-4 / 1.16	9-7 / 1.27	9-10 / 1.37	10-1 / 1.48	10-4 / 1.60	10-7 / 1.71	11-1 / 1.95	11-9 / 2.33		13.7	
7-10 / 0.80	8-1 / 0.89	8-4 / 0.98	8-7 / 1.08	8-10 / 1.17	9-1 / 1.27	9-4 / 1.37	9-7 / 1.48	9-10 / 1.59	10-3 / 1.81	10-10 / 2.16	11-5 / 2.53	16.0	2x4
7-2 / 0.73	7-5 / 0.81	7-8 / 0.90	7-10 / 0.98	8-1 / 1.07	8-4 / 1.16	8-6 / 1.25	8-9 / 1.35	8-11 / 1.45	9-4 / 1.65	9-11 / 1.97	10-5 / 2.31	19.2	
6-5 / 0.66	6-7 / 0.73	6-10 / 0.80	7-0 / 0.88	7-3 / 0.96	7-5 / 1.04	7-8 / 1.12	7-10 / 1.21	8-0 / 1.29	8-4 / 1.48	8-10 / 1.76	9-4 / 2.06	24.0	
14-2 / 0.93	14-8 / 1.03	15-2 / 1.14	15-8 / 1.24	16-1 / 1.36	16-7 / 1.47	17-0 / 1.59	17-5 / 1.71	17-10 / 1.83	18-7 / 2.09	19-9 / 2.49		12.0	
13-3 / 0.87	13-9 / 0.96	14-2 / 1.06	14-8 / 1.16	15-1 / 1.27	15-6 / 1.37	15-11 / 1.48	16-3 / 1.60	16-8 / 1.71	17-5 / 1.95	18-5 / 2.33		13.7	
12-4 / 0.80	12-9 / 0.89	13-2 / 0.98	13-7 / 1.08	13-11 / 1.17	14-4 / 1.27	14-8 / 1.37	15-1 / 1.48	15-5 / 1.59	16-1 / 1.81	17-1 / 2.16	18-0 / 2.53	16.0	2x6
11-3 / 0.73	11-7 / 0.81	12-0 / 0.90	12-4 / 0.98	12-9 / 1.07	13-1 / 1.16	13-5 / 1.25	13-9 / 1.35	14-1 / 1.45	14-8 / 1.65	15-7 / 1.97	16-5 / 2.31	19.2	
10-0 / 0.66	10-5 / 0.73	10-9 / 0.80	11-1 / 0.88	11-5 / 0.96	11-8 / 1.04	12-0 / 1.12	12-4 / 1.21	12-7 / 1.29	13-2 / 1.48	13-11 / 1.76	14-8 / 2.06	24.0	
18-9 / 0.93	19-5 / 1.03	20-0 / 1.14	20-8 / 1.24	21-3 / 1.36	21-10 / 1.47	22-4 / 1.59	22-11 / 1.71	23-6 / 1.83	24-6 / 2.09	26-0 / 2.49		12.0	
17-6 / 0.87	18-2 / 0.96	18-9 / 1.06	19-4 / 1.16	19-10 / 1.27	20-5 / 1.37	20-11 / 1.48	21-5 / 1.60	21-11 / 1.71	22-11 / 1.95	24-4 / 2.33		13.7	
16-3 / 0.80	16-9 / 0.89	17-4 / 0.98	17-10 / 1.08	18-5 / 1.17	18-11 / 1.27	19-5 / 1.37	19-10 / 1.48	20-4 / 1.59	21-3 / 1.81	22-6 / 2.16	23-9 / 2.53	16.0	2x8
14-10 / 0.73	15-4 / 0.81	15-10 / 0.90	16-4 / 0.98	16-9 / 1.07	17-3 / 1.16	17-8 / 1.25	18-2 / 1.35	18-7 / 1.45	19-5 / 1.65	20-7 / 1.97	21-8 / 2.31	19.2	
13-3 / 0.66	13-8 / 0.73	14-2 / 0.80	14-7 / 0.88	15-0 / 0.96	15-5 / 1.04	15-10 / 1.12	16-3 / 1.21	16-7 / 1.29	17-4 / 1.48	18-5 / 1.76	19-5 / 2.06	24.0	
23-11 / 0.93	24-9 / 1.03	25-6 / 1.14	26-4 / 1.24	27-1 / 1.36	27-10 / 1.47	28-7 / 1.59	29-3 / 1.71	29-11 / 1.83	31-3 / 2.09	33-2 / 2.49		12.0	
22-4 / 0.87	23-2 / 0.96	23-11 / 1.06	24-7 / 1.16	25-4 / 1.27	26-0 / 1.37	26-8 / 1.48	27-4 / 1.60	28-0 / 1.71	29-3 / 1.95	31-0 / 2.33		13.7	
20-8 / 0.80	21-5 / 0.89	22-1 / 0.98	22-10 / 1.08	23-5 / 1.17	24-1 / 1.27	24-9 / 1.37	25-4 / 1.48	25-11 / 1.59	27-1 / 1.81	28-9 / 2.16	30-3 / 2.53	16.0	2x10
18-11 / 0.73	19-7 / 0.81	20-2 / 0.90	20-10 / 0.98	21-5 / 1.07	22-0 / 1.16	22-7 / 1.25	23-2 / 1.35	23-8 / 1.45	24-9 / 1.65	26-3 / 1.97	27-8 / 2.31	19.2	
16-11 / 0.66	17-6 / 0.73	18-1 / 0.80	18-7 / 0.88	19-2 / 0.96	19-8 / 1.04	20-2 / 1.12	20-8 / 1.21	21-2 / 1.29	22-1 / 1.48	23-5 / 1.76	24-9 / 2.06	24.0	

Note: The required modulus of elasticity, "E", in 1,000,000 pounds per square inch is shown below each span.

TABLE R-11
MEDIUM OR HIGH SLOPE RAFTERS
No Ceiling Load
Slope over 3 in 12
Live Load - 30 lb. per. sq. ft.
(Heavy roof covering)

DESIGN CRITERIA:
Strength - 15 lbs. per sq. ft. dead load plus 30 lbs. per sq. ft. live load determines required fiber stress.
Deflection - For 30 lbs. per sq. ft. live load. Limited to span in inches divided by 180.

RAFTER SIZE (IN)	SPACING (IN)	Extreme Fiber Stress in Bending, "F_b" (psi).											
		200	300	400	500	600	700	800	900	1000	1100	1200	1300
2x4	12.0	3-0 0.05	3-8 0.09	4-3 0.15	4-9 0.20	5-3 0.27	5-8 0.34	6-0 0.41	6-5 0.49	6-9 0.58	7-1 0.67	7-5 0.76	7-8 0.86
	13.7	2-10 0.05	3-5 0.09	4-0 0.14	4-5 0.19	4-11 0.25	5-3 0.32	5-8 0.39	6-0 0.46	6-4 0.54	6-7 0.62	6-11 0.71	7-2 0.80
	16.0	2-7 0.04	3-2 0.08	3-8 0.13	4-1 0.18	4-6 0.23	4-11 0.29	5-3 0.36	5-6 0.43	5-10 0.50	6-1 0.58	6-5 0.66	6-8 0.74
	19.2	2-5 0.04	2-11 0.08	3-4 0.12	3-9 0.16	4-1 0.21	4-5 0.27	4-9 0.33	5-1 0.39	5-4 0.46	5-7 0.53	5-10 0.60	6-1 0.68
	24.0	2-2 0.04	2-7 0.07	3-0 0.10	3-4 0.14	3-8 0.19	4-0 0.24	4-3 0.29	4-6 0.35	4-9 0.41	5-0 0.47	5-3 0.54	5-5 0.61
2x6	12.0	4-9 0.05	5-10 0.09	6-8 0.15	7-6 0.20	8-2 0.27	8-10 0.34	9-6 0.41	10-0 0.49	10-7 0.58	11-1 0.67	11-7 0.76	12-1 0.86
	13.7	4-5 0.05	5-5 0.09	6-3 0.14	7-0 0.19	7-8 0.25	8-3 0.32	8-10 0.39	9-5 0.46	9-11 0.54	10-5 0.62	10-10 0.71	11-3 0.80
	16.0	4-1 0.04	5-0 0.08	5-10 0.13	6-6 0.18	7-1 0.23	7-8 0.29	8-2 0.36	8-8 0.43	9-2 0.50	9-7 0.58	10-0 0.66	10-5 0.74
	19.2	3-9 0.04	4-7 0.08	5-4 0.12	5-11 0.16	6-6 0.21	7-0 0.27	7-6 0.33	7-11 0.39	8-4 0.46	8-9 0.53	9-2 0.60	9-6 0.68
	24.0	3-4 0.04	4-1 0.07	4-9 0.10	5-4 0.14	5-10 0.19	6-3 0.24	6-8 0.29	7-1 0.35	7-6 0.41	7-10 0.47	8-2 0.54	8-6 0.61
2x8	12.0	6-3 0.05	7-8 0.09	8-10 0.15	9-10 0.20	10-10 0.27	11-8 0.34	12-6 0.41	13-3 0.49	13-11 0.58	14-8 0.67	15-3 0.76	15-11 0.86
	13.7	5-10 0.05	7-2 0.09	8-3 0.14	9-3 0.19	10-1 0.25	10-11 0.32	11-8 0.39	12-5 0.46	13-1 0.54	13-8 0.62	14-4 0.71	14-11 0.80
	16.0	5-5 0.04	6-7 0.08	7-8 0.13	8-7 0.18	9-4 0.23	10-1 0.29	10-10 0.36	11-6 0.43	12-1 0.50	12-8 0.58	13-3 0.66	13-9 0.74
	19.2	4-11 0.04	6-1 0.08	7-0 0.12	7-10 0.16	8-7 0.21	9-3 0.27	9-10 0.33	10-6 0.39	11-0 0.46	11-7 0.53	12-1 0.60	12-7 0.68
	24.0	4-5 0.04	5-5 0.07	6-3 0.10	7-0 0.14	7-8 0.19	8-3 0.24	8-10 0.29	9-4 0.35	9-10 0.41	10-4 0.47	10-10 0.54	11-3 0.61
2x10	12.0	8-0 0.05	9-9 0.09	11-3 0.15	12-7 0.20	13-9 0.27	14-11 0.34	15-11 0.41	16-11 0.49	17-10 0.58	18-8 0.67	19-6 0.76	20-4 0.86
	13.7	7-5 0.05	9-1 0.09	10-6 0.14	11-9 0.19	12-11 0.25	13-11 0.32	14-11 0.39	15-10 0.46	16-8 0.54	17-6 0.62	18-3 0.71	19-0 0.80
	16.0	6-11 0.04	8-5 0.08	9-9 0.13	10-11 0.18	11-11 0.23	12-11 0.29	13-9 0.36	14-8 0.43	15-5 0.50	16-2 0.58	16-11 0.66	17-7 0.74
	19.2	6-4 0.04	7-8 0.08	8-11 0.12	9-11 0.16	10-11 0.21	11-9 0.27	12-7 0.33	13-4 0.39	14-1 0.46	14-9 0.53	15-5 0.60	16-1 0.68
	24.0	5-8 0.04	6-11 0.07	8-0 0.10	8-11 0.14	9-9 0.19	10-6 0.24	11-3 0.29	11-11 0.35	12-7 0.41	13-2 0.47	13-9 0.54	14-4 0.61

Note: The required modulus of elasticity, "E", in 1,000,000 pounds per inch is shown below each span.

TABLE R-11 (cont.)

RAFTERS: Spans are measured along the horizontal projection and loads are considered as applied on the horizontal projection.

_	_	_	_	Extreme Fiber Stress in Bending, "F_b" (psi).	_	_	_	_	_	_	_	RAFTER SPACING (IN)	SIZE (IN)
1400	1500	1600	1700	1800	1900	2000	2100	2200	2400	2700	3000		
8-0 0.96	8-3 1.06	8-6 1.17	8-9 1.28	9-0 1.39	9-3 1.51	9-6 1.63	9-9 1.76	10-0 1.88	10-5 2.15	11-1 2.56		12.0	
7-5 0.89	7-9 0.99	8-0 1.09	8-3 1.20	8-5 1.30	8-8 1.41	8-11 1.53	9-2 1.64	9-4 1.76	9-9 2.01	10-4 2.40		13.7	
6-11 0.83	7-2 0.92	7-5 1.01	7-7 1.11	7-10 1.21	8-0 1.31	8-3 1.41	8-5 1.52	8-8 1.63	9-0 1.86	9-7 2.22	10-1 2.60	16.0	2x4
6-4 0.76	6-6 0.84	6-9 0.92	6-11 1.01	7-2 1.10	7-4 1.20	7-6 1.29	7-9 1.39	7-11 1.49	8-3 1.70	8-9 2.03	9-3 2.37	19.2	
5-8 0.68	5-10 0.75	6-0 0.83	6-3 0.90	6-5 0.99	6-7 1.07	6-9 1.15	6-11 1.24	7-1 1.33	7-5 1.52	7-10 1.81	8-3 2.12	24.0	
12-6 0.96	13-0 1.06	13-5 1.17	13-10 1.28	14-2 1.39	14-7 1.51	15-0 1.63	15-4 1.76	15-8 1.88	16-5 2.15	17-5 2.56		12.0	
11-9 0.89	12-2 0.99	12-6 1.09	12-11 1.20	13-3 1.30	13-8 1.41	14-0 1.53	14-4 1.64	14-8 1.76	15-4 2.01	16-3 2.40		13.7	
10-10 0.83	11-3 0.92	11-7 1.01	11-11 1.11	12-4 1.21	12-8 1.31	13-0 1.41	13-3 1.52	13-7 1.63	14-2 1.86	15-1 2.22	15-11 2.60	16.0	2x6
9-11 0.76	10-3 0.84	10-7 0.92	10-11 1.01	11-3 1.10	11-6 1.20	11-10 1.29	12-2 1.39	12-5 1.49	13-0 1.70	13-9 2.03	14-6 2.37	19.2	
8-10 0.68	9-2 0.75	9-6 0.83	9-9 0.90	10-0 0.99	10-4 1.07	10-7 1.15	10-10 1.24	11-1 1.33	11-7 1.52	12-4 1.81	13-0 2.12	24.0	
16-6 0.96	17-1 1.06	17-8 1.17	18-2 1.28	18-9 1.39	19-3 1.51	19-9 1.63	20-3 1.76	20-8 1.88	21-7 2.15	22-11 2.56		12.0	
15-5 0.89	16-0 0.99	16-6 1.09	17-0 1.20	17-6 1.30	18-0 1.41	18-5 1.53	18-11 1.64	19-4 1.76	20-3 2.01	21-5 2.40		13.7	
14-4 0.83	14-10 0.92	15-3 1.01	15-9 1.11	16-3 1.21	16-8 1.31	17-1 1.41	17-6 1.52	17-11 1.63	18-9 1.86	19-10 2.22	20-11 2.60	16.0	2x8
13-1 0.76	13-6 0.84	13-11 0.92	14-5 1.01	14-10 1.10	15-2 1.20	15-7 1.29	16-0 1.39	16-4 1.49	17-1 1.70	18-2 2.03	19-1 2.37	19.2	
11-8 0.68	12-1 0.75	12-6 0.83	12-10 0.90	13-3 0.99	13-7 1.07	13-11 1.15	14-4 1.24	14-8 1.33	15-3 1.52	16-3 1.81	17-1 2.12	24.0	
21-1 0.96	21-10 1.06	22-6 1.17	23-3 1.28	23-11 1.39	24-6 1.51	25-2 1.63	25-10 1.76	26-5 1.88	27-7 2.15	29-3 2.56		12.0	
19-8 0.89	20-5 0.99	21-1 1.09	21-9 1.20	22-4 1.30	22-11 1.41	23-7 1.53	24-2 1.64	24-8 1.76	25-10 2.01	27-4 2.40		13.7	
18-3 0.83	18-11 0.92	19-6 1.01	20-1 1.11	20-8 1.21	21-3 1.31	21-10 1.41	22-4 1.52	22-10 1.63	23-11 1.86	25-4 2.22	26-8 2.60	16.0	2x10
16-8 0.76	17-3 0.84	17-10 0.92	18-4 1.01	18-11 1.10	19-5 1.20	19-11 1.29	20-5 1.39	20-10 1.49	21-10 1.70	23-2 2.03	24-5 2.37	19.2	
14-11 0.68	15-5 0.75	15-11 0.83	16-5 0.90	16-11 0.99	17-4 1.07	17-10 1.15	18-3 1.24	18-8 1.33	19-6 1.52	20-8 1.81	21-10 2.12	24.0	

Note: The required modulus of elasticity, "E", in 1,000,000 pounds per square inch is shown below each span.

TABLE R-12
MEDIUM OR HIGH SLOPE RAFTERS
No Ceiling Load
Slope over 3 in 12
Live Load - 40 lb. per. sq. ft.
(Heavy roof covering)

DESIGN CRITERIA:
Strength - 15 lbs. per sq. ft. dead load plus 40 lbs. per sq. ft. live load determines required fiber stress.
Deflection - For 40 lbs. per sq. ft. live load. Limited to span in inches divided by 180.

RAFTER SIZE (IN)	SPACING (IN)	Extreme Fiber Stress in Bending, "F_b" (psi).											
		200	300	400	500	600	700	800	900	1000	1100	1200	1300
2x4	12.0	2-9 / 0.05	3-4 / 0.09	3-10 / 0.14	4-4 / 0.20	4-9 / 0.26	5-1 / 0.33	5-5 / 0.41	5-9 / 0.49	6-1 / 0.57	6-5 / 0.66	6-8 / 0.75	6-11 / 0.84
	13.7	2-7 / 0.05	3-1 / 0.09	3-7 / 0.13	4-0 / 0.19	4-5 / 0.25	4-9 / 0.31	5-1 / 0.38	5-5 / 0.46	5-8 / 0.53	6-0 / 0.61	6-3 / 0.70	6-6 / 0.79
	16.0	2-4 / 0.04	2-11 / 0.08	3-4 / 0.12	3-9 / 0.17	4-1 / 0.23	4-5 / 0.29	4-9 / 0.35	5-0 / 0.42	5-3 / 0.49	5-6 / 0.57	5-9 / 0.65	6-0 / 0.73
	19.2	2-2 / 0.04	2-8 / 0.07	3-1 / 0.11	3-5 / 0.16	3-9 / 0.21	4-0 / 0.26	4-4 / 0.32	4-7 / 0.38	4-10 / 0.45	5-1 / 0.52	5-3 / 0.59	5-6 / 0.67
	24.0	1-11 / 0.04	2-4 / 0.07	2-9 / 0.10	3-1 / 0.14	3-4 / 0.19	3-7 / 0.24	3-10 / 0.29	4-1 / 0.34	4-4 / 0.40	4-6 / 0.46	4-9 / 0.53	4-11 / 0.60
2x6	12.0	4-3 / 0.05	5-3 / 0.09	6-1 / 0.14	6-9 / 0.20	7-5 / 0.26	8-0 / 0.33	8-7 / 0.41	9-1 / 0.49	9-7 / 0.57	10-0 / 0.66	10-6 / 0.75	10-11 / 0.84
	13.7	4-0 / 0.05	4-11 / 0.09	5-8 / 0.13	6-4 / 0.19	6-11 / 0.25	7-6 / 0.31	8-0 / 0.38	8-6 / 0.46	8-11 / 0.53	9-5 / 0.61	9-10 / 0.70	10-3 / 0.79
	16.0	3-8 / 0.04	4-6 / 0.08	5-3 / 0.12	5-10 / 0.17	6-5 / 0.23	6-11 / 0.29	7-5 / 0.35	7-10 / 0.42	8-3 / 0.49	8-8 / 0.57	9-1 / 0.65	9-5 / 0.73
	19.2	3-5 / 0.04	4-2 / 0.07	4-9 / 0.11	5-4 / 0.16	5-10 / 0.21	6-4 / 0.26	6-9 / 0.32	7-2 / 0.38	7-7 / 0.45	7-11 / 0.52	8-3 / 0.59	8-8 / 0.67
	24.0	3-0 / 0.04	3-8 / 0.07	4-3 / 0.10	4-9 / 0.14	5-3 / 0.19	5-8 / 0.24	6-1 / 0.29	6-5 / 0.34	6-9 / 0.40	7-1 / 0.46	7-5 / 0.53	7-9 / 0.60
2x8	12.0	5-8 / 0.05	6-11 / 0.09	8-0 / 0.14	8-11 / 0.20	9-9 / 0.26	10-7 / 0.33	11-3 / 0.41	12-0 / 0.49	12-7 / 0.57	13-3 / 0.66	13-10 / 0.75	14-5 / 0.84
	13.7	5-3 / 0.05	6-6 / 0.09	7-6 / 0.13	8-4 / 0.19	9-2 / 0.25	9-11 / 0.31	10-7 / 0.28	11-2 / 0.46	11-10 / 0.53	12-5 / 0.61	12-11 / 0.70	13-6 / 0.79
	16.0	4-11 / 0.04	6-0 / 0.08	6-11 / 0.12	7-9 / 0.17	8-6 / 0.23	9-2 / 0.29	9-9 / 0.35	10-4 / 0.42	10-11 / 0.49	11-6 / 0.57	12-0 / 0.65	12-6 / 0.73
	19.2	4-6 / 0.04	5-6 / 0.07	6-4 / 0.11	7-1 / 0.16	7-9 / 0.21	8-4 / 0.26	8-11 / 0.32	9-6 / 0.38	10-0 / 0.45	10-6 / 0.52	10-11 / 0.59	11-5 / 0.67
	24.0	4-0 / 0.04	4-11 / 0.07	5-8 / 0.10	6-4 / 0.14	6-11 / 0.19	7-6 / 0.24	8-0 / 0.29	8-6 / 0.34	8-11 / 0.40	9-4 / 0.46	9-9 / 0.53	10-2 / 0.60
2x10	12.0	7-2 / 0.05	8-10 / 0.09	10-2 / 0.14	11-5 / 0.20	12-6 / 0.26	13-6 / 0.33	14-5 / 0.41	15-3 / 0.49	16-1 / 0.57	16-11 / 0.66	17-8 / 0.75	18-4 / 0.84
	13.7	6-9 / 0.05	8-3 / 0.09	9-6 / 0.13	10-8 / 0.19	11-8 / 0.25	12-7 / 0.31	13-6 / 0.38	14-3 / 0.46	15-1 / 0.53	15-10 / 0.61	16-6 / 0.70	17-2 / 0.79
	16.0	6-3 / 0.04	7-8 / 0.08	8-10 / 0.12	9-10 / 0.17	10-10 / 0.23	11-8 / 0.29	12-6 / 0.35	13-3 / 0.42	13-11 / 0.49	14-8 / 0.57	15-3 / 0.65	15-11 / 0.73
	19.2	5-8 / 0.04	7-0 / 0.07	8-1 / 0.11	9-0 / 0.16	9-10 / 0.21	10-8 / 0.26	11-5 / 0.32	12-1 / 0.38	12-9 / 0.45	13-4 / 0.52	13-11 / 0.59	14-6 / 0.67
	24.0	5-1 / 0.04	6-3 / 0.07	7-2 / 0.10	8-1 / 0.14	8-10 / 0.19	9-6 / 0.24	10-2 / 0.29	10-10 / 0.34	11-5 / 0.40	11-11 / 0.46	12-6 / 0.53	13-0 / 0.60

Note: The required modulus of elasticity, "E", in 1,000,000 pounds per square inch is shown below each span.

TABLE R-12 (cont.)

RAFTERS: Spans are measured along the horizontal projection and loads are considered as applied on the horizontal projection.

1400	1500	1600	1700	1800	1900	2000	2100	2200	2400	2700	3000	RAFTER SPACING (IN)	SIZE (IN)
7-3 0.94	7-6 1.05	7-8 1.15	7-11 1.26	8-2 1.38	8-5 1.49	8-7 1.61	8-10 1.73	9-0 1.86	9-5 2.12	10-0 2.53		12.0	
6-9 0.88	7-0 0.98	7-3 1.08	7-5 1.18	7-8 1.29	7-10 1.40	8-1 1.51	8-3 1.62	8-5 1.74	8-10 1.98	9-4 2.36		13.7	
6-3 0.82	6-6 0.91	6-8 1.00	6-11 1.09	7-1 1.19	7-3 1.29	7-6 1.40	7-8 1.50	7-10 1.61	8-2 1.83	8-8 2.19	9-2 2.56	16.0	2x4
5-8 0.75	5-11 0.83	6-1 0.91	6-3 1.00	6-6 1.09	6-8 1.18	6-10 1.27	7-0 1.37	7-2 1.47	7-6 1.67	7-11 2.00	8-4 2.34	19.2	
5-1 0.67	5-3 0.74	5-5 0.82	5-7 0.89	5-9 0.97	5-11 1.06	6-1 1.14	6-3 1.23	6-5 1.31	6-8 1.50	7-1 1.79	7-6 2.09	24.0	
11-4 0.94	11-9 1.05	12-1 1.15	12-6 1.26	12-10 1.38	13-2 1.49	13-6 1.61	13-10 1.73	14-2 1.86	14-10 2.12	15-9 2.53		12.0	
10-7 0.88	11-0 0.98	11-4 1.08	11-8 1.18	12-0 1.29	12-4 1.40	12-8 1.51	13-0 1.62	13-3 1.74	13-10 1.98	14-9 2.36		13.7	
9-10 0.82	10-2 0.91	10-6 1.00	10-10 1.09	11-1 1.19	11-5 1.29	11-9 1.40	12-0 1.50	12-4 1.61	12-10 1.83	13-7 2.19	14-4 2.56	16.0	2x6
8-11 0.75	9-3 0.83	9-7 0.91	9-10 1.00	10-2 1.09	10-5 1.18	10-8 1.27	11-0 1.37	11-3 1.47	11-9 1.67	12-5 2.00	13-1 2.34	19.2	
8-0 0.67	8-3 0.74	8-7 0.82	8-10 0.89	9-1 0.97	9-4 1.06	9-7 1.14	9-10 1.23	10-0 1.31	10-6 1.50	11-1 1.79	11-9 2.09	24.0	
14-11 0.94	15-5 1.05	16-0 1.15	16-5 1.26	16-11 1.38	17-5 1.49	17-10 1.61	18-3 1.73	18-9 1.86	19-7 2.12	20-9 2.53		12.0	
14-0 0.88	14-6 0.98	14-11 1.08	15-5 1.18	15-10 1.29	16-3 1.40	16-8 1.51	17-1 1.62	17-6 1.74	18-3 1.98	19-5 2.36		13.7	
12-11 0.82	13-5 0.91	13-10 1.00	14-3 1.09	14-8 1.19	15-1 1.29	15-5 1.40	15-10 1.50	16-3 1.61	16-11 1.83	18-0 2.19	18-11 2.56	16.0	2x8
11-10 0.75	12-3 0.83	12-7 0.91	13-0 1.00	13-5 1.09	13-9 1.18	14-1 1.27	14-6 1.37	14-10 1.47	15-5 1.67	16-5 2.00	17-3 2.34	19.2	
10-7 0.67	10-11 0.74	11-3 0.82	11-8 0.89	12-0 0.97	12-4 1.06	12-7 1.14	12-11 1.23	13-3 1.31	13-10 1.50	14-8 1.79	15-5 2.09	24.0	
19-1 0.94	19-9 1.05	20-4 1.15	21-0 1.26	21-7 1.38	22-2 1.49	22-9 1.61	23-4 1.73	23-11 1.86	24-11 2.12	26-6 2.53		12.0	
17-10 0.88	18-5 0.98	19-1 1.08	19-8 1.18	20-2 1.29	20-9 1.40	21-4 1.51	21-10 1.62	22-4 1.74	23-4 1.98	24-9 2.36		13.7	
16-6 0.82	17-1 0.91	17-8 1.00	18-2 1.09	18-9 1.19	19-3 1.29	19-9 1.40	20-2 1.50	20-8 1.61	21-7 1.83	22-11 2.19	24-2 2.56	16.0	2x10
15-1 0.75	15-7 0.83	16-1 0.91	16-7 1.00	17-1 1.09	17-7 1.18	18-0 1.27	18-5 1.37	18-11 1.47	19-9 1.67	20-11 2.00	22-1 2.34	19.2	
13-6 0.67	13-11 0.74	14-5 0.82	14-10 0.89	15-3 0.97	15-8 1.06	16-1 1.14	16-6 1.23	16-11 1.31	17-8 1.50	18-9 1.79	19-9 2.09	24.0	

Note: The required modulus of elasticity, "E" in 1,000,000 pounds per square inch is shown below each span.

TABLE R-13
MEDIUM OR HIGH SLOPE RAFTERS
No Ceiling Load
Slope over 3 in 12
Live Load - 20 lb. per. sq. ft.
(Light roof covering)

DESIGN CRITERIA:
Strength - 7 lbs. per sq. ft. dead load plus 20 lbs. per sq. ft. live load determines required fiber stress.
Deflection - For 20 lbs. per sq. ft. live load. Limited to span in inches divided by 180.

RAFTER SIZE (IN)	SPACING (IN)	Extreme Fiber Stress in Bending, "F_b" (psi).											
		200	300	400	500	600	700	800	900	1000	1100	1200	1300
2x4	12.0	3-11 0.07	4-9 0.14	5-6 0.21	6-2 0.29	6-9 0.38	7-3 0.49	7-9 0.59	8-3 0.71	8-8 0.83	9-1 0.96	9-6 1.09	9-11 1.23
	13.7	3-8 0.07	4-5 0.13	5-2 0.20	5-9 0.27	6-4 0.36	6-10 0.45	7-3 0.55	7-9 0.66	8-2 0.77	8-6 0.89	8-11 1.02	9-3 1.15
	16.0	3-4 0.06	4-1 0.12	4-9 0.18	5-4 0.25	5-10 0.33	6-4 0.42	6-9 0.51	7-2 0.61	7-6 0.72	7-11 0.83	8-3 0.94	8-7 1.06
	19.2	3-1 0.06	3-9 0.11	4-4 0.17	4-10 0.23	5-4 0.30	5-9 0.38	6-2 0.47	6-6 0.56	6-10 0.65	7-3 0.76	7-6 0.86	7-10 0.97
	24.0	2-9 0.05	3-4 0.10	3-11 0.15	4-4 0.21	4-9 0.27	5-2 0.34	5-6 0.42	5-10 0.50	6-2 0.59	6-5 0.68	6-9 0.77	7-0 0.87
2x6	12.0	6-1 0.07	7-6 0.14	8-8 0.21	9-8 0.29	10-7 0.38	11-5 0.49	12-3 0.59	13-0 0.71	13-8 0.83	14-4 0.96	15-0 1.09	15-7 1.23
	13.7	5-9 0.07	7-0 0.13	8-1 0.20	9-0 0.27	9-11 0.36	10-8 0.45	11-5 0.55	12-2 0.66	12-9 0.77	13-5 0.89	14-0 1.02	14-7 1.15
	16.0	5-4 0.06	6-6 0.12	7-6 0.18	8-4 0.25	9-2 0.33	9-11 0.42	10-7 0.51	11-3 0.61	11-10 0.72	12-5 0.83	13-0 0.94	13-6 1.06
	19.2	4-10 0.06	5-11 0.11	6-10 0.17	7-8 0.23	8-4 0.30	9-0 0.38	9-8 0.47	10-3 0.56	10-10 0.65	11-4 0.76	11-10 0.86	12-4 0.97
	24.0	4-4 0.05	5-4 0.10	6-1 0.15	6-10 0.21	7-6 0.27	8-1 0.34	8-8 0.42	9-2 0.50	9-8 0.59	10-2 0.68	10-7 0.77	11-0 0.87
2x8	12.0	8-1 0.07	9-10 0.14	11-5 0.21	12-9 0.29	13-11 0.38	15-1 0.49	16-1 0.59	17-1 0.71	18-0 0.83	18-11 0.96	19-9 1.09	20-6 1.23
	13.7	7-6 0.07	9-3 0.13	10-8 0.20	11-11 0.27	13-1 0.36	14-1 0.45	15-1 0.55	16-0 0.66	16-10 0.77	17-8 0.89	18-5 1.02	19-3 1.15
	16.0	7-0 0.06	8-7 0.12	9-10 0.18	11-0 0.25	12-1 0.33	13-1 0.42	13-11 0.51	14-10 0.61	15-7 0.72	16-4 0.83	17-1 0.94	17-9 1.06
	19.2	6-4 0.06	7-10 0.11	9-0 0.17	10-1 0.23	11-0 0.30	11-11 0.38	12-9 0.47	13-6 0.56	14-3 0.65	14-11 0.76	15-7 0.86	16-3 0.97
	24.0	5-8 0.05	7-0 0.10	8-1 0.15	9-0 0.21	9-10 0.27	10-8 0.34	11-5 0.42	12-1 0.50	12-9 0.59	13-4 0.68	13-11 0.77	14-6 0.87
2x10	12.0	10-3 0.07	12-7 0.14	14-6 0.21	16-3 0.29	17-10 0.38	19-3 0.49	20-7 0.59	21-10 0.71	23-0 0.83	24-1 0.96	25-2 1.09	26-2 1.23
	13.7	9-7 0.07	11-9 0.13	13-7 0.20	15-2 0.27	16-8 0.36	18-0 0.45	19-3 0.55	20-5 0.66	21-6 0.77	22-7 0.89	23-7 1.02	24-6 1.15
	16.0	8-11 0.06	10-11 0.12	12-7 0.18	14-1 0.25	15-5 0.33	16-8 0.42	17-10 0.51	18-11 0.61	19-11 0.72	20-10 0.83	21-10 0.94	22-8 1.06
	19.2	8-2 0.06	9-11 0.11	11-6 0.17	12-10 0.23	14-1 0.30	15-2 0.38	16-3 0.47	17-3 0.56	18-2 0.65	19-1 0.76	19-11 0.86	20-9 0.97
	24.0	7-3 0.05	8-11 0.10	10-3 0.15	11-6 0.21	12-7 0.27	13-7 0.34	14-6 0.42	15-5 0.50	16-3 0.59	17-1 0.68	17-10 0.77	18-6 0.87

Note: The required modulus of elasticity, "E", in 1,000,000 pounds per square inch is shown below each span.

TABLE R-13 (cont.)

RAFTERS: Spans are measured along the horizontal projection and loads are considered as applied on the horizontal projection.

Extreme Fiber Stress in Bending, "F_b" (psi).											RAFTER SPACING (IN)	SIZE (IN)
1400	1500	1600	1700	1800	1900	2000	2100	2200	2400	2700		
10-3 1.37	10-8 1.52	11-0 1.68	11-4 1.84	11-8 2.00	12-0 2.17	12-4 2.34	12-7 2.52				12.0	
9-7 1.28	10-0 1.42	10-3 1.57	10-7 1.72	10-11 1.87	11-3 2.03	11-6 2.19	11-9 2.36	12-1 2.53			13.7	
8-11 1.19	9-3 1.32	9-6 1.45	9-10 1.59	10-1 1.73	10-5 1.88	10-8 2.03	10-11 2.18	11-2 2.34			16.0	2x4
8-2 1.08	8-5 1.20	8-8 1.33	9-0 1.45	9-3 1.58	9-6 1.71	9-9 1.85	10-0 1.99	10-2 2.14	10-8 2.43		19.2	
7-3 0.97	7-6 1.08	7-9 1.19	8-0 1.30	8-3 1.41	8-6 1.53	8-8 1.66	8-11 1.78	9-1 1.91	9-6 2.18	10-1 2.60	24.0	
16-2 1.37	16-9 1.52	17-3 1.68	17-10 1.84	18-4 2.00	18-10 2.17	19-4 2.34	19-10 2.52				12.0	
15-1 1.28	15-8 1.42	16-2 1.57	16-8 1.72	17-2 1.87	17-7 2.03	18-1 2.19	18-6 2.36	19-0 2.53			13.7	
14-0 1.19	14-6 1.32	15-0 1.45	15-5 1.59	15-11 1.73	16-4 1.88	16-9 2.03	17-2 2.18	17-7 2.34			16.0	2x6
12-9 1.08	13-3 1.20	13-8 1.33	14-1 1.45	14-6 1.58	14-11 1.71	15-3 1.85	15-8 1.99	16-0 2.14	16-9 2.43		19.2	
11-5 0.97	11-10 1.08	12-3 1.19	12-7 1.30	13-0 1.41	13-4 1.53	13-8 1.66	14-0 1.78	14-4 1.91	15-0 2.18	15-11 2.60	24.0	
21-4 1.37	22-1 1.52	22-9 1.68	23-6 1.84	24-2 2.00	24-10 2.17	25-6 2.34	26-1 2.52				12.0	
19-11 1.28	20-8 1.42	21-4 1.57	22-0 1.72	22-7 1.87	23-3 2.03	23-10 2.19	24-5 2.36	25-0 2.53			13.7	
18-5 1.19	19-1 1.32	19-9 1.45	20-4 1.59	20-11 1.73	21-6 1.88	22-1 2.03	22-7 2.18	23-2 2.34			16.0	2x8
16-10 1.08	17-5 1.20	18-0 1.33	18-7 1.45	19-1 1.58	19-8 1.71	20-2 1.85	20-8 1.99	21-1 2.14	22-1 2.43		19.2	
15-1 0.97	15-7 1.08	16-1 1.19	16-7 1.30	17-1 1.41	17-7 1.53	18-0 1.66	18-5 1.78	18-11 1.91	19-9 2.18	20-11 2.60	24.0	
27-2 1.37	28-2 1.52	29-1 1.68	30-0 1.84	30-10 2.00	31-8 2.17	32-6 2.34	33-4 2.52				12.0	
25-5 1.28	26-4 1.42	27-2 1.57	28-0 1.72	28-10 1.87	29-8 2.03	30-5 2.19	31-2 2.36	31-11 2.53			13.7	
23-7 1.19	24-5 1.32	25-2 1.45	25-11 1.59	26-8 1.73	27-5 1.88	28-2 2.03	28-10 2.18	29-6 2.34			16.0	2x10
21-6 1.08	22-3 1.20	23-0 1.33	23-8 1.45	24-5 1.58	25-1 1.71	25-8 1.85	26-4 1.99	26-11 2.14	28-2 2.43		19.2	
19-3 0.97	19-11 1.08	20-7 1.19	21-2 1.30	21-10 1.41	22-5 1.53	23-0 1.66	23-7 1.78	24-1 1.91	25-2 2.18	26-8 2.60	24.0	

Note: The required modulus of elasticity, "E", in 1,000,000 pounds per square inch is shown below each span.

TABLE R-14
MEDIUM OR HIGH SLOPE RAFTERS
No Ceiling Load
Slope over 3 in 12
Live Load - 30 lb. per. sq. ft.
(Light roof covering)

DESIGN CRITERIA:
Strength - 7 lbs. per sq. ft. dead load plus
30 lbs. per sq. ft. live load determines
required fiber stress.
Deflection - For 30 lbs. per sq. ft. live load.
Limited to span in inches divided by 180.

RAFTER SIZE (IN)	SPACING (IN)	Extreme Fiber Stress in Bending, "F_b" (psi).											
		200	300	400	500	600	700	800	900	1000	1100	1200	1300
2x4	12.0	3-4 / 0.07	4-1 / 0.13	4-8 / 0.20	5-3 / 0.27	5-9 / 0.36	6-3 / 0.45	6-8 / 0.55	7-1 / 0.66	7-5 / 0.77	7-9 / 0.89	8-2 / 1.02	8-6 / 1.15
	13.7	3-1 / 0.06	3-10 / 0.12	4-5 / 0.18	4-11 / 0.26	5-5 / 0.34	5-10 / 0.42	6-3 / 0.52	6-7 / 0.62	6-11 / 0.72	7-3 / 0.84	7-7 / 0.95	7-11 / 1.07
	16.0	2-11 / 0.06	3-6 / 0.11	4-1 / 0.17	4-7 / 0.24	5-0 / 0.31	5-5 / 0.39	5-9 / 0.48	6-1 / 0.57	6-5 / 0.67	6-9 / 0.77	7-1 / 0.88	7-4 / 0.99
	19.2	2-8 / 0.05	3-3 / 0.10	3-9 / 0.15	4-2 / 0.22	4-7 / 0.28	4-11 / 0.36	5-3 / 0.44	5-7 / 0.52	5-10 / 0.61	6-2 / 0.71	6-5 / 0.80	6-8 / 0.91
	24.0	2-4 / 0.05	2-11 / 0.09	3-4 / 0.14	3-9 / 0.19	4-1 / 0.25	4-5 / 0.32	4-8 / 0.39	5-0 / 0.47	5-3 / 0.55	5-6 / 0.63	5-9 / 0.72	6-0 / 0.81
2x6	12.0	5-3 / 0.07	6-5 / 0.13	7-5 / 0.20	8-3 / 0.27	9-1 / 0.36	9-9 / 0.45	10-5 / 0.55	11-1 / 0.66	11-8 / 0.77	12-3 / 0.89	12-9 / 1.02	13-4 / 1.15
	13.7	4-11 / 0.06	6-0 / 0.12	6-11 / 0.18	7-9 / 0.26	8-5 / 0.34	9-2 / 0.42	9-9 / 0.52	10-4 / 0.62	10-11 / 0.72	11-5 / 0.84	12-0 / 0.95	12-5 / 1.07
	16.0	4-6 / 0.06	5-6 / 0.11	6-5 / 0.17	7-2 / 0.24	7-10 / 0.31	8-5 / 0.39	9-1 / 0.48	9-7 / 0.57	10-1 / 0.67	10-7 / 0.77	11-1 / 0.88	11-6 / 0.99
	19.2	4-2 / 0.05	5-1 / 0.10	5-10 / 0.15	6-6 / 0.22	7-2 / 0.28	7-9 / 0.36	8-3 / 0.44	8-9 / 0.52	9-3 / 0.61	9-8 / 0.71	10-1 / 0.80	10-6 / 0.91
	24.0	3-8 / 0.05	4-6 / 0.09	5-3 / 0.14	5-10 / 0.19	6-5 / 0.25	6-11 / 0.32	7-5 / 0.39	7-10 / 0.47	8-3 / 0.55	8-8 / 0.63	9-1 / 0.72	9-5 / 0.81
2x8	12.0	6-11 / 0.07	8-5 / 0.13	9-9 / 0.20	10-11 / 0.27	11-11 / 0.36	12-10 / 0.45	13-9 / 0.55	14-7 / 0.66	15-5 / 0.77	16-2 / 0.89	16-10 / 1.02	17-7 / 1.15
	13.7	6-5 / 0.06	7-11 / 0.12	9-1 / 0.18	10-2 / 0.26	11-2 / 0.34	12-1 / 0.42	12-10 / 0.52	13-8 / 0.62	14-5 / 0.72	15-1 / 0.84	15-9 / 0.95	16-5 / 1.07
	16.0	6-0 / 0.06	7-4 / 0.11	8-5 / 0.17	9-5 / 0.24	10-4 / 0.31	11-2 / 0.39	11-11 / 0.48	12-8 / 0.57	13-4 / 0.67	14-0 / 0.77	14-7 / 0.88	15-2 / 0.99
	19.2	5-5 / 0.05	6-8 / 0.10	7-8 / 0.15	8-7 / 0.22	9-5 / 0.28	10-2 / 0.36	10-11 / 0.44	11-6 / 0.52	12-2 / 0.61	12-9 / 0.71	13-4 / 0.80	13-10 / 0.91
	24.0	4-10 / 0.05	6-0 / 0.09	6-11 / 0.14	7-8 / 0.19	8-5 / 0.25	9-1 / 0.32	9-9 / 0.39	10-4 / 0.47	10-11 / 0.55	11-5 / 0.63	11-11 / 0.72	12-5 / 0.81
2x10	12.0	8-9 / 0.07	10-9 / 0.13	12-5 / 0.20	13-11 / 0.27	15-2 / 0.36	16-5 / 0.45	17-7 / 0.55	18-7 / 0.66	19-8 / 0.77	20-7 / 0.89	21-6 / 1.02	22-5 / 1.15
	13.7	8-3 / 0.06	10-1 / 0.12	11-7 / 0.18	13-0 / 0.26	14-3 / 0.34	15-4 / 0.42	16-5 / 0.52	17-5 / 0.62	18-4 / 0.72	19-3 / 0.84	20-1 / 0.95	20-11 / 1.07
	16.0	7-7 / 0.07	9-4 / 0.12	10-9 / 0.19	12-0 / 0.26	13-2 / 0.34	14-3 / 0.43	15-2 / 0.53	16-2 / 0.63	17-0 / 0.74	17-10 / 0.85	18-7 / 0.97	19-5 / 1.09
	19.2	6-11 / 0.05	8-6 / 0.10	9-10 / 0.15	11-0 / 0.22	12-0 / 0.28	13-0 / 0.36	13-11 / 0.44	14-9 / 0.52	15-6 / 0.61	16-3 / 0.71	17-0 / 0.80	17-8 / 0.91
	24.0	6-2 / 0.05	7-7 / 0.09	8-9 / 0.14	9-10 / 0.19	10-9 / 0.25	11-7 / 0.32	12-5 / 0.39	13-2 / 0.47	13-11 / 0.55	14-7 / 0.63	15-2 / 0.72	15-10 / 0.81

Note: The required modulus of elasticity, "E", in 1,000,000 pounds per square inch is shown below each span.

TABLE R-14 (cont.)

RAFTERS: Spans are measured along the horizontal projection and loads are considered as applied on the horizontal projection.

Extreme Fiber Stress in Bending, "F_b" (psi).											RAFTER SPACING (IN)	SIZE (IN)
1400	1500	1600	1700	1800	1900	2000	2100	2200	2400	2700		
8-9 1.28	9-1 1.42	9-5 1.57	9-8 1.72	10-0 1.87	10-3 2.03	10-6 2.19	10-9 2.36	11-0 2.53			12.0	
8-3 1.20	8-6 1.33	8-9 1.47	9-1 1.61	9-4 1.75	9-7 1.90	9-10 2.05	10-1 2.20	10-4 2.36			13.7	
7-7 1.11	7-11 1.23	8-2 1.36	8-5 1.49	8-8 1.62	8-10 1.76	9-1 1.90	9-4 2.04	9-7 2.19	10-0 2.49		16.0	2x4
6-11 1.01	7-2 1.12	7-5 1.24	7-8 1.36	7-11 1.48	8-1 1.60	8-4 1.73	8-6 1.86	8-9 2.00	9-1 2.28		19.2	
6-3 0.91	6-5 1.01	6-8 1.11	6-10 1.21	7-1 1.32	7-3 1.43	7-5 1.55	7-7 1.67	7-9 1.79	8-2 2.04	8-8 2.43	24.0	
13-10 1.28	14-4 1.42	14-9 1.57	15-3 1.72	15-8 1.87	16-1 2.03	16-6 2.19	16-11 2.36	17-4 2.53			12.0	
12-11 1.20	13-4 1.33	13-10 1.47	14-3 1.61	14-8 1.75	15-1 1.90	15-5 2.05	15-10 2.20	16-2 2.36			13.7	
12-0 1.11	12-5 1.23	12-9 1.36	13-2 1.49	13-7 1.62	13-11 1.76	14-4 1.90	14-8 2.04	15-0 2.19	15-8 2.49		16.0	2x6
10-11 1.01	11-4 1.12	11-8 1.24	12-0 1.36	12-5 1.48	12-9 1.60	13-1 1.73	13-4 1.86	13-8 2.00	14-4 2.28		19.2	
9-9 0.91	10-1 1.01	10-5 1.11	10-9 1.21	11-1 1.32	11-5 1.43	11-8 1.55	12-0 1.67	12-3 1.79	12-9 2.04	13-7 2.43	24.0	
18-2 1.28	18-10 1.42	19-6 1.57	20-1 1.72	20-8 1.87	21-3 2.03	21-9 2.19	22-4 2.36	22-10 2.53			12.0	
17-0 1.20	17-8 1.33	18-2 1.47	18-9 1.61	19-4 1.75	19-10 1.90	20-4 2.05	20-10 2.20	21-4 2.36			13.7	
15-9 1.11	16-4 1.23	16-10 1.36	17-4 1.49	17-11 1.62	18-4 1.76	18-10 1.90	19-4 2.04	19-9 2.19	20-8 2.49		16.0	2x8
14-5 1.01	14-11 1.12	15-5 1.24	15-10 1.36	16-4 1.48	16-9 1.60	17-2 1.73	17-8 1.86	18-1 2.00	18-10 2.28		19.2	
12-10 0.91	13-4 1.01	13-9 1.11	14-2 1.21	14-7 1.32	15-0 1.43	15-5 1.55	15-9 1.67	16-2 1.79	16-10 2.04	17-11 2.43	24.0	
23-3 1.28	24-1 1.42	24-10 1.57	25-7 1.72	26-4 1.87	27-1 2.03	27-9 2.19	28-5 2.36	29-1 2.53			12.0	
21-9 1.20	22-6 1.33	23-3 1.47	23-11 1.61	24-8 1.75	25-4 1.90	26-0 2.05	26-7 2.20	27-3 2.36			13.7	
20-1 1.22	20-10 1.35	21-6 1.49	22-2 1.63	22-10 1.78	23-5 1.93	24-1 2.08	24-8 2.24	25-3 2.40			16.0	2x10
18-4 1.01	19-0 1.12	19-8 1.24	20-3 1.36	20-10 1.48	21-5 1.60	21-11 1.73	22-6 1.86	23-0 2.00	24-1 2.28		19.2	
16-5 0.91	17-0 1.01	17-7 1.11	18-1 1.21	18-7 1.32	19-2 1.43	19-8 1.55	20-1 1.67	20-7 1.79	21-6 2.04	22-10 2.43	24.0	

Note: The required modulus of elasticity, "E", in 1,000,000 pounds per square inch is shown below each span.

TABLE R-15
MEDIUM OR HIGH SLOPE RAFTERS
No Ceiling Load
Slope over 3 in 12
Live Load - 40 lb. per. sq. ft.
(Light roof covering)

DESIGN CRITERIA:
Strength - 7 lbs. per sq. ft. dead load plus
  40 lbs. per. sq. ft. live load determines
  required fiber stress.
Deflection - For 40 lbs. per. sq. ft. live load.
  Limited to span in inches divided by 180.

RAFTER SIZE (IN)	SPACING (IN)	Extreme Fiber Stress in Bending, "F$_b$" (psi).											
		200	300	400	500	600	700	800	900	1000	1100	1200	1300
2x4	12.0	2-11 / 0.06	3-7 / 0.12	4-2 / 0.18	4-8 / 0.25	5-1 / 0.34	5-6 / 0.42	5-11 / 0.52	6-3 / 0.62	6-7 / 0.72	6-11 / 0.83	7-3 / 0.95	7-6 / 1.07
	13.7	2-9 / 0.06	3-5 / 0.11	3-11 / 0.17	4-4 / 0.24	4-9 / 0.31	5-2 / 0.40	5-6 / 0.48	5-10 / 0.58	6-2 / 0.67	6-6 / 0.78	6-9 / 0.89	7-0 / 1.00
	16.0	2-7 / 0.06	3-2 / 0.10	3-7 / 0.16	4-0 / 0.22	4-5 / 0.29	4-9 / 0.37	5-1 / 0.45	5-5 / 0.53	5-8 / 0.62	6-0 / 0.72	6-3 / 0.82	6-6 / 0.93
	19.2	2-4 / 0.05	2-10 / 0.09	3-4 / 0.14	3-8 / 0.20	4-0 / 0.26	4-4 / 0.33	4-8 / 0.41	4-11 / 0.49	5-3 / 0.57	5-6 / 0.66	5-8 / 0.75	5-11 / 0.85
	24.0	2-1 / 0.05	2-7 / 0.08	2-11 / 0.13	3-4 / 0.18	3-7 / 0.24	3-11 / 0.30	4-2 / 0.36	4-5 / 0.44	4-8 / 0.51	4-11 / 0.59	5-1 / 0.67	5-4 / 0.76
2x6	12.0	4-8 / 0.06	5-8 / 0.12	6-7 / 0.18	7-4 / 0.25	8-0 / 0.34	8-8 / 0.42	9-3 / 0.52	9-10 / 0.62	10-4 / 0.72	10-10 / 0.83	11-4 / 0.95	11-10 / 1.07
	13.7	4-4 / 0.06	5-4 / 0.11	6-2 / 0.17	6-10 / 0.24	7-6 / 0.31	8-1 / 0.40	8-8 / 0.48	9-2 / 0.58	9-8 / 0.67	10-2 / 0.78	10-7 / 0.89	11-1 / 1.00
	16.0	4-0 / 0.06	4-11 / 0.10	5-8 / 0.16	6-4 / 0.22	6-11 / 0.29	7-6 / 0.37	8-0 / 0.45	8-6 / 0.53	9-0 / 0.62	9-5 / 0.72	9-10 / 0.82	10-3 / 0.93
	19.2	3-8 / 0.05	4-6 / 0.09	5-2 / 0.14	5-9 / 0.20	6-4 / 0.26	6-10 / 0.33	7-4 / 0.41	7-9 / 0.49	8-2 / 0.57	8-7 / 0.66	9-0 / 0.75	9-4 / 0.85
	24.0	3-3 / 0.05	4-0 / 0.08	4-8 / 0.13	5-2 / 0.18	5-8 / 0.24	6-2 / 0.30	6-7 / 0.36	6-11 / 0.44	7-4 / 0.51	7-8 / 0.59	8-0 / 0.67	8-4 / 0.76
2x8	12.0	6-1 / 0.06	7-6 / 0.12	8-8 / 0.18	9-8 / 0.25	10-7 / 0.34	11-5 / 0.42	12-3 / 0.52	12-11 / 0.62	13-8 / 0.72	14-4 / 0.83	14-11 / 0.95	15-7 / 1.07
	13.7	5-9 / 0.06	7-0 / 0.11	8-1 / 0.17	9-0 / 0.24	9-11 / 0.31	10-8 / 0.40	11-5 / 0.48	12-1 / 0.58	12-9 / 0.67	13-5 / 0.78	14-0 / 0.89	14-7 / 1.00
	16.0	5-3 / 0.06	6-6 / 0.10	7-6 / 0.16	8-4 / 0.22	9-2 / 0.29	9-11 / 0.37	10-7 / 0.45	11-3 / 0.53	11-10 / 0.62	12-5 / 0.72	12-11 / 0.82	13-6 / 0.93
	19.2	4-10 / 0.05	5-11 / 0.09	6-10 / 0.14	7-8 / 0.20	8-4 / 0.26	9-0 / 0.33	9-8 / 0.41	10-3 / 0.49	10-10 / 0.57	11-4 / 0.66	11-10 / 0.75	12-4 / 0.85
	24.0	4-4 / 0.05	5-3 / 0.08	6-1 / 0.13	6-10 / 0.18	7-6 / 0.24	8-1 / 0.30	8-8 / 0.36	9-2 / 0.44	9-8 / 0.51	10-2 / 0.59	10-7 / 0.67	11-0 / 0.76
2x10	12.0	7-9 / 0.06	9-6 / 0.12	11-0 / 0.18	12-4 / 0.25	13-6 / 0.34	14-7 / 0.42	15-7 / 0.52	16-6 / 0.62	17-5 / 0.72	18-3 / 0.83	19-1 / 0.95	19-10 / 1.07
	13.7	7-3 / 0.06	8-11 / 0.11	10-4 / 0.17	11-6 / 0.24	12-7 / 0.31	13-8 / 0.40	14-7 / 0.48	15-5 / 0.58	16-4 / 0.67	17-1 / 0.78	17-10 / 0.89	18-7 / 1.00
	16.0	6-9 / 0.06	8-3 / 0.10	9-6 / 0.16	10-8 / 0.22	11-8 / 0.29	12-7 / 0.37	13-6 / 0.45	14-4 / 0.53	15-1 / 0.62	15-10 / 0.72	16-6 / 0.82	17-2 / 0.93
	19.2	6-2 / 0.05	7-7 / 0.09	8-9 / 0.14	9-9 / 0.20	10-8 / 0.26	11-6 / 0.33	12-4 / 0.41	13-1 / 0.49	13-9 / 0.57	14-5 / 0.66	15-1 / 0.75	15-8 / 0.85
	24.0	5-6 / 0.05	6-9 / 0.08	7-9 / 0.13	8-9 / 0.18	9-6 / 0.24	10-4 / 0.30	11-0 / 0.36	11-8 / 0.44	12-4 / 0.51	12-11 / 0.59	13-6 / 0.67	14-1 / 0.76

Note: The required modulus of elasticity, "E", in 1,000,000 pounds per square inch is shown below each span.

TABLE R-15 (cont.)

RAFTERS: Spans are measured along the horizontal projection and loads are considered as applied on the horizontal projection.

| Extreme Fiber Stress in Bending, "F_b" (psi). | | | | | | | | | | | RAFTER SPACING (IN) | SIZE (IN) |
1400	1500	1600	1700	1800	1900	2000	2100	2200	2400	2700		
7-10 / 1.19	8-1 / 1.32	8-4 / 1.46	8-7 / 1.60	8-10 / 1.74	9-1 / 1.89	9-4 / 2.04	9-7 / 2.19	9-9 / 2.35			12.0	
7-4 / 1.12	7-7 / 1.24	7-10 / 1.37	8-0 / 1.50	8-3 / 1.63	8-6 / 1.77	8-9 / 1.91	8-11 / 2.05	9-2 / 2.20	9-7 / 2.51		13.7	
6-9 / 1.03	7-0 / 1.15	7-3 / 1.26	7-5 / 1.38	7-8 / 1.51	7-10 / 1.64	8-1 / 1.77	8-3 / 1.90	8-6 / 2.04	8-10 / 2.32		16.0	2x4
6-2 / 0.94	6-5 / 1.05	6-7 / 1.15	6-10 / 1.26	7-0 / 1.38	7-2 / 1.49	7-4 / 1.61	7-7 / 1.74	7-9 / 1.86	8-1 / 2.12	8-7 / 2.53	19.2	
5-6 / 0.84	5-8 / 0.94	5-11 / 1.03	6-1 / 1.13	6-3 / 1.23	6-5 / 1.34	6-7 / 1.44	6-9 / 1.55	6-11 / 1.66	7-3 / 1.90	7-8 / 2.26	24.0	
12-3 / 1.19	12-8 / 1.32	13-1 / 1.46	13-6 / 1.60	13-11 / 1.74	14-3 / 1.89	14-8 / 2.04	15-0 / 2.19	15-4 / 2.35			12.0	
11-6 / 1.12	11-10 / 1.24	12-3 / 1.37	12-8 / 1.50	13-0 / 1.63	13-4 / 1.77	13-8 / 1.91	14-0 / 2.05	14-4 / 2.20	15-0 / 2.51		13.7	
10-7 / 1.03	11-0 / 1.15	11-4 / 1.26	11-8 / 1.38	12-0 / 1.51	12-4 / 1.64	12-8 / 1.77	13-0 / 1.90	13-4 / 2.04	13-11 / 2.32		16.0	2x6
9-8 / 0.94	10-0 / 1.05	10-4 / 1.15	10-8 / 1.26	11-0 / 1.38	11-3 / 1.49	11-7 / 1.61	11-10 / 1.74	12-2 / 1.86	12-8 / 2.12	13-5 / 2.53	19.2	
8-8 / 0.84	9-0 / 0.94	9-3 / 1.03	9-7 / 1.13	9-10 / 1.23	10-1 / 1.34	10-4 / 1.44	10-7 / 1.55	10-10 / 1.66	11-4 / 1.90	12-0 / 2.26	24.0	
16-2 / 1.19	16-9 / 1.32	17-3 / 1.46	17-10 / 1.60	18-4 / 1.74	18-10 / 1.89	19-4 / 2.04	19-9 / 2.19	20-3 / 2.35			12.0	
15-1 / 1.12	15-8 / 1.24	16-2 / 1.37	16-8 / 1.50	17-2 / 1.63	17-7 / 1.77	18-1 / 1.91	18-6 / 2.05	18-11 / 2.20	19-9 / 2.51		13.7	
14-0 / 1.03	14-6 / 1.15	14-11 / 1.26	15-5 / 1.38	15-10 / 1.51	16-4 / 1.64	16-9 / 1.77	17-2 / 1.90	17-6 / 2.04	18-4 / 2.32		16.0	2x8
12-9 / 0.94	13-3 / 1.05	13-8 / 1.15	14-1 / 1.26	14-6 / 1.38	14-11 / 1.49	15-3 / 1.61	15-8 / 1.74	16-0 / 1.86	16-9 / 2.12	17-9 / 2.53	19.2	
11-5 / 0.84	11-10 / 0.94	12-3 / 1.03	12-7 / 1.13	12-11 / 1.23	13-4 / 1.34	13-8 / 1.44	14-0 / 1.55	14-4 / 1.66	14-11 / 1.90	15-10 / 2.26	24.0	
20-7 / 1.19	21-4 / 1.32	22-0 / 1.46	22-9 / 1.60	23-4 / 1.74	24-0 / 1.89	24-8 / 2.04	25-3 / 2.19	25-10 / 2.35			12.0	
19-3 / 1.12	19-11 / 1.24	20-7 / 1.37	21-3 / 1.50	21-10 / 1.63	22-6 / 1.77	23-1 / 1.91	23-7 / 2.05	24-2 / 2.20	25-3 / 2.51		13.7	
17-10 / 1.03	18-6 / 1.15	19-1 / 1.26	19-8 / 1.38	20-3 / 1.51	20-10 / 1.64	21-4 / 1.77	21-10 / 1.90	22-4 / 2.04	23-4 / 2.32		16.0	2x10
16-4 / 0.94	16-10 / 1.05	17-5 / 1.15	17-11 / 1.26	18-6 / 1.38	19-0 / 1.49	19-6 / 1.61	19-11 / 1.74	20-5 / 1.86	21-4 / 2.12	22-8 / 2.53	19.2	
14-7 / 0.84	15-1 / 0.94	15-7 / 1.03	16-1 / 1.13	16-6 / 1.23	17-0 / 1.34	17-5 / 1.44	17-10 / 1.55	18-3 / 1.66	19-1 / 1.90	20-3 / 2.26	24.0	

Note: The required modulus of elasticity, "E", in 1,000,000 pounds per square inch is shown below each span.

FLOOR JOIST SPANS FOR 30 AND 40 PSF LIVE LOADS, 10 PSF DEAD LOAD AND 1/360 DEFLECTION LIMITATIONS

Size (inches)	Spacing (inches)	No 1 Dense KD Load psf		No. 1 KD Load psf		No. 2 Dense KD Load psf		No. 2 KD Load psf		No. 3 KD Load psf	
		30	40	30	40	30	40	30	40	30	40
2 × 6	12	12-6	11-4	12-3	11-2	12-0	10-11	11-10	10-9	10-5	9-4
	16	11-4	10-4	11-2	10-2	10-11	9-11	10-9	9-9	9-0	8-1
	24	9-11	9-0	9-9	8-10	9-7	8-8	9-4	8-6	7-4	6-7
2 × 8	12	16-6	15-0	16-2	14-8	15-10	14-5	15-7	14-2	13-9	12-4
	16	15-0	13-7	14-8	13-4	14-5	13-1	14-2	12-10	11-11	10-8
	24	13-1	11-11	12-10	11-8	12-7	11-5	12-4	11-3	9-9	8-8
2 × 10	12	21-0	19-1	20-8	18-9	20-3	18-5	19-10	18-0	17-6	15-8
	16	19-1	17-4	18-9	17-0	18-5	16-9	18-0	16-5	15-2	13-7
	24	16-8	15-2	16-5	14-11	16-1	14-7	15-9	14-4	12-5	11-1
2 × 12	12	25-7	23-3	25-1	22-10	24-8	22-5	24-2	21-11	21-4	19-1
	16	23-3	21-1	22-10	20-9	22-5	20-4	21-11	19-11	18-6	16-6
	24	20-3	18-5	19-11	18-1	19-7	17-9	19-2	17-5	15-1	13-6

CEILING JOIST SPANS FOR 10[a] AND 20[b] PSF LIVE LOADS, 10 PSF DEAD LOADS AND 1/240 DEFLECTION LIMITATION

Size (inches)	Spacing (inches)	10	20	10	20	10	20	10	20	10	20
2 × 4	12	13-2	10-5	12-11	10-3	12-8	10-0	12-5	9-10	11-6	8-2
	16	11-11	9-6	11-9	9-4	11-6	9-1	11-3	8-11	10-0	7-1
	24	10-5	8-3	10-3	8-1	10-0	8-0	9-10	7-9	8-2	5-9
2 × 6	12	20-8	16-4	20-3	16-1	19-11	15-9	19-6	15-6	17-0	12-0
	16	18-9	14-11	18-5	14-7	18-1	14-4	17-8	13-9	14-9	10-5
	24	16-4	13-0	16-1	12-5	15-9	12-3	15-6	11-2	12-0	8-6
2 × 8	12	27-2	21-7	26-9	21-2	26-2	20-10	25-8	20-5	22-5	15-10
	16	24-8	19-7	24-3	19-3	23-10	18-11	23-4	18-1	19-5	13-9
	24	21-7	17-2	21-2	16-5	20-10	16-2	20-5	14-9	15-10	11-3
2 × 10	12	34-8	27-6	34-1	27-1	33-5	26-6	32-9	26-0	28-8	20-3
	16	31-6	25-0	31-0	24-7	30-5	24-1	29-9	23-1	24-10	17-6
	24	27-6	21-10	27-1	20-11	26-6	20-7	26-0	18-10	20-3	14-4

[a] Load contemplates no attic storage and no future sleeping rooms.

[b] Load contemplates limited access for attic storage and no future sleeping rooms.

Lumber in the S-Dry grade category which is dried to 19% or less moisture content will have spans several inches less than the KD grade which is dried to 15% or less moisture content.

Spans are determined on the same basis as those given in the nationally recognized Span Tables for Joists and Rafters, published by the National Forest Products Association. More detailed span information for Southern Pine can be found in the SFPA Bulletin No. 2 on maximum spans for joists and rafters.

The conditions of loading are those recognized by HUD, the Model Building Codes, and National Association of Home Builders.

Grademarked lumber is recommended and should be identified by the grademark of an agency certified by the Board of Review of the American Lumber Standards Committee.

LUMBER SIZES

Nominal Sizes (inches)	Dressed Sizes (inches—19% Max. MC)
2 × 4	1½ × 3½
2 × 6	1½ × 5½
2 × 8	1½ × 7¼
2 × 10	1½ × 9¼
2 × 12	1½ × 11¼

RAFTER SPANS FOR ANY SLOPE, DRYWALL CEILING, 20 AND 30 PSF LIVE LOAD—15 PSF DEAD LOAD

		Grade									
		No 1 Dense KD		No. 1 KD		No. 2 Dense KD		No. 2 KD		No. 3 KD	
		Load psf		Load psf		Load psf		Load psf		Load psf	
Size (inches)	Spacing (inches)	20	30	20	30	20	30	20	30	20	30
2 × 6	12	16–4	14–4	16–1	14–1	15–9	13–9	16–3	13–9	12–5	10–6
	16	14–11	13–0	15–8	13–3	15–5	13–1	14–2	11–11	10–9	9–1
	24	13–9	11–8	12–9	10–10	12–7	10–8	11–7	9–9	8–9	7–4
2 × 8	12	21–7	18–10	21–2	18–6	20–10	18–2	21–6	18–2	16–3	13–9
	16	19–7	17–2	20–8	17–6	20–4	17–3	18–7	15–9	14–2	12–0
	24	18–2	15–4	16–10	14–3	16–7	14–0	15–2	12–10	11–7	9–9
2 × 10	12	27–6	24–1	27–1	23–8	26–6	23–2	27–5	23–2	20–10	17–7
	16	25–0	21–10	26–4	22–3	25–11	21–11	23–8	20–1	18–0	15–3
	24	23–1	19–6	21–6	18–2	21–2	17–11	19–4	16–5	14–8	12–5
2 × 12	12	33–6	29–3	32–11	28–9	32–3	28–2	33–4	28–3	25–4	21–5
	16	30–5	26–7	32–0	27–1	31–7	26–8	28–10	24–5	21–11	18–7
	24	28–1	23–9	26–2	22–1	25–9	21–9	23–7	20–0	17–10	15–2

[a]Where there is no snow load of any consequence a minimum of 20 psf live load is anticipated to occur occasionally from construction loads of short duration (7 days) and for design purposes building codes permit a 25% increase in the allowable fiber stress in bending. Such an increase is reflected in the spans.

[b]The 30 psf loading is based on snow loading conditions generally considered by building codes to be of short duration and warrant a 15% increase in allowable fiber stress in bending. Such an increase is reflected in the spans.

**RAFTER SPANS FOR LOW AND HIGH SLOPES, NO CEILING,[a] 20 PSF LIVE LOAD,
10 AND 7 PSF DEAD LOAD RESPECTIVELY**

		Low	High	Low	High	Low	High	Low	High	Low	High
2 × 4	12		11–6		11–3		11–1		10–10		9–6
	16		10–5		10–3		10–0		9–10		8–3
	24		9–1		8–11		8–9		9–2		6–9
2 × 6	12	16–4	18–0	16–1	17–8	15–9	17–4	15–6	18–6	13–4	14–1
	16	14–11	16–4	14–7	17–10	14–4	15–9	15–3	16–1	11–7	12–3
	24	13–0	15–8	13–9	14–6	13–7	14–4	12–5	13–2	9–5	10–0
2 × 8	12	21–7	23–9	21–2	23–4	20–10	22–11	20–5	24–5	17–7	18–7
	16	19–7	21–7	19–3	23–6	18–11	20–10	20–1	21–2	15–3	16–1
	24	17–2	20–8	18–3	19–2	17–11	18–10	16–4	17–4	12–6	13–2
2 × 10	12	27–6	30–4	27–1	29–9	26–6	29–2	26–0	31–2	22–6	23–8
	16	25–0	27–6	24–7	30–0	24–1	26–6	25–7	27–0	19–5	20–6
	24	21–10	26–3	23–3	24–6	22–10	24–1	20–11	22–0	15–11	16–9
2 × 12	12	33–6		32–11		32–3		31–8		27–5	
	16	30–5		29–11		29–4		31–2		23–8	
	24	26–7		28–3		27–9		25–5		19–4	

[a]Low slope: 3 in 12 or less; high slope: over 3 in 12. Reflects construction load increase (see footnote *a* from Table "Rafter Spans for Any Slope").

D-1 NAIL WEIGHTS AND SIZES

NUMBER PER POUND OR KILO

Size	Weight Unit	Common	Casing	Box	Finishing
2d	Pound	876	1010	1010	1351
	Kilo	1927	2222	2222	2972
3d	Pound	586	635	635	807
	Kilo	1289	1397	1397	1775
4d	Pound	316	473	473	548
	Kilo	695	1041	1041	1206
5d	Pound	271	406	406	500
	Kilo	596	893	893	1100
6d	Pound	181	236	236	309
	Kilo	398	519	519	680
7d	Pound	161	210	210	238
	Kilo	354	462	462	524
8d	Pound	106	145	145	189
	Kilo	233	319	319	416
9d	Pound	96	132	132	172
	Kilo	211	290	290	398
10d	Pound	69	94	94	121
	Kilo	152	207	207	266
12d	Pound	64	88	88	113
	Kilo	141	194	194	249
16d	Pound	49	71	71	90
	Kilo	108	156	156	198
20d	Pound	31	52	52	62
	Kilo	68	114	114	136
30d	Pound	24	46	46	
	Kilo	53	101	101	
40d	Pound	18	35	35	
	Kilo	37	77	77	
50d	Pound	14			
	Kilo	31			
60d	Pound	11			
	Kilo	24			

Source: Federal Housing Administration.

LENGTH AND DIAMETER
IN INCHES AND CENTIMETERS

Size	Inches	Length Centimeters	Diameter Inches	Centimeters
2d	1	2.5	0.068	0.17
3d	1/2	3.2	0.102	0.26
4d	1/4	3.8	0.102	0.26
5d	1/6	4.4	0.102	0.26
6d	2	5.1	0.115	0.29
7d	2/2	5.7	0.115	0.29
8d	2/4	6.4	0.131	0.33
9d	2/6	7.0	0.131	0.33
10d	3	7.6	0.148	0.38
12d	3/2	8.3	0.148	0.38
16d	3/4	8.9	0.148	0.38
20d	4	10.2	0.203	0.51
30d	4/4	11.4	0.220	0.58
40d	5	12.7	0.238	0.60
50d	5/4	14.0	0.257	0.66
60d	6	15.2	0.277	0.70

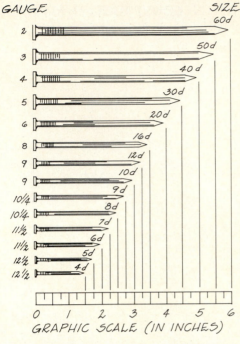

COMMON WIRE NAILS USED IN RESIDENTIAL CONSTRUCTION

D-2 HINGES[1]

Some hinges or butts are left-handed or right-handed. To determine the hand of a loose joint or live butt hinge, hold the hinge open with the hinge face toward you. If you can hold the hinge by the right hand leaf in a vertical position without the hinge falling apart, it is a right-handed hinge. If you must hold the left-hand leaf so the hinge will not fall apart, it is a left-handed hinge.

Hinges should be selected according to door size and weight. The following tables suggest hinge heights and widths.

HINGE HEIGHT

Door Thickness	Door Width	Height of Hinge
3/4" to 1⅛" cabinet	to 24"	2½"
⅞" to 1⅛" screen or combination	to 36"	3"
1⅜"	to 32"	3½" to 4"
	over 32"	4" to 4½"
1¾"	to 36"	4½"ᵃ
	36" to 48"	5"ᵃ
	over 48"	6"ᵃ
2", 2¼", and 2½"	to 42"	5" extra heavy

[a]Heavy-weight hinges should be specified for heavy doors and doors where high-frequency service is expected.

HINGE WIDTH

Door Thickness	Clearance Required	Open Width of Hinge
1⅜"	1¼"	3½"
	1¾"	4"
1¾"	1"	4"
	1½"	4½"
	2"	5"
	3"	6"
2"	1"	4½"
	1½"	5"
	2½"	6"
2¼"	1"	5"
	2"	6"
2½"	¾"	5"
	1¾"	6"
3"	¾"	6"
	2¾"	8"
	4¾"	10"

When specifying hinges, give the knuckle length (excluding tips), then the width open.

To determine hinge width under normal conditions use this equation:

Hinge width = Clearance required
+ [(door thickness − bracket) × 2] + inset (if any)

If this does not come to a standard width, use the next larger standard hinge size.

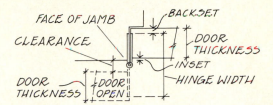

D-3 WOOD SIDING

The table indicates the maximum width and minimum thickness of strip siding that have proved acceptable for the stud spacings shown.

| Maximum Width (inches) | Over Sheathing | | Direct to Framing | |
	Minimum Thickness (inches)[a]	Maximum Stud Spacing (inches)	Minimum Thickness (inches)[a]	Maximum Stud Spacing (inches)
Bevel siding, lapped or rabbeted				
6	7/16		9/16	
8	7/16	24	11/16	16
10	9/16		11/16	
12	11/16		—	—
Drop, shiplap, rustic, and novelty siding				
8	¾	24	¾	16
Vertical siding				
12	¾	b	¾	b
Plywood square-edge lap siding				
12	5/16	16	—	—
24	3/8	24	3/8	16
24	½	24	½	24
Hardboard lap siding				
24	¼	24	—	—

[a]Butt thickness is shown for bevel siding. Tip thickness should be 3/16" for beveled wood siding.

[b]Except when sheathing is nominal ¾" board or ½" plywood, vertical siding should be installed over blocking 24" o.c.

Source: Federal Housing Administration.

D-4   WOOD SHINGLES AND SHAKES

				Maximum Exposure (inches)		
					Double Coursing	
					Face Grade	
Type	Length	Minimum Thickness[a]	Single Coursing	No. 1	No. 2
Shingles and rebutted	16″	5 in 2″	7½	12	10
and rejointed shingles	18″	5 in 2¼″	8½	14	11
	24″	4 in 2″	11½	16	14
Machine-grooved shakes	16″	b	—	12	—
	18″	b	—	14	—
	24″	b	—	16	—
Hand-split and resawn	18″	½″ to 1¼″	8½	14	—
shakes	24″	½″ to 1¼″	11½	20	—
	32″	¾″ to 1¼″	15		
Taper-split shakes	24″	½″ to ⅝″	11½	20	—
Straight-split	18″	⅜″	8½	16	—
shakes	24″	⅜″	11½	22	—

[a]Thickness of shingles is number of butts in specified thickness.
[b]Machined from standard shingle thickness.
Source: Federal Housing Administration.

D-5   WATERPROOFING

Membrane waterproofing used to prevent the passage of water through the exterior walls of a building should be in accordance with the following

Hydrostatic Head	Number of Plies	Laps of Plies	Moppings
Under 6 feet	3	⅓	4 coats
6 to 9 feet	4	¼	5 coats
9 to 12 feet	5	⅕[b]	6 coats
Over 12 feet[a]			

[a]Use elastomeric or equivalent membrane waterproofing in accordance with Public Building Service Interim Guide Specification 4-0711.
[b]Side laps of each successive ply should overlap side joints of preceding ply by ⅕ of membrane width. Each ply should lap at least 3 inches at side joints and 6 inches at end joints with adjacent end joints staggered at least 24 inches.
Source: Federal Housing Administration.

D-6 CONCRETE SLAB REINFORCEMENT

REINFORCEMENT FOR 4-INCH SLAB SUPPORTED AT EDGE AND ON INTERMEDIATE PIERS

	Maximum Spacing of Piers		
Material	*6 Feet o.c.*	*7 Feet o.c.*	*8 Feet o.c.*
Minimum pier size			
Concrete (round)	10 in.	12 in.	14 in.
Concrete or masonry (square)	8 in. × 8 in.	10 in. × 10 in.	12 in. × 12 in.
Minimum footing area	115 sq. in.	130 sq. in.	175 sq. in.
Welded wire fabric	6 × 6-6/6	6 × 6-6/6	6 × 6-6/6

Source: Federal Housing Administration.

REINFORCEMENT FOR 4-INCH SLAB SUPPORTED AT EDGE AND ON INTERMEDIATE WALLS

	Span		
Type of Reinforcement	*8 Feet*	*10 Feet*	*12 Feet*
Welded wire fabric plus bars[a]	6 × 6-6/6 plus no. 3 bars at 18 in. o.c.	6 × 6-6/6 plus no. 3 bars at 18 in. o.c.	6 × 6-6/6 plus no. 3 bars at 12 in. o.c.
Welded wire fabric[b]	3 × 6-8/6	3 × 6-7/6	2 × 6-7/6

[a]Place bars in direction of span.

[b]Place fabric with narrow mesh in direction of span.

Source: Federal Housing Administration.

D-7 VAPOR BARRIERS

MINIMUM NOMINAL THICKNESS OF POLYETHYLENE VAPOR BARRIERS

Warm side walls and ceilings	2 mils (0.002 in.)
Crawl spaces or on sand or tamped earth under nonstructural slabs	4 mils (0.004 in.)
Over gravel or under slabs with reinforcing steel	6 mils (0.006 in.)

Source: Federal Housing Administration.

D-8 MORTAR

MORTAR MIX PROPORTIONS BY VOLUME

Mortar Type	Portland Cement or Portland Blast-Furnace-Slag Cement	Masonry Cement	Hydrated Lime or Lime Putty[a]	Aggregates[b]
M	1	1 (Type II)	—	Not less than 2¼ and not more than 3 times the sum of the volumes of the cement and lime used
	1	—	¼	
S	½	1 (Type II)	—	
	1	—	¼ to ½	
N	—	1 (Type II)	—	
	1	—	½ to 1¼	
O	—	1 (Type I or II)	—	
	1	—	1¼ to 2½	

[a]Quicklime, ASTM C-5, except that lime shall contain not more than 8% magnesium oxide (MgO). Quicklime should be used in plant mixing operations only and shall be thoroughly slaked before mixing.

[b]Measured in a damp, loose condition.

Source: Federal Housing Administration.

MORTAR USES

Kind of Masonry	Type of Mortar Required[a]
Foundation walls or piers	M or S[b]
Exterior walls above grade	S or N[c]
Exterior cavity walls	M or S
Reinforced masonry	M or S[d]
Grouted or filled cell masonry	M or S[d]
Chimneys	S or N
Interior bearing walls	S or N
Interior nonbearing walls	S, N, or O
Retaining walls	M
Sewers and manholes	M
Ceramic veneer	S or N[e]
Glass block	N
Gypsum block	Gypsum mortar
Fire brick	Refractory mortar

[a]Mortar shall comply with ASTM C-270 requirements for unreinforced masonry and ASTM C-476 for reinforced masonry.

[b]Type N may be used with solid units when parged.

[c]Type S may be required in high-wind areas. Type O may be used for exterior walls where facing and backing are solid units.

[d]As required by design strength.

[e]Aggregate shall not exceed 4 times the portland cement content.

Source: Federal Housing Administration.

D-9 PIPING AND CONNECTORS

BRANCH WATER SUPPLY PIPING AND
DRAINAGE AND VENT CONNECTORS (inches)

| | Fixture Supply | | | |
| | Hot Water | Cold Water | Soil or Waste Connections | Vent Connections |
Fixture				
Water closest	—	⅜	3 × 4	2
Lavatory	⅜	⅜	1¼	1¼
Bathtub	½	½	1½	1¼
Sink	½	½	1½	1¼
Laundry tray	½	½	1½	1¼
Sink-and-tray combination	½	½	1½	1¼
Shower	½	½	2	1¼
Dishwasher	½	½	1½	1¼
Washing machine	½	½	2	1¼

Note: Fixtures supplies shall be brass pipe or copper tube, chrome-plated when exposed. All hot- and cold-water branch supply lines shall be not less than ½ in. Exposed openings for fixture supply or branch piping shall be covered with chrome-plated or stainless-steel escutcheons.

Source: Federal Housing Administration.

D-10 WATER HEATERS

The table gives typical water heater storage capacity, input, and recovery requirements; they may vary with individual manufacturers. Any combination of these requirements to produce the 1-hour draw stated will be satisfactory. Recovery is based on a 100° F rise in water temperature.

For example, for a three-bedroom, two-bath residence, there are three choices: a 40-gallon storage/30-gallon-per-hour recovery gas heater, a 50-gallon storage/22-gallon-per-hour recovery electric heater, or a 30-gallon storage/59-gallon-per-hour recovery oil heater, or an equivalent combination that will produce at least 70 gallons per hour total draw.

DIRECT-FIRED WATER HEATER CAPACITIES

Fuel		Gas	Elect.	Oil	Gas	Elect.	Oil	Gas	Elect.	Oil	Gas	Elect.	Oil
Number of Bedrooms			1			2			3			—	
1 to 1½ baths	Storage (gallons)	20	20	30	30	30	30	30	40	30	—	—	—
	Input (BTUs per hour *or* kilowatts)	27K	2.5	70K	36K	3.5	70K	36K	4.5	70K	—	—	—
	Draw (gallons per hour)	43	30	89	60	44	89	60	58	89	—	—	—
	Recovery (gallons per hour)	23	10	59	30	14	59	30	18	59	—	—	—
Number of Bedrooms			2			3			4			5	
2 to 2½ baths	Storage (gallons)	30	40	30	40	50	30	40	50	30	50	66	30
	Input (BTUs per hour *or* kilowatts)	36K	4.5	70K	36K	5.5	70K	38K	5.5	70K	47K	5.5	70K
	Draw (gallons per hour)	60	58	89	70	72	89	72	72	89	90	88	89
	Recovery (gallons per hour)	30	18	59	30	22	59	32	22	59	40	22	59
Number of Bedrooms			3			4			5			6	
3 to 3½ baths	Storage (gallons)	40	50	30	50	66	30	50	66	30	50	80	40
	Input (BTUs per hour *or* kilowatts)	38K	5.5	70K	38K	5.5	70K	47K	5.5	70K	50K	5.5	70K
	Draw (gallons per hour)	72	72	89	82	88	89	90	88	89	92	102	99
	Recovery (gallons per hour)	32	22	59	32	22	59	40	22	59	42	22	59

Source: Federal Housing Administration.

WAGE RATE DISTRIBUTION BY TRADE: UNITED STATES

Trade	Average Rate per Hour ($)	Corresponding Percentages for Various Rates per Hour of Journeymen												
		Under $11.00	$11.00-11.40	$11.40-11.80	$11.80-12.20	$12.20-12.60	$12.60-13.00	$13.00-13.40	$13.40-13.80	$13.80-14.20	$14.20-14.60	$14.60-15.00	$15.00-15.40	$15.40 and over
Journeymen	12.72	10.0	7.9	12.2	10.2	7.9	9.2	9.3	6.6	11.1	4.0	4.1	3.9	3.4
Asbestos workers	12.59	7.1	7.8	13.4	19.5	10.6	11.7	7.8	6.9	1.3	1.9	2.9	6.6	2.6
Boilermakers	13.45	—a	—	12.7	—	—	22.8	9.4	26.1	—	7.2	5.8	16.0	—
Bricklayers	12.64	7.2	4.8	21.3	11.3	9.2	8.1	6.6	9.3	6.6	6.2	9.2	—	—
Carpenters	12.42	11.4	10.0	18.7	11.3	7.2	8.1	5.2	3.2	15.5	2.9	.4	5.7	.4
Cement finishers	12.16	18.4	8.9	13.1	12.2	6.4	2.9	22.1	9.0	2.0	2.8	2.3	—	—
Drywall tapers	12.42	5.1	30.7	9.4	10.0	5.0	2.7	2.3	10.2	8.8	14.2	1.1	.5	—
Electricians	13.46	2.8	1.2	12.5	9.0	8.6	2.5	15.5	8.0	10.8	2.2	10.4	8.7	7.7
Elevator constructors	13.06	5.5	11.1	4.0	7.0	15.0	9.0	19.1	3.1	12.8	1.1	.7	2.4	9.2
Glaziers	11.91	24.5	7.8	18.4	14.4	7.4	9.0	6.0	5.7	—	1.1	—	2.8	2.9
Lathers	12.45	13.4	17.8	5.1	3.0	8.5	11.4	16.2	10.8	8.9	1.2	—	—	3.8
Machinists	13.96	—	2.4	5.1	13.1	15.5	1.0	5.5	2.1	7.7	26.7	25.7	6.7	1.6
Marble setters	12.43	20.1	6.2	14.9	13.1	6.6	1.0	2.5	12.9	.4	1.4	21.0	—	—
Mosaic and terrazzo workers	11.90	13.5	10.1	27.9	5.1	11.4	21.9	1.1	5.0	1.2	2.3	—	.6	—
Painters	12.00	26.6	16.8	5.0	10.4	5.0	4.4	7.7	8.9	5.1	8.8	.1	—	1.2
Paperhangers	12.09	25.0	18.1	6.6	4.9	5.7	7.5	13.8	4.6	5.6	—	7.2	.6	.3
Pipefitters	13.54	1.7	5.2	2.0	12.6	5.5	11.2	12.5	8.4	13.8	4.5	4.8	.9	16.8
Plasterers	12.00	18.1	17.0	8.7	11.6	6.0	12.5	11.8	3.6	4.6	3.0	1.6	—	1.4
Plumbers	12.98	10.5	2.2	8.2	8.1	7.7	16.6	11.9	5.5	9.4	4.5	6.9	—	8.5
Reinforcing iron workers	12.55	8.8	13.5	8.0	8.8	14.1	14.9	7.1	4.3	12.3	1.2	2.6	4.3	—
Roofers, composition	12.08	24.5	2.5	13.2	10.2	11.4	4.9	4.0	5.5	9.7	13.0	1.3	—	—
Roofers, slate and tile	11.70	23.5	1.5	12.0	16.5	22.2	6.4	5.8	.8	1.5	.8	8.0	—	.9
Sheet-metal workers	13.07	5.3	6.1	5.2	6.7	8.0	18.0	9.7	10.6	9.0	6.7	8.2	3.1	3.5
Stonemasons	12.31	12.6	5.7	15.3	15.9	9.8	6.5	5.5	7.1	6.5	1.6	13.4	—	—
Structural iron workers	12.73	6.7	11.2	6.1	10.7	12.0	11.9	8.5	3.6	22.6	.7	3.7	2.5	—
Tile layers	12.25	17.4	13.8	14.3	9.2	5.7	6.2	6.6	6.1	9.3	2.5	—	8.9	—

	Average Rate per Hour ($)	Corresponding Percentages for Various Rates per Hour of Helpers and Laborers													
		Under $7.20	$7.20-7.60	$7.60-8.00	$8.00-8.40	$8.40-8.80	$8.80-9.20	$9.20-9.60	$9.60-10.00	$10.00-10.40	$10.40-10.80	$10.80-11.20	$11.20-11.60	$11.60-12.00	$12.00 and over
Helpers and laborers	9.80	4.6	4.1	5.7	2.0	7.6	8.6	7.6	15.9	9.4	6.7	13.6	2.2	5.3	6.6
Bricklayers' tenders	10.04	4.9	5.1	5.3	1.1	3.7	1.1	5.1	9.3	14.3	13.8	27.4	–	2.2	6.8
Building laborers	9.81	4.3	2.9	5.3	1.2	8.8	10.3	8.4	18.0	9.3	4.8	12.8	2.1	5.7	6.2
Composition roofers' helpers	8.03	28.9	5.3	9.0	24.2	–	8.3	–	3.3	1.0	11.2	–	8.9	–	–
Elevator constructors' helpers	9.24	2.1	6.8	10.6	8.9	11.8	12.3	11.4	20.7	.2	6.1	–	1.2	–	7.9
Marble setters' helpers	9.67	2.4	–	1.0	18.0	.9	4.6	27.6	10.9	16.5	2.1	5.3	5.4	1.8	3.5
Plasterers' laborers	10.12	2.6	7.8	5.4	6.8	8.2	9.4	.9	9.0	8.9	8.6	5.1	3.0	8.9	15.4
Plumbers' laborers	9.21	7.3	26.2	11.2	–	.6	2.1	1.7	2.7	4.4	23.3	–	4.6	12.0	3.9
Terrazzo workers' laborers	9.68	11.7	5.4	7.2	7.9	.7	5.4	4.3	13.1	5.1	1.9	6.7	16.1	9.3	5.2
Tile layers' helpers	10.08	5.0	–	7.7	5.6	3.1	4.4	8.7	17.6	5.0	5.7	10.8	5.6	6.4	14.3

[a]Dash indicates no data reported.

Union building trades, July 1, 1980

Source: U.S. Department of Labor Bureau of Labor Statistics.

AVERAGE WAGE RATES BY TRADE: REGIONS

Trade	United States ($)	New England ($)	Middle Atlantic ($)	Border States ($)	Southeast ($)	Southwest ($)	Great Lakes ($)	Middle West ($)	Mountain ($)	Pacific ($)
All building trades	12.21	11.46	12.09	11.06	10.58	11.06	12.98	12.06	11.88	13.10
Journeymen	12.72	11.94	12.59	11.73	11.25	11.47	13.43	12.47	12.45	13.65
Asbestos workers	12.59	11.82	12.50	12.07	11.68	12.32	13.07	11.93	12.43	14.74
Boilermakers	13.45	13.40	13.49	12.96	11.79	12.70	14.13	13.01	13.66	14.39
Bricklayers	12.64	11.37	12.14	11.65	11.64	11.80	13.29	11.89	12.79	14.14
Carpenters	12.42	11.64	12.24	11.69	10.82	11.10	13.41	12.40	11.66	12.81
Cement finishers	12.16	11.46	11.37	10.76	10.68	11.46	12.59	11.90	11.80	12.88
Drywall tapers	12.42	11.18	11.48	12.40	10.53	11.38	11.76	11.92	11.84	13.68
Electricians (inside wirers)	13.46	12.08	13.88	12.46	12.05	11.99	13.92	13.09	13.32	15.16
Elevator constructors	13.06	12.02	12.87	12.16	11.29	11.57	13.70	12.60	12.50	16.56
Glaziers	11.91	10.89	12.10	11.24	9.78	10.58	12.30	11.85	11.90	12.68
Lathers	12.45	_b	12.21	11.36	10.64	11.12	12.71	12.47	11.95	13.44
Machinists	13.96	12.50	12.50	13.33	—	12.58	14.22	13.88	—	14.58
Marble setters	12.43	11.34	11.69	11.39	10.94	11.71	13.32	11.60	12.12	14.30
Mosaic and terrazzo workers	11.90	11.37	12.27	10.81	11.53	10.67	12.30	12.45	12.27	13.12
Painters	12.00	11.10	11.55	10.56	9.68	11.22	12.07	12.02	11.46	13.59
Paperhangers	12.09	11.03	12.30	11.67	10.29	11.09	11.63	13.24	11.41	13.91
Pipefitters	13.54	12.54	12.99	11.99	12.16	12.34	14.32	13.66	13.57	15.24
Plasterers	12.00	11.23	11.39	11.37	10.34	11.35	12.43	11.45	12.38	13.14
Plumbers	12.98	12.08	12.96	11.45	11.90	11.04	13.96	12.88	13.21	14.98
Reinforcing iron workers	12.55	13.45	12.85	11.73	11.38	11.01	13.45	11.44	12.78	13.02
Roofers, composition	12.08	11.51	12.15	9.94	10.05	9.60	13.44	11.38	12.05	12.74
Roofers, slate and tile	11.70	12.56	12.37	8.26	10.18	9.32	12.99	11.80	11.44	12.66
Sheet-metal workers	13.07	12.14	12.97	11.78	11.39	11.48	13.64	13.07	12.95	14.38
Stonemasons	12.31	11.36	11.29	11.67	11.58	12.34	12.90	11.61	12.65	14.15
Structural iron workers	12.73	13.26	12.47	11.70	11.34	11.38	13.54	11.42	12.94	13.36
Tile layers	12.25	11.36	11.08	10.81	11.15	11.20	12.75	11.28	12.22	13.43

Region[a]

AVERAGE WAGE RATES BY TRADE: REGIONS (CONT.)

	New England	Middle Atlantic	Border States	Southeast	Southwest	Great Lakes	Middle West	Mountain	Pacific
Helpers and laborers	9.80	9.26	10.29	8.27	7.57	8.24	10.77	9.85	10.31
Bricklayers' tenders	10.04	9.63	11.01	7.29	7.68	8.11	10.62	9.84	10.76
Building laborers	9.81	9.18	10.17	8.43	7.56	8.33	10.84	9.70	10.13
Composition roofers' helpers	8.03	9.12	7.25	6.55	6.27	3.84	9.44	—	9.14
Elevator constructors' helpers	9.24	8.27	9.37	8.51	7.73	8.03	9.47	8.85	8.80
Marble setters' helpers	9.67	9.88	9.84	9.83	8.77	8.49	11.69	11.03	10.07
Plasterers' laborers	10.12	9.52	10.40	9.18	7.84	7.96	10.96	10.13	12.15
Plumbers' laborers	9.21	—	10.41	7.90	7.36	—	10.91	11.38	9.71
Terrazzo workers' laborers	9.68	10.63	11.09	7.81	7.89	8.17	11.33	11.55	10.30
Tile layers' helpers	10.08	9.92	10.10	7.74	7.78	8.56	11.56	9.94	10.93

[a]The regions referred to in this study include:: *New England*—Connecticut, Maine, Massachusetts, New Hampshire, Rhode Island, and Vermont; *Middle Atlantic*—New Jersey, New York, and Pennsylvania; *Border States*—Delaware, District of Columbia, Kentucky, Maryland, Virginia, and West Virginia; *Southeast*—Alabama, Florida, Georgia, Mississippi, North Carolina, South Carolina, and Tennessee; *Southwest*—Arkansas, Louisiana, Oklahoma, and Texas; *Great Lakes*—Illinois, Indiana, Michigan, Minnesota, Ohio, and Wisconsin; *Middle West*—Iowa, Kansas, Missouri, Nebraska, North Dakota, and South Dakota; *Mountain*—Arizona, Colorado, Idaho, Montana, New Mexico, Utah, and Wyoming; *Pacific*—Alaska, California, Hawaii, Nevada, Oregon, and Washington.

[b]Dash indicates no data reported.

[c]Union building trades, July 1, 1980.

Source: U.S. Department of Labor Bureau of Labor Statistics.

AVERAGE WAGE RATES AND EMPLOYER CONTRIBUTION
FOR SELECTED BENEFITS BY TRADE: UNITED STATES

Trade	Average Rate per Hour ($)	Average Employer Contribution per Hour[a] ($)	Average Rate Plus Employer Contribution per Hour[a] ($)
All building trades	12.21	2.83	15.05
Journeymen	12.72	2.97	15.69
Asbestos workers	12.59	3.14	15.74
Boilermakers	13.45	3.04	16.50
Bricklayers	12.64	2.56	15.19
Carpenters	12.42	2.80	15.22
Cement finishers	12.16	3.03	15.19
Drywall tapers	12.42	2.41	14.84
Electricians (inside wirers)	13.46	3.21	16.67
Elevator constructors	13.06	3.20	16.26
Glaziers	11.91	2.86	14.77
Lathers	12.45	2.70	15.16
Machinists	13.96	1.82	15.78
Marble setters	12.43	2.57	15.00
Mosaic and terrazzo workers	11.90	2.02	13.91
Painters	12.00	2.23	14.22
Paperhangers	12.09	2.06	14.15
Pipefitters	13.54	3.44	16.98
Plasterers	12.00	2.49	14.50
Plumbers	12.98	3.22	16.21
Reinforcing iron workers	12.55	3.26	15.80
Roofers, composition	12.08	2.40	14.48
Roofers, slate and tile	11.70	2.38	14.08
Sheet-metal workers	13.07	3.19	16.25
Stonemasons	12.31	2.61	14.92
Structural iron workers	12.73	3.78	16.52
Tile layers	12.25	2.69	14.93
Helpers and laborers	9.80	2.22	12.02
Bricklayers' tenders	10.04	1.99	12.03
Building laborers	9.81	2.28	12.09
Composition roofers' helpers	8.03	1.80	9.84
Elevator constructors' helpers	9.24	2.87	12.11
Marble setters' helpers	9.67	1.94	11.61
Plasterers' laborers	10.12	2.24	12.36
Plumbers' laborers	9.21	1.55	10.77
Terrazzo workers' laborers	9.68	1.49	11.17
Tile layers' helpers	10.08	2.07	12.15

[a]Includes employer contributions to insurance (life insurance, hospitalization, medical, surgical, and other similar types of health and welfare programs); pension funds; vacation payments; supplemental unemployment benefits; savings funds; and paid holidays, as provided in labor-management contracts. Averages are for a straight-line hour; in actual practice, however, some employer payments are calculated for total hours worked or gross payroll.

Average refers to all workers in the classification, including those for whom employer contributions were not specified in their particular contracts. Such situations were included in the average computation as zero contributions.

Some contracts also provide additional payments to other funds such as for education and promotion. Information on payments to these funds was not included in these tabulations.

Union building trades, July 1, 1980.

Note: Because of rounding, sums of individual items may not equal totals.

Source: U.S. Department of Labor Bureau of Labor Statistics.

WEEKLY HOURS: UNITED STATES

Trade	Average Weekly Hours	Workweek Length (Percent of Members)					
		Under 35 Hours	35 Hours	36 Hours	37.5 Hours	Over 37.5 and Under 40 Hours	40 Hours
All building trades	39.4	A[a]	9.2	1.8	0.7	A	88.3
Journeymen	39.4	A	9.9	2.1	.8	A	87.1
Asbestos workers	39.5	—[b]	10.7	—	—	—	89.3
Boilermakers	39.9	—	.8	—	2.8	—	96.3
Bricklayers	39.4	—	12.1	—	.7	—	87.2
Carpenters	39.4	—	10.5	2.6	—	—	86.9
Cement finishers	39.9	—	3.0	—	—	—	97.0
Drywall tapers	39.9	—	1.1	1.6	—	—	97.3
Electricians (inside wirers)	39.2	—	14.1	2.4	2.0	—	81.5
Elevator constructors	38.6	2.3	11.3	—	—	—	86.4
Glaziers	39.7	—	5.4	1.1	—	1.5	92.0
Lathers	39.3	—	9.7	5.4	—	—	84.8
Machinists	40.0	—	—	—	—	—	100.0
Marble setters	39.4	—	12.1	—	—	—	87.9
Mosaic and terrazzo workers	39.4	—	12.3	—	—	—	87.7
Painters	38.9	—	18.9	3.4	—	—	77.7
Paperhangers	39.4	—	6.5	6.6	—	—	86.9
Pipefitters	39.3	—	8.5	6.1	.3	—	85.1
Plasterers	39.5	1.4	5.6	2.2	—	—	90.9
Plumbers	39.2	—	12.3	4.0	2.5	—	81.2
Reinforcing iron workers	39.9	—	2.3	—	—	—	97.7
Roofers, composition	39.6	—	7.8	—	—	—	92.2
Roofers, slate and tile	39.3	—	14.1	—	—	—	85.9
Sheet-metal workers	39.6	—	7.5	—	3.0	—	89.5
Stonemasons	39.1	—	17.6	—	2.5	—	79.9
Structural iron workers	39.9	—	1.9	—	—	—	98.1
Tile layers	40.0	—	—	—	—	—	100.0
Helpers and laborers	39.7	.1	5.7	.2	—	—	94.1
Bricklayers' tenders	38.8	—	24.1	.6	—	—	75.3
Building laborers	39.9	—	2.4	—	—	—	97.6
Composition roofers, helpers	40.0	—	.6	—	—	—	99.4
Elevator constructors, helpers	39.4	—	11.3	—	—	—	88.7
Marble setters' helpers	39.8	—	4.3	—	—	—	95.7
Plasterers' laborers	39.6	1.6	2.8	2.4	—	—	93.2
Plumbers' laborers	40.0	—	—	—	—	—	100.0
Terrazzo workers laborers	39.3	—	13.8	—	—	—	86.2
Tile layers' helpers	40.0	—	—	—	—	—	100.0

[a]Less than 0.05 percent.

[b]Dash indicates no members reported for specified interval.

Note: Because of rounding, sums of individual items may not equal totals.

Source: U.S. Department of Labor Bureau of Labor Statistics.

APPENDIX F
BIDDING INFORMATION

There are a number of ways contractors find out about pending jobs. Contractors are often contacted directly by architects, engineers, owners, and others and asked to bid on a certain project. Also, owners may wish to advertise broadly to gain the best competitive bids, or they may choose to negotiate privately with one or more contractors.

Public institutions are often legally required to advertise projects in newspapers, trade journals, and magazines and to post notices in public places. The "invitation for bids" notice or "instructions to bidders" notice usually contains a description of the project, states the project location, names the owner, tells where the bid documents may be obtained and what deposit or charge is required, states bond and similar protective requirements of the owner, and gives the time, the manner, and the place where the bids are to be received. The following is a typical example of a newspaper "invitation for bids" notice.

INVITATION FOR BIDS

The City of Centerville will receive bids for Landscape Improvements and Rehabilitation of the Centerville Town Square and Park until 10:00 a.m. on the 31st day of August, 1985, at City Hall, 101 Walnut Street, Centerville, Georgia 30817, by the Mayor and City Board of Alderman, at which time and place all Bids will be publicly opened and read aloud. Bids are invited upon the following:

Contract 2—Supply and planning of landscape materials in town square area. The Instructions to Bidders, Form of Bid, Form of Contract, Plans, Specifications, and Forms of Bid Bond, Performance and Payment Bond, and other Contract Documents may be examined at the following:

1. Centerville City Hall, 101 Walnut Street, Centerville, Georgia 30817
2. Environmental Planning and Engineering Co., Inc., 3035 Directors Street, Suite 901, Atlanta, Georgia 30301 (Telephone 404-345-0721)

Copies of the Contract Documents may be obtained from Environmental Planning and Engineering Co., Inc., located at 3035 Directors Street, Suite 901, Atlanta, Georgia 30301 upon payment of $100.00, non-refundable, per set of documents.

Performance Bond

KNOW ALL MEN BY THESE PRESENTS: that

(Here insert full name and address or legal title of Contractor)

as Principal, hereinafter called Contractor, and,

(Here insert full name and address or legal title of Surety)

as Surety, hereinafter called Surety, are held and firmly bound unto

(Here insert full name and address or legal title of Owner)

as Obligee, hereinafter called Owner, in the amount of

Dollars ($),

for the payment whereof Contractor and Surety bind themselves, their heirs, executors, administrators, successors and assigns, jointly and severally, firmly by these presents.

WHEREAS,

Contractor has by written agreement dated   19   , entered into a contract with Owner for

in accordance with Drawings and Specifications prepared by

(Here insert full name and address or legal title of Architect)

which contract is by reference made a part hereof, and is hereinafter referred to as the Contract.

NOW, THEREFORE, THE CONDITION OF THIS OBLIGATION is such that, if Contractor shall promptly and faithfully perform said Contract, then this obligation shall be null and void; otherwise it shall remain in full force and effect.

The Surety hereby waives notice of any alteration or extension of time made by the Owner.

Whenever Contractor shall be, and declared by Owner to be in default under the Contract, the Owner having performed Owner's obligations thereunder, the Surety may promptly remedy the default, or shall promptly

1) Complete the Contract in accordance with its terms and conditions, or

2) Obtain a bid or bids for completing the Contract in accordance with its terms and conditions, and upon determination by Surety of the lowest responsible bidder, or, if the Owner elects, upon determination by the Owner and the Surety jointly of the lowest responsible bidder, arrange for a contract between such bidder and Owner, and make available as Work progresses (even though there should be a default or a succession of defaults under the contract or contracts of completion arranged under this paragraph) sufficient funds to pay the cost of completion less the balance of the contract price; but not exceeding, including other costs and damages for which the Surety may be liable hereunder, the amount set forth in the first paragraph hereof. The term "balance of the contract price," as used in this paragraph, shall mean the total amount payable by Owner to Contractor under the Contract and any amendments thereto, less the amount properly paid by Owner to Contractor.

Any suit under this bond must be instituted before the expiration of two (2) years from the date on which final payment under the Contract falls due.

No right of action shall accrue on this bond to or for the use of any person or corporation other than the Owner named herein or the heirs, executors, administrators or successors of the Owner.

Signed and sealed this   day of   19

_____
(Witness)

{ _____ (Principal)   (Seal)

  _____ (Title)

_____
(Witness)

{ _____ (Surety)   (Seal)

  _____ (Title)

PERFORMANCE BOND

BOND NUMBER...

KNOW ALL MEN BY THESE PRESENTS:

That ....

.... as Principal,

hereinafter called Contractor, and   , a corporation organized and existing under the laws

of the State of Maryland, Baltimore, Maryland, as Surety, hereinafter called Surety, are held and firmly bound unto ....

....

as Obligee, hereinafter called Owner, in the amount of ....

.... Dollars ($...............................),

for the payment whereof Contractor and Surety bind themselves, their heirs, executors, administrators, successors and assigns, jointly and severally, firmly by these presents.

WHEREAS,. Contractor has by written agreement dated .... 19......, entered into a contract with Owner for

in accordance with drawings and specifications prepared by ....
                                                                         (Here insert full name, title and address)

...., which contract is by reference made a part hereof, and is hereinafter referred to as the Contract.

NOW, THEREFORE, THE CONDITION OF THIS OBLIGATION is such that, if Contractor shall promptly and faithfully perform said Contract, then this obligation shall be null and void; otherwise it shall remain in full force and effect.

The Surety hereby waives notice of any alteration or extension of time made by the Owner.

Whenever Contractor shall be, and declared by Owner to be in default under the Contract, the Owner having performed Owner's obligations thereunder, the Surety may promptly remedy the default, or shall promptly

(1) Complete the Contract in accordance with its terms and conditions, or
(2) Obtain a bid or bids for completing the Contract in accordance with its terms and conditions, and upon determination by Surety of the lowest responsible bidder, or, if the Owner elects, upon determination by the Owner and the Surety jointly of the lowest responsible bidder, arrange for a contract between such bidder and Owner, and make available as Work progresses (even though there should be a default or a succession of defaults under the contract or contracts of completion arranged under this paragraph) sufficient funds to pay the cost of completion less the balance of the contract price; but not exceeding, including other costs and damages for which the Surety may be liable hereunder, the amount set forth in the first paragraph hereof. The term "balance of the contract price," as used in this paragraph, shall mean the total amount payable by Owner to Contractor under the Contract and any amendments thereto, less the amount properly paid by Owner to Contractor.

Any suit under this bond must be instituted before the expiration of two (2) years from the date on which final payment under the Contract falls due.

No right of action shall accrue on this bond to or for the use of any person or corporation other than the Owner named herein or the heirs, executors, administrators or successors of the Owner.

Signed and sealed this.. day of.., 19....

In the presence of:

....

....   By .... (Seal)
            (Witness)   Principal

....   By .... (Seal)
            (Witness)

Labor and Material Payment Bond

THIS BOND IS ISSUED SIMULTANEOUSLY WITH PERFORMANCE BOND IN FAVOR OF THE
OWNER CONDITIONED ON THE FULL AND FAITHFUL PERFORMANCE OF THE CONTRACT

KNOW ALL MEN BY THESE PRESENTS: that

(Here insert full name and address or legal title of Contractor)

SPECIMEN COPY

as Principal, hereinafter called Principal, and,

(Here insert full name and address or legal title of Surety)

as Surety, hereinafter called Surety, are held and firmly bound unto

(Here insert full name and address or legal title of Owner)

as Obligee, hereinafter called Owner, for the use and benefit of claimants as hereinbelow defined, in the

amount of

(Here insert a sum equal to at least one-half of the contract price)   Dollars ($),

for the payment whereof Principal and Surety bind themselves, their heirs, executors, administrators,
successors and assigns, jointly and severally, firmly by these presents.

WHEREAS,

Principal has by written agreement dated   19   , entered into a contract with Owner for

in accordance with Drawings and Specifications prepared by

(Here insert full name and address or legal title of Architect)

which contract is by reference made a part hereof, and is hereinafter referred to as the Contract.

NOW, THEREFORE, THE CONDITION OF THIS OBLIGATION is such that, if Principal shall promptly make payment to all claimants as hereinafter defined, for all labor and material used or reasonably required for use in the performance of the Contract, then this obligation shall be void; otherwise it shall remain in full force and effect, subject, however, to the following conditions:

1. A claimant is defined as one having a direct contract with the Principal or with a Subcontractor of the Principal for labor, material, or both, used or reasonably required for use in the performance of the Contract, labor and material being construed to include that part of water, gas, power, light, heat, oil, gasoline, telephone service or rental of equipment directly applicable to the Contract.

2. The above named Principal and Surety hereby jointly and severally agree with the Owner that every claimant as herein defined, who has not been paid in full before the expiration of a period of ninety (90) days after the date on which the last of such claimant's work or labor was done or performed, or materials were furnished by such claimant, may sue on this bond for the use of such claimant, prosecute the suit to final judgment for such sum or sums as may be justly due claimant, and have execution thereon. The Owner shall not be liable for the payment of any costs or expenses of any such suit.

3. No suit or action shall be commenced hereunder by any claimant:

a) Unless claimant, other than one having a direct contract with the Principal, shall have given written notice to any two of the following: the Principal, the Owner, or the Surety above named, within ninety (90) days after such claimant did or performed the last of the work or labor, or furnished the last of the materials for which said claim is made, stating with substantial

accuracy the amount claimed and the name of the party to whom the materials were furnished, or for whom the work or labor was done or performed. Such notice shall be served by mailing the same by registered mail or certified mail, postage prepaid, in an envelope addressed to the Principal, Owner or Surety, at any place where an office is regularly maintained for the transaction of business, or served in any manner in which legal process may be served in the state in which the aforesaid project is located, save that such service need not be made by a public officer.

b) After the expiration of one (1) year following the date on which Principal ceased Work on said Contract, it being understood, however, that if any limitation embodied in this bond is prohibited by any law controlling the construction hereof such limitation shall be deemed to be amended so as to be equal to the minimum period of limitation permitted by such law.

c) Other than in a state court of competent jurisdiction in and for the county or other political subdivision of the state in which the Project, or any part thereof, is situated, or in the United States District Court for the district in which the Project, or any part thereof, is situated, and not elsewhere.

4. The amount of this bond shall be reduced by and to the extent of any payment or payments made in good faith hereunder, inclusive of the payment by Surety of mechanics' liens which may be filed of record against said improvement, whether or not claim for the amount of such lien be presented under and against this bond.

Signed and sealed this   day of   19

_____
(Witness)

     (Principal)   (Seal)

_____
(Title)

_____
(Witness)

     (Surety)   (Seal)

_____
(Title)

LABOR AND MATERIAL PAYMENT BOND

KNOW ALL MEN BY THESE PRESENTS: BOND NUMBER ....

That ....

.... as Principal,

hereinafter called Principal, and   a corporation organized and existing under the laws

of the State of Maryland, Baltimore, Maryland as Surety, hereinafter called Surety, are held and firmly bound unto ....

....

as Obligee, hereinafter called Owner, for the use and benefit of claimants as hereinbelow defined, in the amount of ....

.... Dollars ($ ....),

for the payment whereof Principal and Surety bind themselves, their heirs, executors, administrators, successors and assigns, jointly and severally, firmly by these presents.

      WHEREAS, Principal has by written agreement dated .... 19 , entered into a contract with Owner for

in accordance with drawings and specifications prepared by ....

                                       (Here insert full name, title and address)

.... which contract is by reference made a part hereof, and is hereinafter referred to as the Contract.

      NOW, THEREFORE, THE CONDITION OF THIS OBLIGATION is such that if the Principal shall promptly make payment to all claimants as hereinafter defined, for all labor and material used or reasonably required for use in the performance of the Contract, then this obligation shall be void; otherwise it shall remain in full force and effect, subject, however, to the following conditions:

(1) A claimant is defined as one having a direct contract with the Principal or with a sub-contractor of the Principal for labor, material, or both, used or reasonably required for use in the performance of the contract, labor and material being construed to include that part of water, gas, power, light, heat, oil, gasoline, telephone service or rental of equipment directly applicable to the Contract.

(2) The above-named Principal and Surety hereby jointly and severally agree with the Owner that every claimant as herein defined, who has not been paid in full before the expiration of a period of ninety (90) days after the date on which the last of such claimant's work or labor was done or performed, or materials were furnished by such claimant, may sue on this bond for the use of such claimant, prosecute the suit to final judgment for such sum or sums as may be justly due claimant, and have execution thereon. The Owner shall not be liable for the payment of any costs or expenses of any such suit.

(3) No suit or action shall be commenced hereunder by any claimant,

     (a) Unless claimant, other than one having a direct contract with the Principal, shall have given written notice to any two of the following: The Principal, the Owner, or the Surety above named, within ninety (90) days after such claimant did or performed the last of the work or labor, or furnished the last of the materials for which said claim is made, stating with substantial accuracy the amount claimed and the name of the party to whom the materials were furnished, or for whom the work or labor was done or performed. Such notice shall be served by mailing the same by registered mail or certified mail, postage prepaid, in an envelope addressed to the Principal, Owner or Surety, at any place where an office is regularly maintained for the transaction of business, or served in any manner in which legal process may be served in the state in which the aforesaid project is located, save that such service need not be made by a public officer.

     (b) After the expiration of one (1) year following the date on which Principal ceased work on said Contract, it being understood, however, that if any limitation embodied in this bond is prohibited by any law controlling the construction hereof such limitation shall be deemed to be amended so as to be equal to the minimum period of limitation permitted by such law.

     (c) Other than in a state court of competent jurisdiction in and for the county or other political subdivision of the state in which the project, or any part thereof, is situated, or in the United States District Court for the district in which the project, or any part thereof, is situated, and not elsewhere.

(4) The amount of this bond shall be reduced by and to the extent of any payment or payments made in good faith hereunder, inclusive of the payment by Surety of mechanics' liens which may be filed of record against said improvement, whether or not claim for the amount of such lien be presented under and against this bond.

Signed and sealed this .... day of ...., 19....

....   By .... (Seal)
        (Witness)   Principal

....   By .... (Seal)
        (Witness)

This bond is issued simultaneously with performance bond in favor of the Owner conditioned on the full and faithful performance of the Contract.

Contract 211-A   (2-70)

REALTOR©

DAVIES/SOWELL INC.
•REALTORS•
RESIDENTIAL & INVESTMENT PROPERTIES

54 SOUTH COOPER
MEMPHIS, TENNESSEE 38104

(Total number of executed copies made............)

MEMPHIS, TENN., ...19..........

RECEIVED OF...

the sum of..**Dollars**
as earnest money and in part payment for the purchase of the following described real estate (called "Property")
situated in the County of Shelby, and State of Tennessee:

Seller covenants and agrees to sell and convey Property, with all improvements thereon, or cause it to be
conveyed, by good and sufficient warranty deed, to Purchaser, or to such person or persons as Purchaser may
designate; Purchaser, however shall not be released from any of Purchaser's agreements and undertakings as set
forth herein, unless otherwise stated; and Purchaser covenants and agrees to purchase and accept Property for the

total price of ($.................................)..

....

Dollars, upon terms as follows:

* Seller agrees to have the above residence inspected by a reliable State Licensed and Bonded Termite Control Operator; have treated if termite infestation is found, and also remedy and repair any structural insecurities in the visible foundation timbers caused by termite damage.

* Seller agrees to pay the undersigned Agent a commission of _____ of the sale price.

Rents, if any, and all taxes for the current year are to be prorated as of date of closing, and all prior unpaid taxes or liens including front foot assessments are to be paid by Seller, unless otherwise specified. Fire and any additional hazard insurance premiums on the improvements on Property are to be.. (cancelled) (prorated) as of date of closing. If prorated, Purchaser is to pay Seller the unearned premiums for such insurance. (It is recommended that Seller notify his insurance company of the existence of this contract of sale.)

IF THIS CONTRACT REQUIRES FHA OR VA FINANCING, THE SELLERS AGREE TO PAY THE DISCOUNT ON THE NEW LOAN, NOT TO EXCEED_____%.

Title is to be conveyed subject to all restrictions, easements and covenants of record, and subject to zoning o-dinances or laws of any governmental authority. Possession of premises is to be given...

CONTINUED

The improvements on Property are to be delivered in as good condition as they are as of the date of this contract, ordinary wear and tear excepted, and if not in such condition when final settlement is made, Seller is obligated to put them in such condition, or to compensate Purchaser for his failure to do so, but in the event of destruction by fire, or otherwise, Seller's liability shall in no event be more than the appraised value of the improvements so destroyed.

Deferred payments, if any, are to be evidenced by promissory note(s) of Purchaser payable on or before maturity, bearing interest at .... per cent per annum, and secured by a deed of trust on Property in the form generally used by banks and title insurance companies in Memphis, Tennessee. Settlement and payment of balance, if any, of cash payment shall be made upon presentation of a good and valid warranty deed with the usual covenants and conveying a good and merchantable title, after allowing fifteen days from completion of title search or the delivery of abstracts for examination of title. At the election of Purchaser, Seller agrees promptly to furnish, for examination only, either title search or adequate abstracts of title, taxes, and judgments, covering Property, or at Seller's option, a policy of title insurance by one of the title insurance companies with offices in Memphis for the amount of the above purchase price, insuring marketability of title and paid for by Seller. Adequate abstracts of title, taxes and judgments are those required by a title insurance company with an office in Memphis as the basis for the issuance of a policy of title insurance. In the event of controversy regarding title, a title insurance policy covering Property, issued by any local title insurance company for the above purchase price, shall constitute and be accepted by Purchaser as conclusive evidence of good and merchantable title.

If the title is not good and cannot be made good within a reasonable time after written notice has been given that the title is defective, specifically pointing out the defects, then the above earnest money shall be returned to Purchaser and the usual commission shall be paid to the undersigned Agent by Seller. If the title is good and Purchaser shall fail to pay for Property as specified herein, Seller shall have the right to elect to declare this contract cancelled, and upon such election, the earnest money shall be retained by and divided equally between Seller and Agent, as liquidated damages and commission respectively, but in no event shall Agent's share exceed the regular commission. The right given Seller to make the above election shall not be Seller's exclusive remedy, and either party shall have the right to elect to affirm this contract and enforce its specific performance or recover full damages for its breach. Seller's retention of such earnest money shall not be evidence of an election to declare this contract cancelled, as Seller shall have the right to retain his portion of earnest money to be credited against damages actually sustained.

Unless otherwise specified herein Agents commission is to be paid in cash out of the net proceeds of the sale at time of closing this transaction. Failure to close shall not relieve Seller of his obligation to pay a commission as provided herein. If property is being exchanged, each party hereto agrees to furnish either title search or adequate abstracts of title and pay the Agent the commission on the real estate each contracts herein to convey, and otherwise fulfill obligations incumbant upon Seller as outlined above. Any abstracts covering Property only will become the property of Purchaser subject to rights of mortgage holder.

Seller is to pay for preparation of deed, recording of purchase money trust deed, if any, title search or abstract, state tax and Register's fee on trust deed, and notary fee on deed. Seller authorizes Agent to order title search or abstract for which Seller agrees to pay. Purchaser is to pay for preparation of note, or notes, and trust deed, notary fee on trust deed, recording of deed, state tax and Register's fee on deed, and expense of title examination or title insurance, if any. Seller and Purchaser are to share equally in paying closing fee and loan transfer fee, if any, in connection with transaction. If Purchaser obtains a loan on Property, he is to pay all expenses incident thereto.

Should there be any tax, insurance or other accrual items on deposit with the holder of any debt secured by Property and assumed by Purchaser, at the time of closing Purchaser shall reimburse Seller therefor.

This instrument when signed only by the prospective Purchaser shall constitute an offer which shall not be withdrawable in less than 48 hours from the date hereof.

Purchaser accepts Property in its existing condition, no warranties or representations having been made by Seller or Agent which are not expressly stated herein.

As used herein, where applicable: "Seller" and "Purchaser" include the plural; the masculine includes the feminine or neuter gender.

WITNESS the signatures of all parties the day and year above written.

Subject to clearance of any check given, the undersigned Agent acknowledges receipt of the above mentioned earnest money which is held in trust subject to the terms of this contract.

DAVIES/SOWELL INC
•REALTORS•

....
....
Purchaser
....

By...

....

M.L.S. number...
Seller

Purchaser's address..Telephone................

Seller's address..Telephone................

DAVIES/SOWELL INC.
•REALTORS•

REAL ESTATE SALE
CONTRACT

PROPERTY

FROM

TO

DATE CLOSED

Mid South Title Insurance Corporation

Mid South Title
Memphis
Tenn.

Centrum Bldg.
6363 Poplar Ave.
685-2500

•

One Commerce Square
Suite 1200
523-8121

•

Civic Center
100 N. Main
523-8121

Equity Contract

MEMPHIS, TENN., ... 19 ....

RECEIVED OF ....

the sum of ....Dollars
as earnest money and in part payment for the purchase of the following described real estate (called "Property")
situated in the County of Shelby, State of Tennessee:

Seller covenants and agrees to sell and convey property, with all improvements thereon, or cause it to be conveyed,
by good and sufficient warranty deed, to Buyer, or to such person or persons as Buyer may designate; Buyer,
however, shall not be released from any of Buyer's agreements and undertakings as set forth herein, unless
otherwise stated; and Buyer covenants and agrees to purchase and accept property upon terms as follows:

1. Buyer agrees to pay, for the equity, ($ ....) ....
..............................Dollars cash at closing of which the earnest money is a part (unless otherwise provided
herein) and assume the existing indebtedness as follows:

First Mortgage holder .... Approx. Bal. $ ....
Address .... Mo. Payment $ ....
Second Mortgage holder .... Approx. Bal. $ ....
Address .... Mo. Payment $ ....

2. Seller agrees to bring all indebtedness current to date of closing. Any escrow shortages shall be paid by the Seller. Any escrow surplus, reported by the holder, shall be refunded to the Seller.

3. Buyer agrees to pay the ....payment and all subsequent payments on the loan (s) assumed in the approximate amount of $ .... (or the amount agreed to in Item 8 below) including principal, interest, taxes, insurance and assessments, if any.

4. There shall be no proration of current taxes, insurance, interest or assessments. All prior taxes, liens or assessments are to be paid by the Seller.

5. Tax, insurance and other accrual items on deposit with the mortgage holder, after adjustments provided for in Section 2, and any unexpired hazard insurance shall be transferred to the Buyer without charge.

6. $ .... of the cash payment provided for in Item 1 above is to be deferred and shall be evidenced by a promissory note payable to the Seller in monthly payments of $ .... including interest at ....% per annum beginning .... with privilege to prepay at anytime. The note shall be secured by a deed of trust on the property .

7. Other terms agreed upon:

8. It is understood and agreed that, if the mortgage assumed contains an option to accelerate the debt upon the sale of the property, then this contract shall be contingent upon Buyer's ability to assume the loan at its present interest rate or at a rate not to exceed ....% without changing the term of the loan. If necessary to assume the loan, the Buyer covenants and agrees (1) to furnish any credit information reasonably requested by the mortgage holder; (2) to execute any assumption and/or modification agreements; (3) to pay, in addition to one-half of the usual loan transfer fee, all other necessary expense incurred in the assumption.

9. Possession of premises to be given ....

CONTINUED ON REVERSE SIDE

10. Seller agrees to pay a real estate sales commission to ....
.... in the amount of .... % of the total sales price. Unless otherwise agreed, the commission is to be paid in cash out of the proceeds of the sale at the time of closing. Failure to close shall not relieve the Seller of the obligation to pay the commission provided for herein.

11. Seller agrees to furnish either title search or adequate abstracts of title, taxes and judgements as soon as same can be prepared covering property or, at Seller's option, an owner's title policy for the amount of the purchase price to be issued by one of the title insurance companies having offices in Memphis, Tennessee, insuring a good and marketable title which shall constitute and be accepted by the buyer as conclusive evidence of a good and merchantable title.

12. Seller agrees to pay the following: abstracts or title search, attorney's fee for preparation of warranty deed, notary fee, transfer tax and recording fee on purchase money trust deed, if any.

13. Buyer agrees to pay the following: title insurance, expense of title examination, if any, transfer tax and recording fee on warranty deed, attorney's fee for preparation of note and trust deed and notary fee on trust deed. Also any expense incident to a new loan, if any.

14. Buyer and Seller agree to share equally the expense of the escrow closing fee and the usual loan transfer fee.

15. Settlement and payment of balance, if any, of cash payment shall be made upon presentation of a good and valid warranty deed with the usual covenants and conveying a good and merchantable title, after allowing fifteen days from completion of title search or the delivery of abstracts for examination of title.

16. Title is to be conveyed subject to all restrictions, easements and covenants of record and subject to zoning ordinances or laws of any governmental authority.

17. If the title is not good and cannot be made good within a reasonable time after written notice has been given that title is defective, specifically pointing out the defects, then the above mentioned earnest money shall be returned to buyer and the commission shall be paid to the undersigned Agent by Seller.

18. However, if the title is good and Buyer shall fail to pay for property as specified herein, Seller shall have the right to elect to declare this contract cancelled, and upon such election, the earnest money shall be retained by and divided equally between Seller and Agent, as liquidated damages and commission respectively, but in no event shall Agent's share exceed the regular commission. The right given Seller to make the above election shall not be Seller's exclusive remedy, and either party shall have the right to elect to affirm this contract and enforce its specific performance or recover full damages for its breach. Seller's retention of such earnest money shall not be evidence of an election to declare this contract cancelled, as Seller shall have the right to retain his portion of earnest money to be credited against damages actually sustained.

19. The improvements on property are to be delivered in as good condition as they are as of the date of this contract, ordinary wear and tear excepted, and if not in such condition when final settlement is made, Seller is obligated to put them in such condition, or to compensate Purchaser for his failure to do so, but in the event of destruction by fire, or otherwise, Seller's liability shall in no event be more than the appraised value of the improvements so destroyed.

20. Buyer accepts property in its existing condition, no warranties or representations having been made by Seller or Agent which are not expressly stated herein.

21. As used herein, where applicable, ''Seller'' and ''Buyer'' include the plural; the masculine includes the feminine or neuter gender.

22. This instrument when signed only by the prospective Buyer shall constitute an offer which shall not be withdrawable in less than forty-eight hours from the date hereof.

WITNESS the signature of all parties the day and year above written.

Subject to clearance of any check given, the undersigned Agent acknowledges receipt of the above mentioned earnest money which is held in trust subject to the terms of this contract.

....

....
                                                                    Buyer

By ....
                                                                    Seller

Buyer's address .... Telephone ....

Seller's address .... Telephone ....

FORM NO. 63   Rev. 71080

Courtesy Mid South Title Insurance Corp., Memphis, Tenn.

Security Title Company, Inc.

5865 RIDGEWAY PARKWAY, SUITE 104, PHONE 761-2030
MEMPHIS, TENNESSEE 38119

(Total number of executed copies made _____)

MEMPHIS, TENNESSEE _____ , 19 _____

RECEIVED OF _____

the sum of _____ Dollars
as earnest money and in part payment for the purchase of the following described real estate (called ''Property'')
situated in the County of Shelby, and State of Tennessee:

Property known by street address _____ ,
together with all permanent improvements thereon.

Included in the sales price as part of the consideration are:

Seller covenants and agrees to sell and convey Property, with all improvements thereon, or cause it to be conveyed, by good and sufficient warranty deed, to Purchaser, or to such person or persons as Purchaser may designate; Purchaser, however, shall not be released from any of Purchaser's agreements and undertakings as set
forth herein, unless otherwise stated; and Purchaser covenants and agrees to purchase and accept Property for

the total price of ($ _____) _____ Dollars,
upon terms as follows:
   All cash at closing of which the earnest money is a part.

This contract is contingent upon the Purchaser's applying for and obtaining an FHA loan in the amount of

$ _____ from _____ Mortgage Company, amortized for 30 years at _____ %

interest, or whatever the prevailing rate. Application for this loan is to be made within _____ days of this date.

Purchaser to pay all loan and sale expenses legally chargeable to them including origination fee.

Seller warrants all appliances, plumbing, heating, electrical and air conditioning systems are in working order at the time of closing.(Any certifications to be paid for by the purchaser.) Purchaser shall have the right to inspect property prior to closing.

Seller agrees to have the above described property inspected by a licensed and bonded termite control operator, have treated if infestation is found, remedy and repair any insecurities in the visible foundation timbers caused by termites or other wood destroying insects.

Title search, abstract and/or title insurance to be prepared by Security Title Co. of Memphis,

Closing to be on or before _____ .

"It is expressly agreed that, notwithstanding any other provisions of this contract, the Purchaser shall not be obligated to complete the Purchase of the property described herein or to incur any penalty by forfeiture of earnest money deposits or otherwise unless the Seller has delivered to the Purchaser a written statement issued by the Federal Housing Commissioner setting forth the appraised value of the property (excluding closing costs) of not less than $_____ which statement the Seller hereby agrees to deliver to the Purchaser promptly after such appraised values statement is made available to the Seller. The Purchaser shall, however, have the privilege and option of proceeding with the consummation of the contract without regard to the amount of the appraised valuation made by the Federal Housing Commissioner. The appraised valuation is arrived at to determine the maximum mortgage the Department of Housing and Urban Development will insure. HUD does not warrant the value or the condition of the property. The Purchaser should satisfy himself/herself that the price and condition of the property are acceptable."

Rents, if any, and all taxes for the current year and interest (and FHA Mortgage Insurance Premium, if any) upon any debt secured by Property and assumed by Purchaser are to be prorated as of date of closing, and all prior unpaid taxes or liens including front foot assessments are to be paid by Seller, unless otherwise specified. Fire and any additional hazard insurance premiums on the improvements on Property are to be _____(canceled) (prorated) as of date of closing. If prorated, Purchaser is to pay Seller the unearned premiums for such insurance. (Its is recommended that Seller notify his insurance company or the existence of this contract of sale.)

IF THIS CONTRACT REQUIRES FHA OR VA FINANCING, THE SELLERS AGREE TO PAY THE DISCOUNT

ON THE NEW LOAN, NOT TO EXCEED _____ % .

Title is to be conveyed subject to all restrictions, easements and covenants of record, and subject to zoning ordinances or laws of any governmental authority. Possession of premises is to be given _____ .

Security Title/Hulsey Form 103

The improvements on Property are to be delivered in as good condition as they were as of the date of this contract, ordinary wear and tear excepted, and if not in such condition when final settlement is made, Seller is obligated to put them in such condition, or to compensate Purchaser for his failure to do so, but in the event of destruction by fire, or otherwise, Seller's liability shall in no event be more than the appraised value of the improvements so destroyed.

Deferred payments, if any, are to be evidenced by promissory note(s) of Purchaser payable on or before maturity, bearing interest at _____ per cent per annum, and secured by a deed of trust on Property in the form generally used by banks and title insurance companies in Memphis, Tennessee. Settlement and payment of balance, if any, of cash payment shall be made upon presentation of a good and valid warranty deed with the usual covenants and conveying a good and merchantable title, after allowing fifteen days from completion of title search or the delivery of abstracts for examination of title. At the election of Purchaser, Seller agrees to furnish promptly, for examination only, either title search or adequate abstracts of title, taxes, and judgments, covering Property, or at Seller's option, a policy of title insurance by one of the title insurance companies with offices in Memphis for the amount of the above purchase price, insuring marketability of title and paid for by Seller. Adequate abstracts of title, taxes and of a policy of title insurnace. In the event of controversy regarding title, a title insurance policy covering Property, issued by any local title insurance company for the above purchase price, shall constitute and be accepted by Purchaser as conclusive evidence of good and merchantable title.

If the title is not good and cannot be made good within a reasonable time after written notice has been given that the title is defective, specifically pointing out the defects then the above earnest money shall be returned to Purchaser and the usual commission shall be paid to the undersigned Agent by Seller. If the title is good and Purchaser shall fail to pay for Property as specified herein, Seller shall have the right to elect to declare this contract cancelled, and upon such election, the earnest money shall be retained by and divided equally between Seller and Agent, as liquidated damages and commission respectively, but in no event shall Agent's share exceed the regular commission. The right given Seller to make the above election shall not be Seller's exclusive remedy, and either party shall have the right to elect to affirm this contract and enforce its specific performance or recover full damages for its breach. Seller's retention of such earnest money shall not be evidence of an election to declare this contract cancelled, as Seller shall have the right to retain his portion of earnest money to be credited against damages actually sustained. Seller agrees to pay the undersigned Agent a commission of _____ % of the sale price. Unless otherwise specified herein, such commission is to be paid in cash out of the net proceeds of the sale at time of closing this transaction. Failure to close shall not relieve Seller of his obligation to pay a commission as provided herein. If property is being exchanged, each party hereto agrees to furnish either title search or adequate abstracts of title and pay the Agent the commission on the real estate each contracts herein to convey, and otherwise fulfill obligations incumbent upon Seller as outlined above. Any abstracts covering Property only will become the property of Purchaser subject to rights of mortgage holder.

Seller is to pay for preparation of deed, recording of purchase money trust deed, if any, title search or abstract, state tax and Register's fee on trust deed, and notary fee on deed. Seller authorizes Agent to order title search or abstract for which Seller agrees to pay. Purchaser is to pay for preparation of note, or notes, and trust deed, notary fee on trust deed, recording of deed, state tax and Register's fee on deed, and expense of title examination or title insurance, if any. Seller and Purchaser are to share equally in paying closing fee and loan transfer fee, if any, in connection with transaction. If Purchaser obtains a loan on Property, he is to pay all expenses incident thereto.

Should there be any tax, insurance or other accrual items on deposit with the holder of any debt secured by Property and assumed by Purchaser, at the time of closing Purchaser shall reimburse Seller therefor.

This instrument when signed only by the prospective Purchaser shall constitute an offer which shall not be withdrawable in less than 48 hours from the date hereof.

Purchaser accepts Property in its existing condition, no warranties or representations having been made by Seller or Agent which are not expressly stated herein.

As used herein, where applicable: "Seller" and "Purchaser" include the plural; the masculine includes the feminine or neuter gender.

Witness the signatures of all parties the day and year above written.

Subject to clearance of any check given, the undersigned Agent acknowledges receipt of the above mentioned earnest money which is held in trust by _____ *, subject to the terms of this contract.*

_____

_____
                                                                    Purchaser

By _____

_____

Cooperating Broker _____
                                                                    Seller

Cooperating Agent _____

Purchaser's address _____
                                                                    Telephone

Seller's address _____
                                                                    Telephone

Courtesy Security Title Co., Inc., Memphis, Tenn.

Security Title Company, Inc.

5865 RIDGEWAY PARKWAY, SUITE 104, PHONE 761-2030
MEMPHIS, TENNESSEE 38119

(Total number of executed copies made _____)

MEMPHIS, TENNESSEE _____ , 19 _____

RECEIVED OF _____

the sum of _____ Dollars
as earnest money and in part payment for the purchase of the following described real estate (called "Property")
situated in the County of Shelby, and State of Tennessee:

Property known by street address _____ ,
together with all permanent improvements thereon.

Included in the sales price as part of the consideration are:

 Seller Covenants and agrees to sell and convey Property, with all improvements thereon, or cause it to be conveyed, by good and sufficient warranty deed, to Purchaser, or to such person or persons as Purchaser may designate; Purchaser, however shall not be released from any of Purchaser's agreements and undertakings as set
forth herein, unless otherwise stated; and Purchaser covenants and agrees to purchase and accept Property for

the total price of ($ _____) _____
Dollars, upon terms as follows:
 All cash at closing of which the earnest money is a part.

This contract is contingent upon the Purchaser's applying for and obtaining a V. A. loan in the amount of

$ _____ from _____ **Mortgage Company,**

amortized for 30 years at _____ % interest, or whatever the prevailing rate. Application for this loan is to be

made within _____ days of this date.

Purchaser to pay all loan and sale expenses legally chargeable to him/her including origination fee.

Seller warrants all appliances, Plumbing heating, electrical and air conditioning systems are in working order at the time of closing. (Any certifications to be paid for by the purchaser.) Purchaser shall have the right to inspect property prior to closing.

The Seller shall furnish the veteran-purchaser at no cost to the veteran prior to settlement a written statement from a recognized exterminator that based on careful inspection of accessible areas and on sounding of accessible structural members, there is no evidence or termite or other wood destroying insect infestation in the subject property, and, if such infestation previously existed, it has been corrected and any damage due to such infestation has also been corrected, or alternatively, has been fully disclosed as follows—NONE.

Title search, abstract and/or title insurance to be prepared by Security Title Co. of Memphis, phone:761-2030.

Closing to be on or before _____ .

It is expressly agreed that, notwithstanding any other provisions of this contract, the Purchaser shall not incur any penalty by forfeiture of earnest money or otherwise be obligated to complete the purchase or the property described herein, if the contract purchase price or cost exceeds the reasonable value of the property established by V. A. The Purchaser, shall, however have the privilege and option of proceeding with the consummation or this contract without regard to the amount of reasonable value established by the V. A.

Rents, if any, and all taxes for the current year and interest (and FHA Mortgage Insurance Premium, if any) upon debt secured by Property and assumed by Purchaser are to be prorated as of date of closing, and all prior unpaid taxes or liens including front foot assessments are to be paid by Seller, unless otherwise specified. Fire and any additional hazard insurance premiums on the improvements on Property are to be _____(canceled) (prorated) as of date of closing. If prorated, Purchaser is to pay Seller the unearned premiums for such insurance. (It is recommended the Teller notify his insurance company of the existence of this contract of sale.)

IF THIS CONTRACT REQUIRES FHA OR VA FINANCING, THE SELLER AGREES TO PAY THE DISCOUNT ON THE NEW LOAN, NOT TO EXCEED _____ %.

Title is to be conveyed subject to all restrictions, easements and covenants of record, and subject to zoning

ordinances or laws of any governmental authority. Possession of premises is to be given _____

Security Title/Hulsey Form 104

The improvements on Property are to be delivered in as good condition as they were as of the date of this contract, ordinary wear and tear excepted, and if not in such condition when final settlement is made, Seller is obligated to put them in such condition, or to compensate Purchaser for his failure to do so, but in the event of destruction by fire, or otherwise, Seller's liability shall in no event be more than the appraised value of the improvements so destroyed.

Deferred payments, if any, are to be evidenced by promissory note(s) of Purchaser payable on or before maturity, bearing interest at _____ per cent per annum, and secured by a deed of trust on Property in the form generally used by banks and title insurance companies in Memphis, Tennessee. Settlement and payment of balance, if any, of cash payment shall be made upon presentation of a good and valid warranty deed with the usual covenants and conveying a good and merchantable title, after allowing fifteen days from completion of title search or the delivery of abstracts for examination of title. At the election of Purchaser, Seller agrees to furnish promptly, for examination only, either title search or adequate abstracts of title, taxes, and judgments, covering Property, or at Seller's option, a policy of title insurance by one of the title insurance companies with offices in Memphis for the amount of the above purchase price, insuring marketability of title and paid for by Seller. Adequate abstracts of title, taxes and of a policy of title insurnace. In the event of controversy regarding title, a title insurance policy covering Property, issued by any local title insurance company for the above purchase price, shall constitute and be accepted by Purchaser as conclusive evidence of good and merchantable title.

If the title is not good and cannot be made good within a reasonable time after written notice has been given that the title is defective, specifically pointing out the defectsc then the above earnest money shall be returned to Purchaser and the usual commission shall be paid to the undersigned Agent by Seller. If the title is good and Purchaser shall fail to pay for Property as specified herein, Seller shall have the right to elect to declare this contract cancelled, and upon such election, the earnest money shall be retained by and divided equally between Seller and Agent, as liquidated damages and commission respectively, but in no event shall Agent's share exceed the regular commission. The right given Seller to make the above election shall not be Seller's exclusive remedy, and either party shall have the right to elect to affirm this contract and enforce its specific performance or recover full damages for its breach. Seller's retention of such earnest money shall not be evidence of an election to declare this contract cancelled, as Seller shall have the right to retain his portion of earnest money to be credited against damages actually sustained. Seller agrees to pay the undersigned Agent a commission of _____ % of the sale price. Unless otherwise specified herein, such commission is to be paid in cash out of the net proceeds of the sale at time of closing this transaction. Failure to close shall not relieve Seller of his obligation to pay a commission as provided herein. If property is being exchanged, each party hereto agrees to furnish either title search or adequate abstracts of title and pay the Agent the commission on the real estate each contracts herein to convey, and otherwise fulfill obligations incumbent upon Seller as outlined above. Any abstracts covering Property only will become the property of Purchaser subject to rights of mortgage holder.

Seller is to pay for preparation of deed, recording of purchase money trust deed, if any, title search or abstract, state tax and Register's fee on trust deed, and notary fee on deed. Seller authorizes Agent to order title search or abstract for which Seller agrees to pay. Purchaser is to pay for preparation of note, or notes, and trust deed, notary fee on trust deed, recording of deed, state tax and Register's fee on deed, and expense of title examination or title insurance, if any. Seller and Purchaser are to share equally in paying closing fee and loan transfer fee, if any, in connection with transaction. If Purchaser obtains a loan on Property, he is to pay all expenses incident thereto.

Should there be any tax, insurance or other accrual items on deposit with the holder of any debt secured by Property and assumed by Purchaser, at the time of closing Purchaser shall reimburse Seller therefor.

This instrument when signed only by the prospective Purchaser shall constitute an offer which shall not be withdrawable in less than 48 hours from the date hereof.

Purchaser accepts Property in its existing condition, no warranties or representations having been made by Seller or Agent which are not expressly stated herein.

As used herein, where applicable: "Seller" and "Purchaser" include the plural; the masculine includes the feminine or neuter gender.

Witness the signatures of all parties the day and year above written.

Subject to clearance of any check given, the undersigned Agent acknowledges receipt of the above mentioned earnest money which is held in trust by _____ *, subject to the terms of this contract.*

_____

_____ *Purchaser*

By _____

_____

Cooperating Broker _____ *Seller*

Cooperating Agent _____

Purchaser's address _____ *Telephone*

Seller's address _____ *Telephone*

NOTICE:

Protect your Investment with OWNER'S Title Insurance

The real estate investment that you have just contracted to make is probably one of the largest individual investments that you will make in your lifetime. Doesn't it make sense to protect it?

There is only one way purchasers can safeguard their land title against defects that may render their investments unsafe or invalid.

OWNER'S TITLE INSURANCE

Owner's title insurance pays for a legal defense against an attack on a title as insured, and for all claims proved to be valid. All for a modest one-time premium.

Now that you have found the home of your choice, make sure you enjoy it with peace of mind by asking for owner's title insurance by name. It's the only title coverage that protects the home buyer.

TITLE INSURANCE is your best

Security — and vice versa!!

Security Title Company, Inc.

5865 RIDGEWAY PARKWAY, SUITE 104, PHONE 761-2030
MEMPHIS, TENNESSEE 38119

Courtesy Security Title Co., Inc., Memphis, Tenn.

APPENDIX H
MORTGAGE PAYMENT TABLES

Amounts shown include monthly payments of interest and principal but not taxes and insurance.

20-YEAR MORTGAGE
(monthly payments, interest and principal)

Amount	9%	9½%	10%	10½%	11%	12%	13%	14%	15%
$16,000	144.00	149.15	154.41	159.71	165.15	176.17	187.45	198.96	210.69
20,000	180.00	186.43	193.01	199.68	206.44	220.22	234.32	248.70	263.36
24,000	216.00	223.72	231.61	239.62	247.73	264.26	281.18	298.44	316.03
28,000	252.00	261.00	270.21	279.55	289.01	308.30	328.04	348.19	368.70
30,000	270.00	279.64	289.51	299.52	309.66	330.33	351.47	373.06	395.04
36,000	324.00	335.57	347.41	359.42	371.58	396.39	421.76	447.67	474.04
40,000	360.00	372.86	386.01	399.36	412.88	440.43	468.63	497.41	526.72
44,000	396.00	410.14	424.61	439.29	454.16	484.48	515.49	547.15	579.39
50,000	450.00	466.07	482.52	499.19	516.10	550.54	585.78	621.76	658.40
54,000	486.00	503.36	521.12	539.13	557.38	594.58	632.64	671.50	711.07
60,000	540.00	559.28	579.02	599.03	619.32	660.66	702.93	746.11	790.08
64,000	576.00	596.57	617.62	638.97	660.61	704.70	749.79	795.85	842.75
72,000	647.80	671.14	694.82	718.84	743.16	792.78	843.52	895.33	948.09
80,000	719.78	745.72	772.02	798.72	825.76	880.86	937.25	994.81	1053.43
85,000	764.77	792.33	820.27	848.64	877.37	935.91	995.83	1056.99	1119.24

25-YEAR MORTGAGE
(monthly payments, interest and principal)

Amount	9%	9½%	10%	10½%	11%	12%	13%	14%	15%
$16,000	134.27	139.80	145.40	151.07	156.82	168.52	180.45	192.60	204.93
20,000	167.84	174.74	181.75	188.84	196.02	210.64	225.57	240.75	256.17
24,000	201.41	209.69	218.09	226.61	235.23	252.77	270.68	288.90	307.40
28,000	234.97	244.64	254.44	264.38	274.43	294.90	315.79	337.05	358.63
30,000	251.76	262.11	272.62	283.26	294.03	315.97	338.35	361.13	384.25
36,000	302.11	314.54	327.14	339.91	352.84	379.16	406.02	433.36	461.10
40,000	335.68	349.48	363.49	377.68	392.05	421.28	451.13	481.50	512.33
44,000	369.25	384.43	399.83	415.44	431.24	463.42	496.24	529.66	563.56
50,000	419.60	436.85	454.36	472.10	490.06	526.62	563.91	601.89	640.41
54,000	453.52	471.80	490.70	509.86	529.26	568.75	609.02	650.04	691.64
60,000	503.52	524.22	545.23	566.51	588.06	631.94	676.69	722.26	768.49
64,000	537.09	559.17	581.57	604.28	627.26	674.07	721.80	770.42	819.72
72,000	604.22	629.08	654.28	679.82	705.68	758.33	812.03	866.72	922.19
80,000	671.36	698.96	726.98	755.36	784.10	842.59	902.26	963.02	1024.66
85,000	713.32	742.64	772.42	802.57	833.11	895.25	958.65	1023.21	1088.78

APPENDIX I
F.W. DODGE COMPANY OFFICES

The F.W. Dodge Company, a division of McGraw-Hill Information Systems Company, offers construction information to owners, architects, engineers, contractors, and others interested in construction. (*Source:* F.W. Dodge, McGraw-Hill Information Systems Co.)

Alabama
Birmingham
Huntsville
Mobile
Montgomery

Arizona
Phoenix
Tucson

Arkansas
Little Rock

California
Long Beach
Los Angeles
San Bernardino
San Diego
San Francisco
Santa Ana
Ventura

Colorado
Denver

Connecticut
Avon
Milford

**District of
Columbia**
Washington

Florida
Cocoa
Ft. Lauderdale
Jacksonville
Miami
Pensacola
St. Petersburg
Sarasota
Tallahassee
Tampa
W. Palm Beach
Winter Park

Georgia
Albany
Atlanta
Augusta
Columbus
Macon
Savannah

Illinois
Chicago
Springfield

Indiana
Evansville
Ft. Wayne
Highland
Indianapolis

Iowa
Bettendorf
W. Des Moines

Kansas
Mission
Wichita

Kentucky
Lexington
Louisville

Louisiana
Baton Rouge
Metairie
Shreveport

Maine
Portland

Maryland
Baltimore
Silver Spring

Massachusetts
Boston
West Springfield

Michigan
Detroit
Flint
Grand Rapids
Kalamazoo
Lansing

Minnesota
Minneapolis

Mississippi
Jackson

Missouri
Manchester
Springfield

Nebraska
Omaha

Nevada
Las Vegas

New Hampshire
Manchester

New Jersey
Clifton

New Mexico
Albuquerque

New York
Albany
East Syracuse
Elmsford
Hicksville
New York
Rochester
Williamsville

North Carolina
Charlotte
Greensboro
Raleigh

Ohio
Cincinnati
Cleveland
Columbus
Dayton
Toledo
Youngstown

Oklahoma
Oklahoma City
Tulsa

Oregon
Portland

Pennsylvania
Erie
Philadelphia
Pittsburgh

Rhode Island
Cranston

South Carolina
Charleston
Columbia
Greenville

Tennessee
Chattanooga
Knoxville
Memphis
Nashville

Texas
Abilene
Amarillo
Austin
Beaumont
Dallas
El Paso
Ft. Worth
Houston
Lubbock
Midland
San Antonio
Tyler
Wichita Falls

Virginia
Norfolk
Richmond
Roanoke

Washington
Seattle

West Virginia
Charleston

Wisconsin
Madison
Milwaukee

APPENDIX J
GLOSSARY[1]

Absorption bed: A pit of relatively large dimensions filled with coarse aggregate, containing a distribution pipe system in which absorption of septic tank effluent through bottom area is obtained.

Absorption field: A system of trenches containing coarse aggregate and distribution pipe through which septic tank effluent may seep or leach into surrounding soil.

Accessory building: A secondary building, the use of which is incidental to that of the main building and which is located on the same plot.

Addition: Any construction that increases the size of a building or adds to the building, such as a porch or an attached garage or carport.

Aggregate: The total or mass of materials, such as sand or stone, used in making concrete; as an underlayment for concrete, or asphalt; and for other similar construction purposes.

A lightweight aggregate as defined by the Federal Housing Administration has a dry, loose weight of 55 to 77 pounds per cubic foot.

Air conditioning: The process of treating air so as to control simultaneously its temperature, humidity, cleanliness, and distribution to meet the comfort requirements of the occupants of the conditioned space. The system may be designed for summer air conditioning, winter air conditioning, or both.

Alley: A service way providing a secondary public means of access to abutting properties.

Alteration: Construction that may change the structural parts, mechanical equipment, or location of openings but does not increase the size of the building.

Apartment: Same as *living unit*. Generally used in connection with multifamily buildings.

Area:

Building area. The total ground area of each building and accessory building, not including uncovered entrance platforms, terraces, and steps.

Floor area. The total area of all stories or floors finished as living accommodations. This area includes bays and dormers but does not include space in garages or carports or in attics. Measurements are taken to the outside of exterior walls.

Areaway: An open subsurface space adjacent to a building used to admit light and air or as a means of access to a basement or crawl space.

Attached garage: See *garage*.

Attic: Accessible space between the top of the uppermost ceiling and the underside of the roof. Inaccessible spaces are considered structural cavities.

Attic room (finished attic): Attic space that is finished as living accommodations but does not qualify as a half story. See *story*.

Backfill: To place earth or selected fill material in an excavated void.

Balcony: An exterior platform, enclosed by a railing or parapet, projecting from the wall of a building for the private use of tenants or for exterior access to the above-grade living units. When a balcony is roofed and enclosed with operating windows, it is considered part of the room it serves.

Basement: A space of full-story height below the first floor that is not designed or used primarily for year-round living accommodations. Space, partly below grade, that is designed and finished as habitable space is not defined as basement space. See *story*.

Basementless space (crawl space): An unfinished, accessible space below the first floor which is usually less than full-story height.

Bearing: The portion of a beam, truss, or other structural member that rests on the supports.

British thermal unit (BTU): A unit of measurement of the quantity of heat required to raise the temperature of 1 pound of water $1°$ F. The mean BTU is usually used, which is $\frac{1}{180}$ of the heat required to raise the temperature of 1 pound of water from $32°$ F to $212°$ F at a constant atmospheric pressure of 14.69 pounds per square inch.

Building area: See *area*.

Building height: Vertical distance measured from the average grade to the highest level of a flat roof or to the average height of a pitched roof. Where a height limitation is set forth in stories, such height shall include each full story as defined herein.

Building line: A line established by law or agreement, usually parallel to property line, beyond which a structure may not extend. This generally does not apply to uncovered entrance platforms, terraces, and steps.

Caisson or pier foundations: A foundation system whereby the building structure is supported upon a system of holes (usually lined) bored in the earth to a stratum that will provide adequate support of design loads and filled with concrete. Borings of 2 feet or larger in diameter permit bottom inspection are usually considered *caissons*, while borings of less than 2 feet in diameter are considered *piers*.

Carport: A roofed space having at least one side open to the weather, primarily designed or used for housing motor vehicles.

Cavity walls: See *walls*.

Cesspool: A covered pit with open-jointed lining into which raw sewage is discharged.

Chase: A groove or shaft in a masonry wall provided for accommodation of pipes, ducts, or conduits.

Cold-rolled temper: Metal formed into a sheet or other

[1]*Source:* Federal Housing Administration

shape by cold-working to the final dimensions with a specific hardness.

Community system: A central water or sewarage system, the rates and service of which are not controlled by a governmental authority.

Concrete:

Cellular concrete. Low-density concrete containing a large volume of air and weighing not more than 55 pounds per cubic foot.

Lightweight structural concrete. Concrete having a density of not over 115 pounds per cubic foot.

Plain concrete. Concrete without reinforcement or with only minimum reinforcement required to meet shrinkage or temperature stresses.

Precast concrete. A plain or reinforced concrete element case in other than its final position in the structure.

Reinforced concrete. Concrete in which reinforcement is embedded in such a manner that the concrete and steel act together in resisting forces.

Conductance, thermal: The time rate of heat flow through a unit area of a body, of given size and shape, per unit temperature difference. Value is expressed in BTU per hour per square foot per degree Fahrenheit (symbol C).

Conductivity, thermal: The time rate of heat flow through a unit area of a homogeneous material under the influence of a unit temperature gradient. Value is expressed in BTU per hour per square foot per degree Fahrenheit per inch (symbol k).

Construction classifications: Classification of buildings into types of construction on the basis of the fire properties of walls, floors, roofs, ceilings, and other elements.

Type 1: Fire-resistive construction. A type of construction in which the walls, columns, floors, roof, ceiling, and other structural members are noncombustible, with sufficient fire resistance to withstand the effects of a fire and prevent its spread from one story to another.

Type 2: Noncombustible construction. A type of construction in which the walls, columns, floors, roof, ceiling, and other structural members are noncombustible but which does not qualify as Type 1. Type 2 construction is further classified as Type 2a (1-hour protected) and Type 2b, which does not require protection for certain members.

Type 3: Exterior-protected construction. A type of construction in which the exterior walls are of noncombustible construction, having a fire-resistance rating as specified and which are structurally stable under fire conditions and in which the interior structural members and roof are of protected combustible construction or of unprotected heavy-timber construction in public areas only. Type 3 construction is divided into two subtypes as follows:

Type 3a. Exterior-protected construction in which the interior exitways, columns, beams, and bearing walls are noncombustible, in combination with the protected-floor-system roof construction and non-load-bearing partitions of combustible construction.

Type 3b. Exterior-protected construction in which the interior structural members are of protected combustible materials or of heavy-timber unprotected construction.

Type 4: Protected wood-frame construction. A type of construction in which the exterior walls, partitions, floors, roof, and other structural members are wholly or partly of wood or other combustible materials.

Control joint: A joint either formed or sawed into concrete or masonry surfaces to accommodate movement caused by shrinkage, heat, or other forces.

Corridor: A passageway or hallway that provides a common way of travel to an exit. A *dead-end corridor* is one that provides only one exit.

Corrosion-resistant: Describing sheet metal, nails, or hardware of aluminum alloys, brass, bronze, copper, galvanized steel, lead, stainless steels, terneplate, or zinc-copper alloys. Nails that are coated with paint, cement, or similar materials are not considered corrosion-resistant.

Court:

Inner court. An open outdoor space enclosed on all sides by exterior walls of a building or by exterior walls and property lines on which walls are allowable.

Outer court. An open outdoor space enclosed on at least two sides by exterior walls of a building or by exterior walls and property lines on which walls are allowable, with one side open to a street, driveway, alley, or yard.

Coverage: The percentage of the plot area covered by buildings, including accessory buildings.

Crawl space: Same as *basementless space.*

Curing compounds: Preparations that are applied to concrete surfaces to form membranes that reduce evaporation of water from the concrete.

Dampproofing: A treatment of a surface or structure that retards the passage of water. See *waterproofing.*

Differential earth movement: A displacement horizontally, vertically, or along some resultant vector of an identifiable soil mass, this displacement being measurably different from that of other immediately adjacent soil masses having at least one common boundary.

Differential settlement: Uneven or variable settlement of building foundations due to nonuniform earth support.

Disposal field: Same as *absorption field.*

Driveway: A private way for the use of vehicles and pedestrians.

Drywall finish: Interior covering material, such as gypsum board or plywood, that is applied in large sheets or panels.

Drywell: A covered pit with open-jointed lining or a covered pit filled with coarse aggregate through which drainage from roofs, basement floors, foundation drain tile, or areaways may seep or leach into the surrounding soil.

Dwelling: A building designed or used as the living quarters for one or more families.

Detached. Completely surrounded by permanent open spaces.

Row. Describing a dwelling, the walls on two sides of which are party or lot-line walls.

Semidetached. Describing a dwelling, one side wall of which is a party or lot-line wall. Also called *end-row*.

Dwelling unit: See *living unit*.

Easement: A vested or acquired right to use land other than as a tenant, for a specific purpose, such right being held by someone other than the owner who holds title to the land.

Elastomer: A rubbery material that returns approximately to its original dimension in a short time after a relatively large amount of deformation.

Existing fill: Typically applied to a tract of land that has been extensively reformed by fill, grading, or other earthwork prior to identification of a HUD interest.

Family: One or more persons occupying a single living unit.

Finish rating: The amount of time needed for the stud or joist to experience an average temperature rise of 250° F or an individual temperature rise of 325° F as measured on the plane of the stud or joist nearest the fire, in accordance with ASTM E-119 test.

Fire door: A door, including its frame, so constructed and assembled in place to prevent or retard the passage of flame or hot gases.

Fire-resistive: Descriptive of the ability of materials and assemblies to resist fire and its spread.

Fire-retardant wood: Wood treated by a recognized impregnation process so as to reduce its combustibility or surface-flame spread.

Fire separation: A construction of specified fire resistance separating parts of a building horizontally or vertically.

Firestopping: A barrier within concealed spaces that is effective against the spread of flames or hot gases.

Fire wall: See *wall*.

Flame spread: The propagation of flame over a surface.

Flashing: Sheet metal or other impervious material used in roof and wall construction to protect a building from seepage of water.

Floor: See *story*.

Floor area: See *area*.

Foundation: Construction, below or partly below grade, that provides support for exterior walls or other structural parts of the building.

Frost line: The depth below finish grade where frost action on footings or foundations is improbable.

Garage: A building or enclosure primarily designed or used for motor vehicles.

Attached. Having all or part of one or more walls common to the dwelling or to a covered porch attached to the dwelling.

Built-in. Located within the exterior walls of a dwelling.

Detached. Completely surrounded by open space. A garage connected to the dwelling by an uncovered terrace is defined as a detached garage.

Gas vent: A chimney designed and approved for use with gas-burning appliances only.

Glazing: The act or art of installing and securing glass or glasslike (plastic) materials in prepared openings of installations such as doors, windows, and enclosures. Such openings are defined as having been *glazed*.

Glazing material: Bedding compound, mastic, and clips used in installing or securing glazing panels in prepared openings such as doors, windows, and enclosures.

Glazing panel: A pane or sheet of transparent or translucent glass or glasslike (plastic) material that is installed in prepared openings such as doors, windows, and enclosures.

Grade:

Finish grade. The top surface elevation of lawns, walks, drives, or other improved surfaces after completion of construction or grading operations.

Natural grade. The elevation of the original or undisturbed natural surface of the ground.

Subgrade. The elevation established to receive top surfacing or finishing materials.

Grade beam: A reinforced concrete beam supporting the exterior-wall construction, in contact with the earth but supported by piers.

Gradient: The slope, or rate of increase or decrease in elevation, of a surface, road, or pipe, usually expressed in percent.

Grounds: Small strips of wood used to maintain the thickness of plaster at floor intersections and at openings and for the attachment of trim, base, and other millwork.

Ground water: The surface water that saturates a zone of earth or soil.

Ground-water drain: A drain designed and constructed to convey groundwater away from a saturated zone of earth.

Grout: Masonry mortar of pouring consistency such as is used to fill voids in masonry units.

Hydronics: The science of heating and cooling with liquids.

Hydrostatic head: Water pressure on a surface due to the height of the water above a specific point.

Impervious soils: Dense, compact, fine-grained soils that restrict the infiltration of water.

Individual system: A water or sewage system serving a single property.

Joists: A series of floor, roof, or ceiling framing members spaced not more than 30 inches on center. Members supporting roofs have slopes over 3 in 12 are referred to as *rafters*.

Listing: A list published by an approved testing agency. Equipment appearing on such a list is referred to as *listed*.

Living unit: A dwelling or portion thereof providing complete living facilities for one family, including permanent provisions for living, sleeping, eating, cooking, and sanitation.

Loads:

Concentrated load. A load concentrated on a specified small area of a floor, roof, wall, or other member.

Dead load. The weight of all permanent construction in a building.

Design load. Total load that a structure or member is

designed to sustain safely without exceeding specified deformation.

Live load. The weight of all moving and variable loads that may be placed on or in a building, such as snow, wind, and occupants.

Uniform load. An average load applied uniformly over a floor, roof, or wall or along a beam or girder.

Lot: A parcel of land that is described by reference to a recorded plat or by metes and bounds.

Corner lot. A lot abutting upon two or more streets at their intersection.

Double-fronted lot. An interior lot bounded by a street on front and back.

Interior lot. A lot bounded by a street on one side only.

Lot coverage: The percentage of the plot area covered by the building area.

Lot line: A line bounding the lot as described in the title to the property.

Lot line wall: See *wall*.

Masonry: A construction of units of such materials as clay, shale, concrete, glass, gypsum, or stone, set in mortar.

Hollow masonry. Units in which the voids exceed 25% of the cross-sectional area.

Solid masonry. Units in which the voids do not exceed 25% of the cross-sectional area at any plane parallel to the bearing surface.

Mastic: A thick, pasty sealant or adhesive.

Means of egress: A continuous and unobstructed way of exit travel from any point in a building or structure to a public space and consisting of three separate and distinct parts: the way of exit access, the exit, and the way of exit discharge. A means of egress comprises the vertical and horizontal ways of exit travel and includes intervening rooms, doors, hallways, corridors, passageways, balconies, ramps, stairs, enclosures, lobbies, escalators, horizontal exits, courts, and yards.

Moisture protection: Safeguarding living units against the penetration or passage of water, water vapor, and dampness.

Natural grade: See *grade*.

Nominal size: The commercial size given a piece of lumber or other material by the trade, other than its actual size. For example, the actual size of a nominal two-by-eight is $1\frac{5}{8}$ by $7\frac{1}{2}$ inches. (See Appendix C-1.)

Noncombustible: Describing a material or a combination of materials that will not ignite or support combustion during a 5-minute exposure to a temperature of 1,200° F.

Open channel: A drainage course designed to carry major storm-sewer flow on the surface (generally used only where no other feasible method of drainage is possible).

Perm: The unit of measurement of the water vapor permeance of a material. Value of one perm is equal to one grain of water vapor per square foot per hour per inch of mercury vapor pressure difference.

Previous soils: Loose, porous, coarse-grained soils that allow the infiltration of water.

Pier: A masonry or concrete column supporting foundations or the floor structure in basementless spaces. A pier may be free-standing or bonded at its sides to other masonry or concrete.

Pilaster: A pier forming part of a masonry or concrete wall, partially projecting from it and bonded to it.

Pile foundations: A foundation system consisting of steel, timber, or concrete members driven vertically into the ground. Piles transfer loads through soft layers to ends bearing on hard soil or rock or by friction to suitable soils.

Plat: A map, plan, or chart of a city, town, section, or subdivision, indicating the location and boundaries of individual properties.

Plates: Horizontal wood members that provide bearing and anchorage for wall, floor, ceiling, and roof framing.

Rafter or *joist plate*. Plate at the top of a masonry or concrete wall, supporting a rafter or roof joist and ceiling framing.

Sill plate. A plate on top of the foundation wall that supports the floor framing.

Wall plate. A plate at the top (*top plate*) or bottom (*bottom plate*) of a wall or partition framing.

Plenum: An air compartment or chamber to which one or more ducts are connected and which forms part of an air-distribution system. Properly constructed, crawl space may be used as a plenum, without ductwork.

Plinth: A solid concrete or masonry block at the base of a column.

Plot: A parcel of land consisting of one or more lots or portions thereof, described by reference to a recorded plat or by metes and bounds.

Pozzuolana: A siliceous or siliceous and aluminous material that forms compounds possessing cementitious properties by reacting with calcium hydroxide in the presence of moisture. (Also called *pozzolan*.)

Property: A lot or plot, including all buildings and improvements on it.

Property line: A recorded boundary of a plot.

Public system: A water or sewage system that is owned and operated by a local government authority or by a local utility company adequately controlled by a governmental authority.

Purlin: An intermediate supporting member at right angles to the rafter or truss framing.

R (thermal resistance): A measure of ability to retard heat flow. R is the numerical reciprocal of U. thus R = 1/U.

Raft or mat foundations: A foundation system consisting of a number of spread footings that have been joined to form a monolithic unit foundation.

Rafters: A series of roof-framing members, spaced not more than 30 inches on center. In roofs having slopes over 3 in 12. Members supporting roofs having slopes 3 in 12 or less are referred to as roof *joists*.

Screeds: Strips of plaster of the desired coat thickness laid on a surface to serve as guides for plastering the intervals between them. Also, the intermediate leveling strips in concrete slabs.

Scuppers: Openings in the wall above the roof that allow water to run off the roof.

Sealants: Any material used to seal joints or openings against the intrusion or passage of any foreign substance, such as water, gas, air, or dirt.

Seepage pit: A covered pit with an open-jointed lining through which septic-tank effluent may seep or leach into the surrounding porous soil.

Septic tank: A covered, watertight sewage settling tank intended to retain the solids in the sewage flowing through the tank long enough for satisfactory decomposition by bacterial action to take place.

Service lines: The various conduits that provide channelized space for the collection or distribution of sewage, water, electricity, steam, gas, storm water, or other utility services.

Shall: In contracts, indicates that which is required.

Should: In contracts, indicates that which is recommended but not mandatory.

Site work: Productive effort designed to improve the area or tract of land, a part of this tract being the exact plot upon which a building or service facility is, has been, or is to be located.

Slump: Degree of wetness of concrete related to workability; a measurement of consistency.

Soil saturation: The water content of the soil mass, which may be visualized as a compressible skeleton containing voids that are filled with water.

Sound transmission loss: The reduction in the sound pressure level between two stated points, usually expressed in *decibels*.

Specified density: The measurable control of the voids in a material by identification of the ratio of mass or weight per unit of volume. The specified density should be practicably obtainable in construction and maintainable during the life of the structure.

Stairway: One or more flights of stairs and any landings or platforms connected with them to form a continuous passage from one floor to another.

Storm drainage system: Facilities, structures, appurtenances, pipes, channels, and natural watercourse improvements to collect, convey, and dispose of surface runoff to an outlet.

Storm water runoff: The water from precipitation that flows along the surface due to gravity and is the excess of that water which wets the surface, is absorbed through infiltration, or remains in puddles or pools.

Story: The portion of a building between a floor and the next floor above.

First story (first floor). The lowermost story that has at least half its total floor area designed for and finished as living accommodations. For the purpose of determining this area, the area of halls, closets, and stairs is included. The area of storage, utility, or heating rooms or spaces is not included. The location of the first story as defined

here is based on the use of the space rather than on the location of entrance doors or the finished grade.

Half story. A story finished as living accommodations located wholly or partly within the roof frame and having a floor area at least half as large as the story below. Space with less than 5 feet of clear headroom is not considered as floor area.

Top story. The story between the uppermost floor and the ceiling or roof above.

Street: A public or private way that affords principal means of vehicular access to properties that abut thereon.

Studs: A series of vertical wall or partition framing members spaced not more than 24 inches on center.

Subgrade support strength: The measurable capacity of the prepared earth surface to sustain weight, loading, pressure, strain, and wear without failure in accordance with design assumptions relative to serviceability as a foundation for buildings, structures, pavement bases, or pavements and walks.

Summer cooling hours: Operating hours per season (considered to be all hours when the outdoor temperature is 80° F or higher).

Surface water: Water from any source that appears in a diffused state, with no permanent continuous source of supply or regular course to flow along, as distinguished from *surface water runoff*, which flows along a regular watercourse or drainage facility or has collected into lakes or ponds.

Swale: A drainage channel formed by the convergence of intersecting slopes.

Thickened-edge slab: A type of concrete floor-slab foundation where the slab is constructed integrally with the foundation wall.

Trimmer: A beam or joist into which a header is framed in framing a chimney, stairway, or other opening.

Truss: A structural framework composed of a series of members so arranged and fastened together that external loads applied at the joints will cause only direct stress in the members.

Trussed rafter: A truss whose chord members also serve as rafters and ceiling joists and are subject to bending stress in addition to direct stress.

U (overall coefficient of heat transmission): The combined thermal value of all the materials in a building section, air spaces, and surface-air films. U is expressed in BTU per hour per square foot of area per degree Fahrenheit temperature difference.

Ventilation:

Mechanical ventilation. Supply and removal of air by power-driven devices.

Natural ventilation. Ventilation to outside air through windows, doors, or other openings.

Walls:

Bearing wall. A wall that supports any vertical load in addition to its own weight.

Cavity wall. A masonry or concrete wall consisting of two wythes arranged to provide an air space within the wall in which the inner and outer wythes of the wall are tied together with metal ties.

Common wall. A wall separating two living units free of common utilities except where an agreement is established that allows necessary repairs, replacement, and services of such utilities.

Curtain wall. A wall, usually nonbearing, between piers or columns.

Faced wall. A wall in which the masonry facing and the backing are so bonded as to exert a common reaction under load.

Firewall. A wall with qualities of fire resistance and structural stability that subdivides a building into fire areas and that resists the spread of fire.

Foundation wall. A wall, below or partly below grade, providing support for the exterior or other structural parts of a building.

Lot line wall. A wall adjoining and parallel to the lot line used primarily by the party upon whose lot the wall is located. Lot line walls may share common foundations.

Masonry wall. A bearing or nonbearing wall of hollow or solid masonry units.

Masonry-bonded hollow wall. A cavity wall bonded together with solid masonry units in lieu of metal ties.

Nonbearing wall. A wall that supports no vertical load other than its own weight.

Parapet wall. The part of any wall that extends entirely above the roof.

Party wall. A wall used jointly by two parties under easement, erected upon a line separating two parcels of land, each of which is a separate real estate entity.

Veneered wall. A wall with a masonry face that is attached to but not so bonded to the body of the wall as to exert a common reaction under load.

Waterproofing: A treatment of a surface or structure that prevents the passage of water. See also *dampproofing.*

Water vapor transmission: The passage of moisture into or through a material or construction, in a gaseous form, due to the difference in vapor pressure at the two faces. The unit of measurement is the *perm.*

Way: A street, alley, or other thoroughfare or easement permanently established for the passage of persons or vehicles.

Winter degree-day: A unit based on temperature difference and time, used in estimating fuel consumption and specifying nominal heating loads of a building in winter. A degree-day represents each degree below $65°$ F of the mean temperature for the day.

Wythe: The partition between two chimney flues in the same stack. Also, the inner and outer walls of a cavity wall.

Yard: The open, unoccupied space on the plot between the property line and the front, rear, or side wall of the building.

Front yard. The yard across the full width of the plot facing the street, extending from the front line of the building to the front property line. Either yard facing a street may be selected as the front yard of a corner lot.

Rear yard. The yard across the full width of the plot opposite the front yard, extending from the rear line of building to the rear property line. The rear yard of a corner lot is the yard opposite the selected front yard.

Side yard. The yard between the side line of building and the adjacent side property line, extending from front yard to the rear yard.

APPENDIX K
ABBREVIATIONS[1]

AA	Aluminum Association
AAMA	Architectural Aluminum Manufacturers Association
ABPA	American Board Products Association
ABS	Acrylonitrile-butadiene styrene
ACI	American Concrete Institute
agg	Aggregate
AGA	American Gas Association
AHAM	Association of Home Appliance Manufacturers
AHLI	American Home Lighting Institute
AIA	American Institute of Architects
AIA	American Insurance Association
AISC	American Institute of Steel Construction
AISI	American Iron and Steel Institute
AITC	American Institute of Timber Construction
AMCA	Air Moving and Conditioning Association
ANSI	American National Standards Institute
APA	American Plywood Association
ARI	Air Conditioning and Refrigeration Institute
ASCE	American Society of Civil Engineers
ASHRAE	American Society of Heating, Refrigerating and Air Conditioning Engineers
ASME	American Society of Mechanical Engineers
ASTM	American Society for Testing and Materials
AWPB	American Wood Preservers Bureau
AWPI	American Wood Preservers Institute
AWS	American Welding Society
AWWA	American Water Works Association
BIA	Brick Institute of American (formerly structural Clay Products Institute, SCPI)
BOCA	Building Officials and Code Administrators, International, Inc.
BRAB	Building Research Advisory Board
BTU	British thermal units
cf	Cubic feet
cfm	Cubic feet per minute
CFR	Code of Federal Regulations
CISPI	Cast Iron Soil Pipe Institute
CPVC	Chlorinated polyvinyl chloride
CS	Commercial standard
CSPE	Chlorosulfonated polyethylene
CTI	Ceramic Tile Institute
cts	Coats
cy	Cubic yards
°	Degrees
dBA	Sound pressure level as measured using the A-weighted network (decibels)
dwv	Drain, waste, and vent
EEA	Electric Energy Association
EEI	Edison Electric Institute
EPA	Environmental Protection Agency
ext.	Exterior
F	Fahrenheit (degrees)

FGJA	Flat Glass Jobbers Association
FHA	Federal Housing Administration
FHDA	Fir and Hemlock Door Association
fpm	Feet per minute
FS	Federal Specifications
ft	Feet
GA	Gypsum Association
ga	Gauge
gal	Gallon
galv	Galvanized
GAMA	Gas Appliance Manufacturers Association
gph/kw	Gallons per hour per kilowatt
gpm	Gallons per minute
gpsf/h	Gallons per square foot per hour
HPMA	Hardwood Plywood Manufacturers Association
hr	Hour
HUD	Department of Housing and Urban Development
HVI	Home Ventilating Institute
HYDI	Hydronics Institute
IAPMO	International Association of Plumbing and Mechanical Officials
IBR	Institute of Boiler and Radiator Manufacturers (superceded by HYDI)
ICBO	International Conference of Building Officials
id	Inside diameter
in.	Inches
IPS	International Pipe Standard
int.	Interior
kw	Kilowatts
lb	Pounds
LP	Liquified petroleum
max.	Maximum
MCA	Mechanical Contractors' Association of America
min.	Minimum
mph	Miles per hour
MPS	Minimum Property Standards
NA	Not applicable
NAAMM	National Association of Architectural Metal Manufacturers
NAHB-RF	National Association of Home Builders— Research Foundation
NBBPVI	National Board of Boiler and Pressure Vessels Inspectors
NBS	National Bureau of Standards
NCMA	National Concrete Masonry Association
NEC	National Electric Code
NECA	National Electrical Contractors Association
NEMA	National Electrical Manufacturers Association
NESCA	National Environmental Systems Contractors Association
NFPA	National Fire Protection Association
NFPA	National Forest Products Association
NIMA	National Insulation Manufacturers Association

[1] *Source:* Federal Housing Administration.

NKCA	National Kitchen Cabinet Association	SCS	Soil Conservation Service
NMWIA	National Mineral Wood Insulation Association	SEAOC	Structural Engineers Association of California
NPA	National Particleboard Assocation	SGCC	Safety Glazing Certification Council
NRMCA	National Ready-Mixed Concrete Association	SI	International standard
NSF	National Sanitation Foundation	SMACNA	Sheet Metal and Air Conditioning Contractors National Association, Inc.
NTMA	National Terrazzo and Mosaic Association		
NWMA	National Woodwork Manufacturers Association	SPMA	Sump Pump Manufacturers Association
od	Outside diameter	sq yd	Square yard
oc	On center	SSBC	Southern Standard Building Code
oz	ounces	T&G	Tongue and groove
%	Percent	TCA	Tile Council of America
PCA	Portland Cement Association	TECO	Timber Engineering Company
PCI	Prestressed Concrete Institute	U	Coefficient of thermal transmission
PS	Product standard	UBC	Uniform Building Code (by ICBO)
ppm	Parts per million	UL	Underwriters' Laboratories, Inc.
psf	Pounds per square foot	UM	Use of Materials Bulletin
psi	Pounds per square inch	USDA	United States Department of Agriculture
PVC	Polyvinyl chloride	vcp	Vitrified clay pipe
R	Thermal resistance	wi	Wrought iron
SBI	Steel Boiler Institute (a division of HYDI)	WM	Western Wood Moulding and Millwork Producers
SCACM	Southern California Association of Cabinet Manufacturers	WQA	Water Quality Association (formerly Water Conditioning Foundation)

Index